Progress in Probability
Volume 73

Series Editors
Steffen Dereich
Davar Khoshnevisan
Andreas E. Kyprianou
Sidney I. Resnick

More information about this series at http://www.springer.com/series/4839

AF810112

Progress in Probability

Birkhäuser

Daniel Hernández-Hernández • Juan Carlos Pardo • Victor Rivero
Editors

XII Symposium of Probability and Stochastic Processes

Merida, Mexico, November 16–20, 2015

Birkhäuser

Editors
Daniel Hernández-Hernández
Department of Probability and Statistics
Center for Research in Mathematics
Guanajuato, Mexico

Juan Carlos Pardo
Department of Probability and Statistics
Center for Research in Mathematics
Guanajuato, Mexico

Victor Rivero
Department of Probability and Statistics
Center for Research in Mathematics
Guanajuato, Mexico

ISSN 1050-6977 ISSN 2297-0428 (electronic)
Progress in Probability
ISBN 978-3-030-08501-8 ISBN 978-3-319-77643-9 (eBook)
https://doi.org/10.1007/978-3-319-77643-9

Mathematics Subject Classification (2010): 60J80, 60J25, 60G51, 60F17, 91A15, 91B30, 93E20

© Springer International Publishing AG, part of Springer Nature 2018
Softcover re-print of the Hardcover 1st edition 2018
This work is subject to copyright. All rights are reserved by the Publisher, whether the whole or part of the material is concerned, specifically the rights of translation, reprinting, reuse of illustrations, recitation, broadcasting, reproduction on microfilms or in any other physical way, and transmission or information storage and retrieval, electronic adaptation, computer software, or by similar or dissimilar methodology now known or hereafter developed.
The use of general descriptive names, registered names, trademarks, service marks, etc. in this publication does not imply, even in the absence of a specific statement, that such names are exempt from the relevant protective laws and regulations and therefore free for general use.
The publisher, the authors and the editors are safe to assume that the advice and information in this book are believed to be true and accurate at the date of publication. Neither the publisher nor the authors or the editors give a warranty, express or implied, with respect to the material contained herein or for any errors or omissions that may have been made. The publisher remains neutral with regard to jurisdictional claims in published maps and institutional affiliations.

Printed on acid-free paper

This book is published under the imprint Birkhäuser, www.birkhauser-science.com by the registered company Springer International Publishing AG part of Springer Nature.
The registered company address is: Gewerbestrasse 11, 6330 Cham, Switzerland

Introduction

The present volume contains contributions and lecture notes of the XII Symposium on Probability and Stochastic Processes, held at the Universidad Autónoma de Yucatán (UAdY), Mexico in November 16–20, 2015.

The traces of this symposium reach back to December 1988 at CIMAT, when it was held for the first time. The symposium is one of the main events in the field, and it takes place every 2 years at different academic institutions in Mexico. During these 27 years and up until today, this series of symposia has readily accomplished its main goal of exchanging ideas and discussing the latest developments in the field by gathering both national and international researchers as well as graduate students.

The symposium in 2015 gathered scholars from over seven countries and covered a wide range of topics that highlight the interaction between applied and theoretical probability. The scientific programme included two courses: *Optimality of two-parameter strategies in stochastic control* organized by Kazutoshi Yamazaki, and *Scaling limits of large random trees* organized by Bénédicte Haas. The event also benefited from nine plenary talks that were delivered by José Blanchet, Loïc Chaumont, Alex Cox, Takis Konstantopoulos, Andreas Kyprianou, Hubert Lacoin, Mihai Sirbu, Gerónimo Uribe and Hasnaa Zidani. Another four thematic sessions and fourteen contributed talks completed the outline of the symposium.

This volume is split into two main parts: first the lectures notes of the two courses provided by Bénédicte Haas and Kazutoshi Yamazaki, followed by research contributions of some of the participants. The lecture notes of Bénédicte Haas and Kazutoshi Yamazaki give an overview of the recent progress on describing the large-scale structure of random trees, and on stochastic control problems where the optimal strategies are described by two parameters under a setting that is driven by a spectrally one-sided Lévy process, respectively. The research contributions start with an illustrative article written by Ekaterina Kolkovska and Ehyter Martín-González, in which they investigate a classical risk process with two-sided jumps that is perturbed by a spectrally negative α-stable process, where the gain size distribution has a rational Laplace transform. The contribution of Daniel Hernández-Hernández and Leonel Pérez-Hernández analyses the minimality of the penalty function associated with a convex risk measure. By considering dynamic programming, Laurent series and the study of sensitive discount optimality, Beatris

Escobedo-Trujillo, Héctor Jasso-Fuentes and José Daniel López-Barrientos analyse Blackwell-Nash equilibria for a general class of zero-sum stochastic differential games. Γ-convergence of monotone functionals is discussed in the contribution written by Erick Treviño-Aguilar, where a criterion is presented under which a functional that is defined on vectors of non-decreasing functions is the Γ-limit of a functional that is defined on vectors of continuous non-decreasing functions. A criterion for the blow-up of a system of one-dimensional reaction-diffusion equations in a finite time is proposed by Eugenio Guerrero and José Alfredo López-Mimbela, where the criterion depends on the drift terms of the system of partial differential equations and on some measures which turn out to be invariant distributions of some diffusions that are associated with the system. Finally, Arno Siri-Jégousse and Linglong Yuan study the asymptotic behaviour, for small times, of the largest block size of Beta-n-coalescents as n increases.

In summary, the high quality and variety of these contributions give a broad panorama of the rich academic programme of the symposium and of its impact. It is worth noting that all papers, including the lecture notes of the invited courses, were subject to a strict peer review process with high international standards. We are very grateful to the referees, many of whom are leading experts in their fields, for their diligent and useful reports. Their comments were implemented by the authors and considerably improve the material presented herein.

We would also like to express our gratitude to all the authors whose original contributions are published in this book, as well as to all the speakers and session organizers of the symposium for their stimulating talks and support. Their valuable contributions show the interest and activity in the area of probability and stochastic processes in Mexico.

We hold in high regard the editors of the book series *Progress in Probability*, Steffen Dereich, Davar Khoshnevisan, Andreas E. Kyprianou and Sidney I. Resnick, for giving us the opportunity to publish the symposium volume in this prestigious series.

Special thanks to the symposium venue Universidad Autónoma de Yucatán and its staff for their great hospitality and for providing excellent conference facilities. We are also indebted to Rosy Davalos, whose outstanding organizational work permitted us to focus on the academic aspects of the conference.

The symposium as well as this volume would not have been possible without the generous support of our sponsors: Centro de Investigación en Matemáticas, RED-CONACYT Matemáticas y Desarrollo, Laboratorio Internacional Solomon Lefschetz CNRS-CONACYT, Instituto de Investigaciones en Matemáticas Aplicadas y en Sistemas and Instituto de Matemáticas at UNAM as well as Universidad Autónoma de Yucatán.

Finally, we hope that the reader of this volume will enjoy learning about the various topics that are treated therein, as much as we did editing it.

Guanajuato, Mexico — Daniel Hernández-Hernández
Guanajuato, Mexico — Juan Carlos Pardo
Guanajuato, Mexico — Victor Rivero

Previous Volumes from the Symposium on Probability and Stochastic Processes

- M. E. Caballero and L. G. Gorostiza, editors. *Simposio de Probabilidad y Procesos Estocásticos*, volume 4 of *Aportaciones Matemáticas: Notas de Investigación [Mathematical Contributions: Research Notes]*. Sociedad Matemática Mexicana, México, 1989.
  Held at Universidad de Guanajuato, Guanajuato, December 12–14, 1988.
- M. E. Caballero and L. G. Gorostiza, editors. *II Simposio de Probabilidad y Procesos Estocásticos. I Encuentro México-Chile de Análisis Estocástico*, volume 7 of *Aportaciones Matemáticas: Notas de Investigación [Mathematical Contributions: Research Notes]*. Sociedad Matemática Mexicana, México, 1992.
  Held at Universidad de Guanajuato, Guanajuato, January 27–31, 1992.
- M. E. Caballero and L. G. Gorostiza, editors. *III Simposio de Probabilidad y Procesos Estocásticos*, volume 11 of *Aportaciones Matemáticas: Notas de Investigación [Mathematical Contributions: Research Notes]*. Sociedad Matemática Mexicana, México, 1994.
  Held at Universidad de Sonora, Hermosillo, March 7–11, 1994.
- L. G. Gorostiza, J. A. León and J. A. López-Mimbela, editors. *IV Simposio de Probabilidad y Procesos Estocásticos*, volume 12 of *Aportaciones Matemáticas: Notas de Investigación [Mathematical Contributions: Research Notes]*. Sociedad Matemática Mexicana, México, 1996.
  Held at Centro de Investigación en Matemáticas, Guanajuato, April 1–5, 1996.
- J. M. González Barrios and L. G. Gorostiza, editors. *Modelos estocásticos*, volume 14 of *Aportaciones Matemáticas: Investigación [Mathematical Contributions: Research]*. Sociedad Matemática Mexicana, México, 1998. Papers from the 5th Symposium on Probability and Stochastic Processes.
  Held at Centro de Investigación en Matemáticas, Guanajuato, March 29 to April 3, 1998.
- D. Hernández, J. A. López-Mimbela and R. Quezada, editors. *Modelos estocásticos II*, volume 16 of *Aportaciones Matemáticas: Investigación [Mathematical Contributions: Research]*. Sociedad Matemática Mexicana, México, 2001. Papers from the 6th Symposium on Probability and Stochastic Processes.
  Held at Centro de Investigación en Matemáticas, Guanajuato, May 23–27, 2000.
- J. M. González-Barrios, J. A. León and A. Meda, editors. *Stochastic models*, volume 336 of *Contemporary Mathematics*. American Mathematical Society, Providence, RI; Sociedad Matemática Mexicana, México, 2003. Aportaciones Matemáticas. [Mathematical Contributions]. Proceedings of the 7th Symposium on Probability and Stochastic Processes.
  Held at the National University of Mexico, Mexico City, June 23–28, 2002.
- M. Bladt, J. A. Lopez Mimbela, J. Ruiz de Chavez and R. Navarro, editors. *Special issue: Proceedings of the 8th Symposium on Probability and Stochastic Processes*, volume 22 of *Stochastic Models*, 2006.
  Held at Universidad de las Americas in Puebla, June 20–25, 2004.

- M. E. Caballero, V. Rivero and J. Ruiz de Chávez, editors. *Ninth Symposium on Probability and Stochastic Processes*, volume 24 of *Stochastic Models*, 2008.
  Held at Centro de Investigación en Matemáticas, Guanajuato, November 20–24, 2006.
- M. E. Caballero, L. Chaumont, D. Hernández-Hernández and V. Rivero, editors. *X Symposium on Probability and Stochastic Processes and the First Joint Meeting France-Mexico of Probability*, volume 31 of *ESAIM Proceedings*. EDP Sciences, Les Ulis, 2011.
  Held at Centro de Investigación en Matemáticas, Guanajuato, November 3–7, 2008.
- R. H. Mena, J. C. Pardo, V. Rivero and G. Uribe Bravo, editors. *XI Symposium on Probability and Stochastic Processes*, volume 69 of *Progress in Probability*. Birkhäuser/Springer, Cham, 2015.
  Held at Centro de Investigación en Matemáticas, Guanajuato, México, November 18–22, 2013.

Contents

Part I
Courses

Scaling Limits of Markov-Branching Trees and Applications

Lecture Notes of the XII Simposio de Probabilidad y Procesos Estocásticos 16–20 Novembre 2015, Mérida, Yucatán

Bénédicte Haas

Abstract The goal of these lecture notes is to survey some of the recent progress on the description of large-scale structure of random trees. We use the framework of Markov-Branching sequences of trees and discuss several applications.

Keywords Random trees · Scaling limits · Self-similar fragmentations · self-similar Markov processes

Mathematics Subject Classification 05C05, 60F17, 60J05, 60J25, 60J80

1 Introduction

The goal of these lecture notes is to survey some of the recent progress on the description of large-scale structure of random trees. Describing the structure of large (random) trees, and more generally large graphs, is an important goal of modern probabilities and combinatorics. Beyond the purely probabilistic or combinatorial aspects, motivations come from the study of models from biology, theoretical computer science or mathematical physics.

The question we will typically be interested in is the following. For $(T_n, n \geq 1)$ a sequence of random unordered (i.e. non-planar) trees, where, for each n, T_n is a tree of size n (the size of a tree may be its number of vertices or its number of leaves, for example): does there exist a deterministic sequence $(a_n, n \geq 1)$ and a *continuous*

B. Haas (✉)
Université Paris 13, Sorbonne Paris Cité, LAGA, CNRS (UMR 7539), Villetaneuse, France
e-mail: haas@math.univ-paris13.fr

© Springer International Publishing AG, part of Springer Nature 2018

D. Hernández-Hernández et al. (eds.), *XII Symposium of Probability and Stochastic Processes*, Progress in Probability 73,
https://doi.org/10.1007/978-3-319-77643-9_1

random tree $\mathcal{T}$ such that

$$\frac{T_n}{a_n} \underset{n\to\infty}{\longrightarrow} \mathcal{T}?$$

To make sense of this question, we will view T_n as a metric space by "replacing" its edges with segments of length 1, and then use the notion of Gromov-Hausdorff distance to compare compact metric spaces. When such a convergence holds, the continuous limit highlights some properties of the discrete objects that approximate it, and vice-versa.

As a first example, consider T_n a tree picked uniformly at random in the set of trees with n vertices labelled by $\{1, \ldots, n\}$. The tree T_n has to be understood as a *typical element* of this set of trees. In this case the answer to the previous question dates back to a series of works by Aldous in the beginning of the 1990s [8–10]: Aldous showed that

$$\frac{T_n}{2\sqrt{n}} \overset{\text{(d)}}{\underset{n\to\infty}{\longrightarrow}} \mathcal{T}_{\text{Br}} \tag{1}$$

where the limiting tree is called the Brownian Continuum Random Tree (CRT), and can be constructed from a standard Brownian excursion. This result has various interesting consequences, e.g. it gives the asymptotics in distribution of the diameter, the height (if we consider rooted versions of the trees) and several other statistics related to the tree T_n. Consequently it also gives the asymptotic proportion of trees with n labelled vertices that have a diameter larger than $x\sqrt{n}$ or/and a height larger than $y\sqrt{n}$, etc. Some of these questions on statistics of uniform trees were already treated in previous works, the strength of Aldous's result is that it describes the asymptotics of the *whole* tree T_n.

Aldous has actually established a version of the convergence (1) in a much broader context, that of *conditioned Galton–Watson trees* with finite variance. In this situation, to fit to our context, T_n is an unordered version of the genealogical tree of a Galton–Watson process (with a given, fixed offspring distribution with mean one and finite variance) conditioned on having a total number of vertices equal to $n, n \geq 1$. Multiplied by $1/\sqrt{n}$, this tree converges in distribution to the Brownian CRT multiplied by a constant that only depends on the variance of the offspring distribution. This should be compared with (and is related to) the convergence of rescaled sums of i.i.d. random variables towards the normal distribution and its functional analog, the convergence of rescaled random walks towards the Brownian motion. It turns out that the above sequence of uniform labelled trees can be seen as a sequence of conditioned Galton–Watson trees (when the offspring distribution is a Poisson distribution) and more generally that several sequences of *combinatorial* trees reduce to conditioned Galton–Watson trees. In the early 2000s, Duquesne [44] extended Aldous's result to conditioned Galton–Watson trees with offspring distributions in the domain of attraction of a stable law. We also refer to [46, 70] for related results. In most of these cases the scaling sequences (a_n) are asymptotically much smaller, i.e. $a_n \ll \sqrt{n}$, and other continuous trees arise in the limit, the so-called family of *stable Lévy trees*. All these results on conditioned Galton–Watson trees are now well established, and have a lot of applications in the study of large random graphs (see e.g. Miermont's book [78] for the connections with random maps and Addario-Berry et al. [4] for connections with Erdős–Rényi random graphs in the critical window).

The classical proofs to establish the scaling limits of Galton–Watson trees consist in considering specific ordered versions of the trees and rely on a careful study of their so-called *contour functions*. It is indeed a common approach to encode trees into functions (similarly to the encoding of the Brownian tree by the Brownian excursion), which are more familiar objects. It turns out that for Galton–Watson trees, the contour functions are closely related to random walks, whose scaling limits are well known. Let us also mention that another common approach to study large random combinatorial structures is to use technics of analytic combinatorics, see [54] for a complete overview of the topic. None of these two methods will be used here.

In these lecture notes, we will focus on another point of view, that of sequences of random trees that satisfy a certain *Markov-Branching property*, which appears naturally in a large set of models and includes conditioned Galton–Watson trees. This property is a sort of discrete fragmentation property which roughly says that in each tree of the sequence, the subtrees above a given height are independent with a law that depends only on their total size. Under appropriate assumptions, we will see that Markov-Branching sequences of trees, suitably rescaled, converge to a family of continuous fractal trees, called the *self-similar fragmentation trees*. These continuous trees are related to the self-similar fragmentation processes studied by Bertoin in the 2000s [14], which are models used to describe the evolution of objects that randomly split as time passes. The main results on Markov-Branching trees presented here were developed in the paper [59], which has its roots in the earlier paper [63]. Several applications have been developed in these two papers, and in more recent works [15, 60, 89]: to Galton–Watson trees with arbitrary degree constraints, to several combinatorial trees families, including the Pólya trees (i.e. trees uniformly distributed in the set of rooted, unlabelled, unordered trees with n vertices, $n \geq 1$), to several examples of dynamical models of tree growth and to sequence of *cut-trees*, which describe the genealogy of some deletion procedure of edges in trees. The objective of these notes is to survey and gather these results, as well as further related results.

In Sect. 2 below, we will start with a series of definitions related to discrete trees and then present several classical examples of sequences of random trees. We will also introduce there the Markov-Branching property. In Sect. 3 we set up the topological framework in which we will work, by introducing the notions of real trees and Gromov–Hausdorff topology. We also recall there the classical results of Aldous [9] and Duquesne [44] on large conditioned Galton–Watson trees. Section 4 is the core of these lecture notes. We present there the results on scaling limits of Markov-Branching trees, and give the main ideas of the proofs. The key ingredient is the study of an integer-valued Markov chain describing the sizes of the subtrees containing a typical leaf of the tree. Section 5 is devoted to the applications mentioned above. Last, Sect. 6 concerns further perspectives and related models (multi-type trees, local limits, applications to other random graphs).

All the sequences of trees we will encounter here have a power growth. There is however a large set of random trees that naturally arise in applications that do not have such a behavior. In particular, many models of trees arising in the analysis of algorithms have a logarithmic growth. See e.g. Drmota's book [42] for an overview of the most classical models. These examples do not fit into our framework.

2 Discrete Trees, Examples and Motivations

2.1 *Discrete Trees*

Our objective is mainly to work with unordered trees. We give below a precise definition of these objects and mention nevertheless the notions of ordered or/and labelled trees to which we will sometimes refer.

A **discrete tree** (or graph-theoretic tree) is a finite or countable graph (V, E) that is connected and has no cycle. Here V denotes the set of vertices of the graph and E its set of edges. Note that two vertices are then connected by exactly one path and that $\#V = \#E + 1$ when the tree is finite.

In the following, we will often denote a (discrete) tree by the letter t, and for $\mathsf{t} = (V, E)$ we will use the slight abuse of notation $v \in \mathsf{t}$ to mean $v \in V$.

A tree t can be seen as a metric space, when endowed with the **graph distance** d_{gr}: given two vertices $u, v \in \mathsf{t}$, $d_{\mathrm{gr}}(u, v)$ is defined as the number of edges of the unique path from u to v.

A **rooted tree** (t, ρ) is an ordered pair where t is a tree and $\rho \in \mathsf{t}$. The vertex ρ is then called the root of t. This gives a genealogical structure to the tree. The root corresponds to the generation 0, its neighbors can be interpreted as its children and form the generation 1, the children of its children form the generation 2, etc. We will usually call the **height** of a vertex its generation, and denote it by $\mathsf{ht}(v)$ (the height of a vertex is therefore its distance to the root). The height of the tree is then

$$\mathsf{ht}(\mathsf{t}) = \sup_{v \in \mathsf{t}} \mathsf{ht}(v)$$

and its **diameter**

$$\mathsf{diam}(\mathsf{t}) = \sup_{u, v \in \mathsf{t}} d_{\mathrm{gr}}(u, v).$$

The **degree of a vertex** $v \in \mathsf{t}$ is the number of connected components obtained when removing v (in other words, it is the number of neighbors of v). A vertex v different from the root and of degree 1 is called a **leaf**. In a rooted tree, the **out-degree of a vertex** v is the number of children of v. Otherwise said, out-degree(v)=degree(v)-$\mathbb{1}_{\{v \neq \mathrm{root}\}}$. A (full) **binary** tree is a rooted tree where all vertices but the leaves have out-degree 2. A **branch-point** is a vertex of degree at least 3.

In these lecture notes, we will mainly work with rooted trees. Moreover we will consider, unless specifically mentioned, that two isomorphic trees are equal, or, when the trees are rooted, that **two root-preserving isomorphic trees are equal**. Such trees can be considered as *unordered unlabelled* trees, in opposition to the following definitions.

Ordered or/and Labelled Trees In the context of rooted trees, it may happen that one needs to order the children of the root, and then, recursively, the children of each vertex in the tree. This gives an **ordered** (or planar) tree. Formally, we generally see such a tree as a subset of the infinite Ulam–Harris tree

$$\mathcal{U} = \bigcup_{n=0}^{\infty} \mathbb{N}^n$$

where $\mathbb{N} := \{1, 2, \dots\}$ and $\mathbb{N}^0 = \{\varnothing\}$. The element $\varnothing$ is the root of the Ulam–Harris tree, and any other $u = u_1u_2 \dots u_n \in \mathcal{U}\backslash\{\varnothing\}$ is connected to the root via the unique shortest path

$$\varnothing \to u_1 \to u_1u_2 \to \dots \to u_1 \dots u_n.$$

The height (or generation) of such a sequence u is therefore its length, n. We then say that $\mathrm{t} \subset \mathcal{U}$ is a (finite or infinite) rooted *ordered* tree if:

- $\varnothing \in \mathrm{t}$
- if $u = u_1 \dots u_n \in \mathrm{t}\backslash\{\varnothing\}$, then $u = u_1 \dots u_{n-1} \in \mathrm{t}$ (the parent of an individual in t that is not the root is also in t)
- if $u = u_1 \dots u_n \in \mathrm{t}$, there exists an integer $c_u(\mathrm{t}) \geq 0$ such that the element $u_1 \dots u_n j \in \mathrm{t}$ if and only if $1 \leq j \leq c_u(\mathrm{t})$.

The number $c_u(\mathrm{t})$ corresponds to the number of children of u in t, i.e., its out-degree.

We will also sometimes consider **labelled** trees. In these cases, the vertices are labelled in a bijective way, typically by $\{1, \dots, n\}$ if there are n vertices (whereas in an unlabelled tree, the vertices but the root are indistinguishable). Partial labelling is also possible, e.g. by labelling only the leaves of the tree.

In the following we will always specify when a tree is ordered or/and labelled. When not specified, it is implicitly unlabelled, unordered.

Counting Trees It is sometimes possible, but not always, to have explicit formulæ for the number of trees of a specific structure. For example, it is known that the number of trees with n labelled vertices is

$$n^{n-2} \quad \text{(Cayley formula)},$$

and consequently, the number of rooted trees with n labelled vertices is

$$n^{n-1}.$$

The number of rooted ordered binary trees with $n+1$ leaves is

$$\frac{1}{n+1}\binom{2n}{n}$$

(this number is called the nth Catalan number) and the number of rooted ordered trees with n vertices is

$$\frac{1}{n}\binom{2n-2}{n-1}.$$

On the other hand, there is no explicit formula for the number of rooted (unlabelled, unordered) trees. Otter [79] shows that this number is asymptotically proportional to

$$c\kappa^n n^{-3/2}$$

where $c \sim 0.4399$ and $\kappa \sim 2.9557$. This should be compared to the asymptotic expansion of the nth Catalan number, which is proportional (by Stirling's formula) to $\pi^{-1/2}4^n n^{-3/2}$.

We refer to the book of Drmota [42] for more details and technics, essentially based on generating functions.

2.2 *First Examples*

We now present a first series of classical families of random trees. Our goal will be to describe their scaling limits when the sizes of the trees grow, as discussed in the Introduction. This will be done in Sect. 5. Most of these families (but not all) share a common property, the Markov-Branching property that will be introduced in the next section.

Combinatorial Trees Let $\mathbb{T}_n$ denote a finite set of trees with n vertices, all sharing some structural properties. E.g. $\mathbb{T}_n$ may be the set of all rooted trees with n vertices, or the set of all rooted ordered trees with n vertices, or the set of all binary trees with n vertices, etc. We are interested in the asymptotic behavior of a "typical element" of $\mathbb{T}_n$ as $n \to \infty$. That is, we pick a tree *uniformly* at random in $\mathbb{T}_n$, denote it by T_n and study its scaling limit. The global behavior of T_n as $n \to \infty$ will represent some of the features shared by most of the trees. For example, if the probability that the height of T_n is larger than $n^{\frac{1}{2}+\varepsilon}$ tends to 0 as $n \to \infty$, this means that the proportion of trees in the set that have a height larger than $n^{\frac{1}{2}+\varepsilon}$ is asymptotically negligible, etc. We will more specifically be interested in the following cases:

- T_n is a uniform rooted tree with n vertices
- T_n is a uniform rooted ordered tree with n vertices

- T_n is a uniform tree with n labelled vertices
- T_n is a uniform rooted ordered binary tree with n vertices (n odd)
- T_n is a uniform rooted binary tree with n vertices (n odd),

etc. Many variations are of course possible, in particular one may consider trees picked uniformly amongst sets of trees with a given structure and n *leaves*, or more general degree constraints. Some of these uniform trees will appear again in the next example.

Galton–Watson Trees Galton–Watson trees are random trees describing the genealogical structure of Galton–Watson processes. These are simple mathematical models for the evolution of a population that continue to play an important role in probability theory and in applications. Let η be a probability on $\mathbb{Z}_+$ (η is called the *offspring distribution*) and let $m := \sum_{i \geq 1} i\eta(i) \in [0, \infty]$ denote its mean. Informally, in a Galton–Watson tree with offspring distribution η, each vertex has a random number of children distributed according to η, independently. We will always assume that $\eta(1) < 1$ in order to avoid the trivial case where each individual has a unique child. Formally, an η-Galton–Watson tree T^η is usually seen as an ordered rooted tree and defined as follows (recall the Ulam–Harris notation $\mathcal{U}$):

- $c_\varnothing(T^\eta)$ is distributed according to η
- conditionally on $c_\varnothing(T^\eta) = p$, the p ordered subtrees $\tau_i = \{u \in \mathcal{U} : iu \in T^\eta\}$ descending from $i = 1, \ldots, p$ are independent and distributed as T^η.

From this construction, one sees that the distribution of T^η is given by:

$$\mathbb{P}\left(T^\eta = \mathtt{t}\right) = \prod_{v \in \mathtt{t}} \eta_{c_v(\mathtt{t})} \tag{2}$$

for all rooted ordered tree $\mathtt{t}$. This definition of Galton–Watson trees as ordered trees is the simplest, avoiding any symmetry problems. However in the following we will mainly see these trees up to isomorphism, which roughly means that we can "forget the order".

Clearly, if we call Z_k the number of individuals at height k, then $(Z_k, k \geq 1)$ is a Galton–Watson process starting from $Z_0 = 1$. It is well known that the extinction time of this process,

$$\inf\{k \geq 0 : Z_k = 0\}$$

if finite with probability 1 when $m \leq 1$ and with a probability $\in [0, 1)$ when $m > 1$. The offspring distribution η and the tree T^η are said to be *subcritical* when $m < 1$, *critical* when $m = 1$ and *supercritical* when $m > 1$. From now on, we assume that

$$m = 1$$

and for integers n such that $\mathbb{P}(\#T^\eta = n) > 0$, we let $T_n^{\eta,\mathrm{v}}$ denote a *non-ordered* version of the Galton–Watson tree T^η conditioned to have n vertices. Sometimes, we will need to keep the order and we will let $T_n^{\eta,\mathrm{v,ord}}$ denote this ordered conditioned version. We point out that in most cases, *but not all*, a subcritical or a supercritical Galton–Watson tree conditioned to have n vertices is distributed as a critical Galton–Watson tree conditioned to have n vertices with a different offspring distribution. So the assumption $m = 1$ is not too restrictive. We refer to [66] for details on that point.

It turns out that conditioned Galton–Watson trees are closely related to combinatorial trees. Indeed, one can easily check with (2) that:

- if $\eta = \mathrm{Geo}(1/2)$, $T_n^{\eta,\mathrm{v,ord}}$ is uniform amongst the set of rooted ordered trees with n vertices
- if $\eta = \mathrm{Poisson}(1)$, $T_n^{\eta,\mathrm{v}}$ is uniform amongst the set of rooted trees with n labelled vertices
- if $\eta = \frac{1}{2}(\delta_0 + \delta_2)$, $T_n^{\eta,\mathrm{v,ord}}$ is uniform amongst the set of rooted ordered binary trees with n vertices.

We refer e.g. to Aldous [9] for additional examples.

Hence, studying the large-scale structure of conditioned Galton–Watson trees will also lead to results in the context of combinatorial trees. As mentioned in the Introduction, the scaling limits of large conditioned Galton–Watson trees are now well known. Their study has been initiated by Aldous [8–10] and then expanded by Duquesne [44]. This will be reviewed in Sect. 3. However, there are some sequences of combinatorial trees that *cannot* be reinterpreted as Galton–Watson trees, starting with the example of the uniform rooted tree with n vertices or the uniform rooted binary tree with n vertices. Studying the scaling limits of these trees remained open for a while, because of the absence of symmetry properties. These scaling limits are presented in Sect. 5.2.

In another direction, one may also wonder what happens when considering versions of Galton–Watson trees conditioned to have n leaves, instead of n vertices, or more general degree constraints. This is discussed in Sect. 5.1.2.

Dynamical Models of Tree Growth We now turn to several sequences of finite rooted random trees that are built recursively by adding at each step new edges on the pre-existing tree. We start with a well known algorithm that Rémy [88] introduced to generate uniform binary trees with n leaves.

Rémy's Algorithm The sequence $(T_n(\mathrm{R}), n \geq 1)$ is constructed recursively as follows:

- Step 1: $T_1(\mathrm{R})$ is the tree with one edge and two vertices: one root, one leaf
- Step n: given $T_{n-1}(\mathrm{R})$, choose uniformly at random one of its edges and graft on "its middle" one new edge-leaf. By this we mean that the selected edge is split into two so as to obtain two edges separated by a new vertex, and then a new edge-leaf is glued on the new vertex. This gives $T_n(\mathrm{R})$.

It turns out (see e.g. [88]) that the tree $T_n(\mathrm{R})$, to which has been subtracted the edge between the root and the first branch point, is distributed as a binary critical Galton–Watson tree conditioned to have $2n - 1$ vertices, or equivalently n leaves (after forgetting the order in the GW-tree). As so, we deduce its asymptotic behavior from that of Galton–Watson trees. However this model can be extended in several directions, most of which are not related to Galton–Watson trees. We detail three of them.

Ford's α-Model [55] Let $\alpha \in [0, 1]$. We construct a sequence $(T_n(\alpha), n \geq 1)$ by modifying Rémy's algorithm as follows:

- Step 1: $T_1(\alpha)$ is the tree with one edge and two vertices: one root, one leaf
- Step n: given $T_{n-1}(\alpha)$, give a weight $1 - \alpha$ to each edge connected to a leaf, and α to all other edges (the internal edges). The total weight is $n - 1 - \alpha$. Now, if $n \neq 2$ or $\alpha \neq 1$, choose an edge at random with a probability proportional to its weight and graft on "its middle" one new edge-leaf. This gives $T_n(\alpha)$. When $n = 2$ and $\alpha = 1$ the total weight is 0 and we decide to graft anyway on the middle of the edge of T_1 one new edge-leaf.

Note that when $\alpha = 1/2$ the weights are the same on all edges and we recover Rémy's algorithm. When $\alpha = 0$, the new edge is always grafted uniformly on an edge-leaf, which gives a tree $T_n(0)$ known as the *Yule tree* with n leaves. When $\alpha = 1$, we obtain a deterministic tree called the *comb tree*. This family of trees indexed by $\alpha \in [0, 1]$ was introduced by Ford [55] in order to interpolate between the Yule, the uniform and the comb models. His goal was to propose new models for phylogenetic trees.

k-Ary Growing Trees [60] This is another extension of Rémy's algorithm, where now several edges are added at each step. Consider an integer $k \geq 2$. The sequence $(T_n(k), n \geq 1)$ is constructed recursively as follows:

- Step 1: $T_1(k)$ is the tree with one edge and two vertices: one root, one leaf
- Step n: given $T_{n-1}(k)$, choose uniformly at random one of its edges and graft on "its middle" $k - 1$ new edges-leaf. This gives $T_n(k)$.

When $k = 2$, we recover Rémy's algorithm. For larger k, there is no connection with Galton–Watson trees.

Marginals of Stable Trees: Marchal's Algorithm In [73], Marchal considered the following algorithm, that attributes weights also to the vertices. Fix a parameter $\beta \in (1, 2]$ and construct the sequence $(T_n(\beta), n \geq 1)$ as follows:

- Step 1: $T_1(\beta)$ is the tree with one edge and two vertices: one root, one leaf
- Step n: given $T_{n-1}(\beta)$, attribute the weight
  - $\beta - 1$ on each edge
  - $d - 1 - \beta$ on each vertex of degree $d \geq 3$.

  The total weight is $n\beta - 1$. Then select at random an edge or vertex with a probability proportional to its weight and graft on it a new edge-leaf. This gives $T_n(\beta)$.

The reason why Marchal introduced this algorithm is that $T_n(\beta)$ is actually distributed as the shape of a tree with edge-lengths that is obtained by sampling n leaves at random in the stable Lévy tree with index β. The class of stable Lévy trees plays in important role in the theory of random trees. It is introduced in Sect. 3.2 below.

Note that when $\beta = 2$, vertices of degree 3 are never selected (their weight is 0). So the trees $T_n(\beta), n \geq 1$ are all binary, and we recover Rémy's algorithm.

Of course, several other extensions of trees built by adding edges recursively may be considered, some of which are mentioned in Sects. 5.3.3 and 6.1.

Remark In these dynamical models of tree growth, we build *on a same probability space* the sequence of trees, contrary to the examples of Galton–Watson trees or combinatorial trees that give sequences of *distributions* of trees. In this situation, one may expect to have more than a convergence in distribution for the rescaled sequences of trees. We will see in Sect. 5.3 that it is indeed the case.

2.3 *The Markov-Branching Property*

Markov-Branching trees were introduced by Aldous [11] as a class of random binary trees for phylogenetic models and later extended to non-binary cases in Broutin et al. [30], and Haas et al. [63]. It turns out that many natural models of sequence of trees satisfy the **Markov-Branching property** (**MB-property** for short), starting with the example of conditioned Galton–Watson trees and most of the examples of the previous section.

Consider

$$\left(T_n, n \geq 1\right)$$

a sequence of trees where T_n is a rooted (unordered, unlabelled) tree with n *leaves*. The MB-property is a property of the sequence of *distributions* of $T_n, n \geq 1$. Informally, the MB-property says that for each tree T_n, *given that*

the root of T_n splits it in p subtrees with respectively $\lambda_1 \geq \ldots \geq \lambda_p$ leaves,

then T_n is distributed as the tree obtained by gluing on a common root p independent trees with respective distributions those of $T_{\lambda_1}, \ldots, T_{\lambda_p}$. The way the leaves are distributed in the sub-trees above the root, in each T_n, for $n \geq 1$, will then allow to fully describe the distributions of the $T_n, n \geq 1$.

We now explain rigorously how to build such sequences of trees. We start with a sequence of probabilities $(q_n, n \geq 1)$, where for each n, q_n is a probability on the set of partitions of the integer n. If $n \geq 2$, this set is defined by

$$\mathcal{P}_n := \left\{\lambda = (\lambda_1, \ldots, \lambda_p), \lambda_i \in \mathbb{N}, \lambda_1 \geq \ldots \geq \lambda_p \geq 1 : \sum_{i=1}^{p} \lambda_i = n\right\},$$

whereas if $n = 1$, $\mathcal{P}_1 := \{(1), \emptyset\}$ (we need to have a cemetery point). For a partition $\lambda \in \mathcal{P}_n$, we denote by $p(\lambda)$ its length, i.e. the number of terms in the sequence λ. The probability q_n will determine how the n leaves of T_n are distributed into the subtrees above its root. We call such a probability a *splitting distribution*. In order that effective splittings occur, we will always assume that

$$q_n((n)) < 1, \quad \forall n \geq 1.$$

We need to define a notion of *gluing* of trees. Consider $\mathrm{t}_1, \ldots, \mathrm{t}_p$, p discrete rooted (unordered) trees. Informally, we want to glue them on a same common root in order to form a tree $\langle \mathrm{t}_1, \ldots, \mathrm{t}_p \rangle$ whose root splits into the p subtrees $\mathrm{t}_1, \ldots, \mathrm{t}_p$. Formally, this can e.g. be done as follows. Consider first ordered versions of the trees $\mathrm{t}_1^{\mathrm{ord}}, \ldots, \mathrm{t}_p^{\mathrm{ord}}$ seen as subsets of the Ulam–Harris tree $\mathcal{U}$ and then define a new ordered tree by

$$\langle \mathrm{t}_1^{\mathrm{ord}}, \ldots, \mathrm{t}_p^{\mathrm{ord}} \rangle := \{\varnothing\} \cup_{i=1}^{p} i\mathrm{t}_i^{\mathrm{ord}}.$$

The tree $\langle \mathrm{t}_1, \ldots, \mathrm{t}_p \rangle$ is then defined as the unordered version of $\langle \mathrm{t}_1^{\mathrm{ord}}, \ldots, \mathrm{t}_p^{\mathrm{ord}} \rangle$.

Definition 2.1 For each $n \geq 1$, let q_n be a probability on $\mathcal{P}_n$ such that $q_n((n)) < 1$. From the sequence $\mathbf{q} = (q_n, n \geq 1)$ we construct recursively a sequence of distributions $(\mathcal{L}_n^{\mathbf{q}})$ such that for all $n \geq 1$, $\mathcal{L}_n^{\mathbf{q}}$ is carried by the set of rooted trees with n leaves, as follows:

- $\mathcal{L}_1^{\mathbf{q}}$ is the distribution of a line-tree with $G + 1$ vertices and G edges where G is a geometric distribution:

$$\mathbb{P}(G = k) = q_1(\emptyset)(1 - q_1(\emptyset))^k, \quad k \geq 0,$$

- for $n \geq 2$, $\mathcal{L}_n^{\mathbf{q}}$ is the distribution of

$$\langle T_1, \ldots, T_{p(\Lambda)} \rangle$$

where Λ is a partition of n distributed according to q_n, and given Λ, the trees $T_1, \ldots, T_{p(\Lambda)}$ are independent with respective distributions $\mathcal{L}_{\Lambda_1}^{\mathbf{q}}, \ldots, \mathcal{L}_{\Lambda_{p(\Lambda)}}^{\mathbf{q}}$.

A sequence $(T_n, n \geq 1)$ of random rooted trees such that $T_n \sim \mathcal{L}_n^{\mathbf{q}}$ for each $n \in \mathbb{N}$ is called a ***MB-sequence of trees* indexed by the leaves, *with splitting distributions* $(q_n, n \geq 1)$**.

This construction may be re-interpreted as follows: we start from a collection of n *indistinguishable* balls, and with probability $q_n(\lambda_1, \ldots, \lambda_p)$, split the collection into p sub-collections with $\lambda_1, \ldots, \lambda_p$ balls. Note that there is a chance $q_n((n)) < 1$ that the collection remains unchanged during this step of the procedure. Then, re-iterate the splitting operation independently for each sub-collection using this time the probability distributions $q_{\lambda_1}, \ldots, q_{\lambda_p}$. If a sub-collection consists of a single

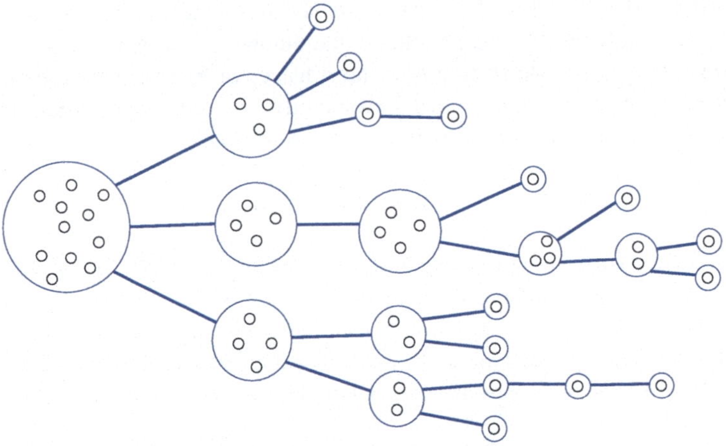

Fig. 1 A sample tree T_{11}. The first splitting arises with probability $q_{11}(4, 4, 3)$

ball, it can remain single with probability $q_1((1))$ or get wiped out with probability $q_1(\emptyset)$. We continue the procedure until all the balls are wiped out. The tree T_n is then the genealogical tree associated with this process: it is rooted at the initial collection of n balls and its n leaves correspond to the n isolated balls just before they are wiped out, See Fig. 1 for an illustration.

We can define similarly MB-sequences of (distributions of) trees indexed by their number of *vertices*. Consider here a sequence $(p_n, n \geq 1)$ such that p_n is a probability on $\mathcal{P}_n$ with no restriction but

$$p_1((1)) = 1.$$

Mimicking the previous balls construction and starting from a collection of n indistinguishable balls, we first remove a ball, split the $n-1$ remaining balls in sub-collections with $\lambda_1, \ldots, \lambda_p$ balls with probability $p_{n-1}((\lambda_1, \ldots, \lambda_p))$, and iterate independently on sub-collections until no ball remains. Formally, this gives:

Definition 2.2 For each $n \geq 1$, let p_n be a probability on $\mathcal{P}_n$, such that $p_1((1)) = 1$. From the sequence $(p_n, n \geq 1)$ we construct recursively a sequence of distributions $(\mathcal{V}_n^{\mathbf{p}})$ such that for all $n \geq 1$, $\mathcal{V}_n^{\mathbf{p}}$ is carried by the set of trees with n vertices, as follows:

- $\mathcal{V}_1^{\mathbf{p}}$ is the deterministic distribution of the tree reduced to one vertex,
- for $n \geq 2$, $\mathcal{V}_n^{\mathbf{p}}$ is the distribution of

$$\langle T_1, \ldots, T_{p(\Lambda)} \rangle$$

where Λ is a partition of $n-1$ distributed according to p_{n-1}, and given Λ, the trees $T_1, \ldots, T_{p(\Lambda)}$ are independent with respective distributions $\mathcal{V}_{\Lambda_1}^{\mathbf{p}}, \ldots, \mathcal{V}_{\Lambda_{p(\Lambda)}}^{\mathbf{p}}$.

A sequence $(T_n, n \geq 1)$ of random rooted trees such that $T_n \sim \mathcal{V}_n^{\mathbf{p}}$ for each $n \in \mathbb{N}$ is called a ***MB-sequence of trees indexed by the vertices, with splitting distributions*** $(p_n, n \geq 1)$.

More generally, the MB-property can be extended to sequences of trees $(T_n, n \geq 1)$ with arbitrary degree constraints, i.e. such that for all n, T_n has n vertices in A, where A is a given subset of $\mathbb{Z}_+$. We will not develop this here and refer the interested reader to [89] for more details.

Some Examples

1. **A deterministic example.** Consider the splitting distributions on $\mathcal{P}_n$

$$q_n(\lceil n/2 \rceil, \lfloor n/2 \rfloor) = 1, \quad n \geq 2,$$

as well as $q_1(\emptyset) = 1$. Let $(T_n, n \geq 1)$ the corresponding MB-sequence indexed by leaves. Then T_n is a deterministic discrete binary tree, whose root splits in two subtrees with both $n/2$ leaves when n is even, and respectively $(n+1)/2$, $(n-1)/2$ leaves when n is odd. Clearly, when $n = 2^k$, the height of T_n is exactly k, and more generally for large n, $\mathrm{ht}(T_n) \sim \ln(n)/\ln(2)$.

2. **A basic example.** For $n \geq 2$, let q_n be the probability on $\mathcal{P}_n$ defined by

$$q_n((n)) = 1 - \frac{1}{n^\alpha} \quad \text{and} \quad q_n(\lceil n/2 \rceil, \lfloor n/2 \rfloor) = \frac{1}{n^\alpha} \quad \text{for some } \alpha > 0,$$

and let $q_1(\emptyset) = 1$. Let $(T_n, n \geq 1)$ be an MB-sequence indexed by leaves with splitting distributions (q_n). Then T_n is a discrete tree with vertices with degrees $\in \{1, 2, 3\}$ where the distance between the root and the first branch point (i.e. the first vertex of degree 3) is a Geometric distribution on $\mathbb{Z}_+$ with success parameter $n^{-\alpha}$. The two subtrees above this branch point are independent subtrees, independent of the Geometric r.v. just mentioned, and whose respective distances between the root and first branch point are Geometric distributions with respectively $(\lceil n/2 \rceil)^{-\alpha}$ and $(\lfloor n/2 \rfloor)^{-\alpha}$ parameters. Noticing the weak convergence

$$\frac{\mathrm{Geo}(n^{-\alpha})}{n^\alpha} \xrightarrow[n \to \infty]{(\mathrm{d})} \mathrm{Exp}(1)$$

one may expect that $n^{-\alpha} T_n$ has a limit in distribution. We will later see that it is indeed the case.

3. **Conditioned Galton–Watson trees**. Let $T_n^{\eta,1}$ be a Galton–Watson tree with offspring distribution η, conditioned on having n leaves, for integers n for which this is possible. The branching property is then preserved by conditioning and the sequence $(T_n^{\eta,1}, n : \mathbb{P}(\#_{\mathrm{leaves}} T^\eta) > 0)$ is Markov-Branching, with splitting

distributions

$$q_n^{\mathrm{GW},\eta}(\lambda) = \eta(p) \times \frac{p!}{\prod_{i=1}^{p} m_i(\lambda)!} \times \frac{\prod_{i=1}^{p} \mathbb{P}(\#_{\mathrm{leaves}} T^{\eta} = \lambda_i)}{\mathbb{P}(\#_{\mathrm{leaves}} T^{\eta} = n)}$$

for all $\lambda \in \mathcal{P}_n, n \geq 2$, where $\#_{\mathrm{leaves}} T^{\eta}$ is the number of leaves of the unconditioned Galton–Watson tree T^{η}, and $m_i(\lambda) = \#\{1 \leq j \leq p : \lambda_j = i\}$. The probability $q_1^{\mathrm{GW},\eta}$ is given by $q_1^{\mathrm{GW},\eta}((1)) = \eta(1)$.

Similarly, if $T_n^{\eta,\mathrm{v}}$ denotes a Galton–Watson tree with offspring distribution η, conditioned on having n *vertices*, the sequence $(T_n^{\eta,\mathrm{v}}, \mathbb{P}(\#_{\mathrm{vertices}} T^{\eta}) > 0)$ is MB, with splitting distributions

$$p_{n-1}^{\mathrm{GW},\eta}(\lambda) = \eta(p) \times \frac{p!}{\prod_{i=1}^{p} m_i(\lambda)!} \times \frac{\prod_{i=1}^{p} \mathbb{P}(\#_{\mathrm{vertices}} T^{\eta} = \lambda_i)}{\mathbb{P}(\#_{\mathrm{vertices}} T^{\eta} = n)} \tag{3}$$

for all $\lambda \in \mathcal{P}_{n-1}, n \geq 3$ where $\#_{\mathrm{vertices}} T^{\eta}$ is the number of leaves of the unconditioned GW-tree T^{η}. Details can be found in [59, Section 5].

4. **Dynamical models of tree growth.** Rémy's, Ford's, Marchal's and the k-ary algorithms all lead to MB-sequences of trees indexed by leaves. To be precise, we have to remove in each of these trees the edge adjacent to the root to obtain MB-sequences of trees (the roots have all a unique child). The MB-property can be proved by induction on n. By construction, the distribution of the leaves in the subtrees above the root is closely connected to urns models. We have the following expressions for the splitting distributions:

   Ford's α-Model For $k \geq \frac{n}{2}, n \geq 2$,

$$q_n^{\mathrm{Ford},\alpha}(k, n-k) = \left(1 + \mathbb{1}_{k \neq \frac{n}{2}}\right) \frac{\Gamma(k-\alpha)\Gamma(n-k-\alpha)}{\Gamma(n-\alpha)\Gamma(1-\alpha)} \left(\frac{\alpha}{2}\binom{n}{k} + (1-2\alpha)\binom{n-2}{k-1}\right),$$

   and $q_1(\emptyset) = 1$. See [55] for details. In particular, taking $\alpha = 1/2$ one sees that

$$q_n^{\text{Rémy}}(k, n-k) = \frac{1}{4}\left(1 + \mathbb{1}_{k \neq \frac{n}{2}}\right) \frac{\Gamma(k-1/2)\Gamma(n-k-1/2)}{\Gamma(n-1/2)\Gamma(1-1/2)} \binom{n}{k}, \quad k \geq \frac{n}{2}, n \geq 2.$$

   k-Ary Growing Trees Note that in these models, there are $1+(k-1)(n-1)$ leaves in the tree $T_n(k)$, so that the indices do not exactly correspond to the definitions of the Markov-Branching properties seen in the previous section. However, by relabelling, defining for $m = 1 + (k-1)(n-1)$ the tree $\overline{T}_m(k)$ to be the tree $T_n(k)$ to which the edge adjacent to the root has been removed, we obtain an MB-sequence $(\overline{T}_m(k), m \in (k-1)\mathbb{N} + 2 - k)$. The splitting distributions are

defined for $m = 1 + (k-1)(n-1)$, $n \geq 2$ and $\lambda = (\lambda_1, \ldots, \lambda_k) \in \mathcal{P}_m$ such that $\lambda_i = 1 + (k-1)\ell_i$, for some $\ell_i \in \mathbb{Z}_+$ for all i (note that $\sum_{i=1}^k \ell_i = n-2$) by

$$q_m^k(\lambda) = \sum_{\mathbf{n}=(n_1,\ldots,n_k)\in\mathbb{N}^k : \mathbf{n}^{\downarrow}=\lambda} \overline{q}_m(\mathbf{n})$$

where $\mathbf{n}^{\downarrow}$ is the decreasing rearrangement of the elements of $\mathbf{n}$ and

$$\overline{q}_m(\mathbf{n}) = \frac{1}{k(\Gamma(\frac{1}{k}))^{k-1}} \left(\prod_{i=1}^{k} \frac{\Gamma(\frac{1}{k}+n_i)}{n_i!} \right) \times \frac{(n-2)!}{\Gamma(\frac{1}{k}+n-1)} \left(\sum_{j=1}^{n_1+1} \frac{n_1!}{(n_1-j+1)!} \frac{(n-j-1)!}{(n-2)!} \right).$$

See [60, Section 3].

Marchal's Algorithm For $\lambda = (\lambda_1, \ldots, \lambda_p) \in \mathcal{P}_n$, $n \geq 2$,

$$q_n^{\text{Marchal},\beta}(\lambda) = \frac{n!}{\lambda_1! \ldots \lambda_p! m_1(\lambda)! \ldots m_n(\lambda)!} \frac{\beta^{2-p}\Gamma(2-\beta^{-1})\Gamma(p-\beta)}{\Gamma(n-\beta^{-1})\Gamma(2-\beta)} \prod_{i=1}^{p} \frac{\Gamma(\lambda_j - \beta^{-1})}{\Gamma(1-\beta^{-1})}$$

where $m_i(\lambda) = \#\{1 \leq j \leq p : \lambda_j = i\}$. This is a consequence of [46, Theorem 3.2.1] and [75, Lemma 5].

5. **Cut-trees.** *Cut-tree of a uniform Cayley tree.* Consider C_n a uniform Cayley tree of size n, i.e. a tree picked uniformly at random amongst the set of rooted tree with n labelled vertices. This tree has the following *recursive property* (see Pitman [85, Theorem 5]): removing an edge uniformly at random in C_n gives two trees, which given their numbers of vertices, $k, n-k$ say, are independent uniform Cayley trees of respective sizes $k, n-k$. Now, consider the following deletion procedure: remove in C_n one edge uniformly at random, then remove another edge in the remaining set of $n-2$ edges uniformly at random and so on until all edges have been removed. It was shown by Janson [65] and Panholzer [83] that the number of steps needed to isolate the root divided by $\sqrt{n}$ converges in distribution to a Rayleigh distribution (i.e. with density $x\exp(-x^2/2)$ on $\mathbb{R}_+$). Bertoin [15] was more generally interested in the number of steps needed to isolate ℓ distinguished vertices, and in that aim he introduced the *cut-tree* T_n^{cut} of C_n. The tree T_n^{cut} is the genealogical tree of the above deletion procedure, i.e. it describes the genealogy of the connected components, see Fig. 2 for an illustration and [15] for a precise construction of T_n^{cut}. Let us just mention here that T_n^{cut} is a rooted binary tree with n leaves, and that Pitman's recursive property

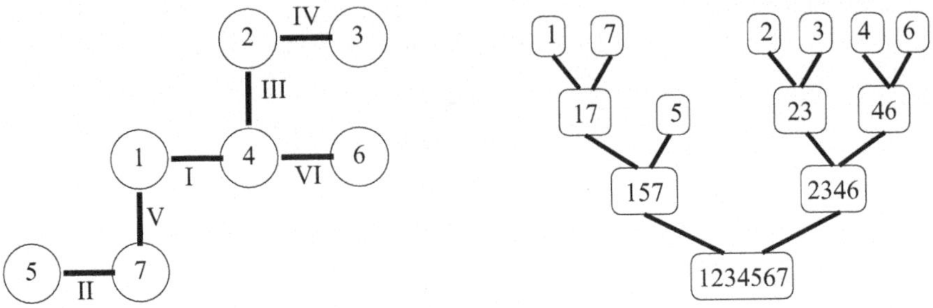

Fig. 2 On the left, a version of the tree C_7, with edges labelled in order of deletion. On the right the associated cut-tree T_7^{cut}, whose vertices are the different connected components arising in the deletion procedure

implies that $(T_n^{\text{cut}}, n \geq 1)$ is MB. The corresponding splitting probabilities are:

$$q_n^{\text{Cut,Cayley}}(k, n-k) = \frac{(n-k)^{n-k-1}}{(n-k)!} \frac{k^{k-1}}{k!} \frac{(n-2)!}{n^{n-3}}, \quad n/2 < k \leq n-1,$$

the calculations are detailed in [15, 84].

Cut-tree of a uniform recursive tree. A recursive tree with n vertices is a tree with vertices labelled by $1, \ldots, n$, rooted at 1, such that the sequence of labels of vertices along any branch from the root to a leaf is increasing. It turns out that the cut-tree of a uniform recursive tree is also MB and with splitting probabilities

$$q_n^{\text{Cut,Recursive}}(k, n-k) = \frac{n}{(n-1)} \left(\frac{1}{k(k+1)} + \frac{1}{(n-k)(n-k+1)} \right) \quad n/2 < k \leq n-1,$$

see [16].

Remark The first example is a simple example of models where *macroscopic branchings are frequent*, unlike the second example where *macroscopic branchings are rare* (they occur with probability $n^{-\alpha} \to 0$). By macroscopic branchings, we mean that the way that the n leaves (or vertices) are distributed above the root gives at least two subtrees with a size proportional to n. Although it is not completely obvious yet, nearly all other examples above have rare macroscopic branchings (in a sense that will be specified later) and this is typically the context in which we will study the scaling limits of MB-trees. Typically the tree T_n will then grow as a power of n. When macroscopic branchings are frequent, there is no scaling limit in general for the Gromov–Hausdorff topology, a topology introduced in the next section. However it is known that the height of the tree T_n is then often of order $c \ln(n)$. This case has been studied in [30].

3 The Example of Galton–Watson Trees and Topological Framework

We start with an informal version of the prototype result of Aldous on the description of the scaling limits of conditioned Galton–Watson trees. Let η be a critical offspring distribution with *finite variance* $\sigma^2 \in (0, \infty)$, and let $T_n^{\eta,\mathrm{v}}$ denote a Galton–Watson tree with offspring distribution η, conditioned to have n vertices (in the following it is implicit that we only consider integers n such that this conditioning is possible). Aldous [10] showed that

$$\frac{\sigma}{2} \times \frac{T_n^{\eta,\mathrm{v}}}{\sqrt{n}} \xrightarrow[n\to\infty]{(\mathrm{d})} \mathcal{T}_{\mathrm{Br}} \tag{4}$$

where the continuous tree $\mathcal{T}_{\mathrm{Br}}$ arising in the limit is the Brownian Continuum Random Tree, sometimes simply called the Brownian tree. Note that the limit only depends on η via its variance σ^2.

This result by Aldous was a breakthrough in the study of large random trees, since it was the first to describe the behavior of the tree as a whole. We will discuss this in more details in Sect. 3.2. Let us first introduce the topological framework in order to make sense of this convergence.

3.1 Real Trees and the Gromov–Hausdorff Topology

Since the pioneering works of Evans et al. [52] in 2003 and Duquesne and Le Gall [47] in 2005, the theory of *real trees* (or $\mathbb{R}$-trees) has been intensively used in probability. These trees are metric spaces having a "tree property" (roughly, this means that for each pair of points x, y in the metric space, there is a unique path going from x to y—see below for a precise definition). This point of view allows behavior such as infinite total length of the tree, vertices with infinite degree, and density of the set of leaves.

In these lecture notes, all the real trees we will consider are compact metric spaces. For this reason, we restrict ourselves to the theory of compact real trees. We now briefly recall background on real trees and the Gromov–Hausdorff and Gromov–Hausdorff–Prokhorov distances, and refer to [51, 71] for more details on this topic.

Real Trees A *real tree* is a metric space $(\mathcal{T}, d)$ such that, for any points x and y in $\mathcal{T}$,

- there is an isometry $\varphi_{x,y} : [0, d(x, y)] \to \mathcal{T}$ such that $\varphi_{x,y}(0) = x$ and $\varphi_{x,y}(d(x, y)) = y$
- for every continuous, injective function $c : [0, 1] \to \mathcal{T}$ with $c(0) = x$, $c(1) = y$, one has $c([0, 1]) = \varphi_{x,y}([0, d(x, y)])$.

Note that a discrete tree may be seen as a real tree by "replacing" its edges by line segments. Unless specified, it will be implicit throughout these notes that these line segments are all of length 1.

We denote by $[[x, y]]$ the line segment $\varphi_{x,y}([0, d(x, y)])$ between x and y. A *rooted* real tree is an ordered pair $((\mathcal{T}, d), \rho)$ such that $(\mathcal{T}, d)$ is a real tree and $\rho \in \mathcal{T}$. This distinguished point ρ is called the root. The height of a point $x \in \mathcal{T}$ is defined by

$$\mathsf{ht}(x) = d(\rho, x)$$

and the height of the tree itself is the supremum of the heights of its points, while the diameter is the supremum of the distance between two points:

$$\mathrm{ht}(\mathcal{T}) = \sup_{x \in \mathcal{T}} d(\rho, x) \qquad \mathrm{diam}(\mathcal{T}) = \sup_{x,y \in \mathcal{T}} d(x, y).$$

The *degree* of a point x is the number of connected components of $\mathcal{T}\backslash\{x\}$. We call *leaves* of $\mathcal{T}$ all the points of $\mathcal{T}\backslash\{\rho\}$ which have degree 1. Given two points x and y, we define $x \wedge y$ as the unique point of $\mathcal{T}$ such that $[[\rho, x]] \cap [[\rho, y]] = [[\rho, x \wedge y]]$. It is called the *branch point* of x and y if its degree is larger than or equal to 3. For $a > 0$, we define the rescaled tree $a\mathcal{T}$ as $(\mathcal{T}, ad)$ (the metric d thus being implicit and dropped from the notation).

As mentioned above, we will only consider compact real trees. We now want to measure how close two such metric spaces are. We start by recalling the definition of Hausdorff distance between compact subsets of a metric space.

Hausdorff Distance If A and B are two nonempty compact subsets of a metric space (E, d), the Hausdorff distance between A and B is defined by

$$d_{E,\mathrm{H}}(A, B) = \inf\left\{\varepsilon > 0\,;\; A \subset B^{\varepsilon} \text{ and } B \subset A^{\varepsilon}\right\},$$

where A^{ε} and B^{ε} are the closed ε-enlargements of A and B, i.e. $A^{\varepsilon} = \{x \in E : d(x, A) \leq \varepsilon\}$ and similarly for B^{ε}.

The Gromov–Hausdorff extends this concept to compact real trees (or more generally compact metric spaces) that are not necessarily compact subsets of a single metric space, by considering embeddings in an arbitrary common metric space.

Gromov–Hausdorff Distance Given two compact rooted trees $(\mathcal{T}, d, \rho)$ and $(\mathcal{T}', d', \rho')$, let

$$d_{\mathrm{GH}}(\mathcal{T}, \mathcal{T}') = \inf\left\{\max\left(d_{\mathcal{Z},\mathrm{H}}(\phi(\mathcal{T}), \phi'(\mathcal{T}')), d_{\mathcal{Z}}(\phi(\rho), \phi'(\rho'))\right)\right\},$$

where the infimum is taken over all pairs of isometric embeddings ϕ and ϕ' of $\mathcal{T}$ and $\mathcal{T}'$ in the same metric space $(\mathcal{Z}, d_{\mathcal{Z}})$, for all choices of metric spaces $(\mathcal{Z}, d_{\mathcal{Z}})$.

We will also be concerned with *measured* trees, that are real trees equipped with a probability measure on their Borel sigma-field. To this effect, recall first the definition of the Prokhorov distance between two probability measures μ and μ' on a metric space (E, d):

$$d_{E,\mathrm{P}}(\mu, \mu') = \inf\left\{\varepsilon > 0\,;\ \forall A \in \mathcal{B}(E),\, \mu(A) \leq \mu'(A^{\varepsilon}) + \varepsilon \text{ and } \mu'(A) \leq \mu(A^{\varepsilon}) + \varepsilon\right\}.$$

This distance metrizes the weak convergence on the set of probability measures on (E, d).

Gromov–Hausdorff–Prokhorov Distance Given two measured compact rooted trees $(\mathcal{T}, d, \rho, \mu)$ and $(\mathcal{T}', d', \rho', \mu')$, we let

$$d_{\mathrm{GHP}}(\mathcal{T}, \mathcal{T}') = \inf\left\{ \max\left(d_{\mathcal{Z},\mathrm{H}}(\phi(\mathcal{T}), \phi'(\mathcal{T}')), d_{\mathcal{Z}}(\phi(\rho), \phi'(\rho')), d_{\mathcal{Z},\mathrm{P}}(\phi_*\mu, \phi'_*\mu')\right)\right\},$$

where the infimum is taken on the same space as before and $\phi_*\mu$, $\phi'_*\mu'$ are the push-forwards of μ, μ' by ϕ, ϕ'.

The Gromov–Hausdorff distance d_{GH} indeed defines a distance on the set of compact rooted real trees taken up to root-preserving isomorphisms. Similarly, The Gromov–Hausdorff–Prokhorov distance d_{GHP} is a distance on the set of compact measured rooted real trees taken up to root-preserving and measure-preserving isomorphisms. Moreover these two metric spaces are Polish, see [52] and [3]. We will always identify two (measured) rooted $\mathbb{R}$-trees when they are isometric and still use the notation $(\mathcal{T}, d)$ (or $\mathcal{T}$ when the choice of the metric is clear) to design their isometry class.

Statistics It is easy to check that the function that associates with a compact rooted tree its diameter is continuous (with respect to the GH-topology on the set of compact rooted real trees and the usual topology on $\mathbb{R}$). Similarly, the function that associates with a compact rooted tree its height is continuous. The function that associates with a compact rooted measured tree the distribution of the height of a leaf chosen according to the probability on the tree is continuous as well (with respect to the GHP-topology on the set of compact rooted measured real trees and the weak topology on the set of probability measures on $\mathbb{R}$). Consequently, the existence of scaling limits with respect to the GHP-topology will directly imply scaling limits for the height, the diameter and the height of a typical vertex of the trees.

3.2 *Scaling Limits of Conditioned Galton–Watson Trees*

We can now turn to rigorous statements on the scaling limits of conditioned Galton–Watson trees. We reformulate the above result (4) by Aldous in the finite variance case and also present the result by Duquesne [44] when the offspring distribution η is heavy tailed, in the domain of attraction of a stable distribution. In the following,

η always denotes a critical offspring distribution, $T_n^{\eta,\mathrm{v}}$ is an η-GW tree conditioned to have n vertices, and $\mu_n^{\eta,\mathrm{v}}$ is the uniform probability on its vertices. The following convergences hold for the Gromov–Hausdorff–Prokhorov topology.

Theorem 3.1

(i) (Aldous [10]) *Assume that η has a finite variance σ^2. Then, there exists a random compact real tree, called the Brownian tree and denoted $\mathcal{T}_{\mathrm{Br}}$, endowed with a probability measure μ_{Br} supported by its set of leaves, such that as $n \to \infty$*

$$\left(\frac{\sigma T_n^{\eta,\mathrm{v}}}{2\sqrt{n}}, \mu_n^{\eta,\mathrm{v}}\right) \xrightarrow[\mathrm{GHP}]{\mathrm{(d)}} (\mathcal{T}_{\mathrm{Br}}, \mu_{\mathrm{Br}}).$$

(ii) (Duquesne [44]) *If $\eta_k \sim \kappa k^{-1-\alpha}$ as $k \to \infty$ for $\alpha \in (1,2)$, then there exists a random compact real tree $\mathcal{T}_\alpha$, called the* stable Lévy tree with index α, *endowed with a probability measure μ_α supported by its set of leaves, such that as $n \to \infty$*

$$\left(\frac{T_n^{\eta,\mathrm{v}}}{n^{1-1/\alpha}}, \mu_n^{\eta,\mathrm{v}}\right) \xrightarrow[\mathrm{GHP}]{\mathrm{(d)}} \left(\left(\frac{\alpha(\alpha-1)}{\kappa\Gamma(2-\alpha)}\right)^{1/\alpha} \alpha^{1/\alpha-1} \cdot \mathcal{T}_\alpha, \mu_\alpha\right).$$

The result by Duquesne actually extends to cases where the offspring distribution η is in the domain of attraction of a stable distribution with index $\alpha \in (1,2]$. See [44] for details.

The Brownian tree was first introduced by Aldous in the early 1990s in the series of papers [8–10]. This tree can be constructed in several ways, the most common being the following. Let $(\mathbf{e}(t), t \in [0,1])$ be a normalized Brownian excursion, which, formally, can be defined from a standard Brownian motion B by letting

$$\mathbf{e}(t) = \frac{|B_{g+t(d-g)}|}{\sqrt{d-g}}, \quad 0 \leq t \leq 1,$$

where $g := \sup\{s \leq 1 : B_s = 0\}$ and $d = \inf\{s \geq 1 : B_s = 0\}$ (note that $d - g > 0$ a.s. since $B_1 \neq 0$ a.s.). Then consider for $x, y \in [0,1], x \leq y$, the non-negative quantity

$$d_{\mathbf{e}}(x,y) = \mathbf{e}(x) + \mathbf{e}(y) - 2 \inf_{z \in [x,y]} \{\mathbf{e}(z)\},$$

and then the equivalent relation $x \sim_{\mathbf{e}} y \Leftrightarrow d_{\mathbf{e}}(x,y) = 0$. It turns out that the quotient space $[0,1]/\sim_{\mathbf{e}}$ endowed with the metric induced by $d_{\mathbf{e}}$ (which indeed gives a true metric) is a compact real tree. The Brownian excursion $\mathbf{e}$ is called the *contour function* of this tree. Equipped with the measure $\mu_{\mathbf{e}}$, which is the push-forward of the Lebesgue measure on $[0,1]$, this gives a version $([0,1]/\sim_{\mathbf{e}}, d_{\mathbf{e}}, \mu_{\mathbf{e}})$ of the measured tree $(\mathcal{T}_{Br}, \mu_{Br})$. To get a better intuition of what this means, as well as more details and other constructions of the Brownian tree, we refer to the three papers by Aldous [8–10] and to the survey by Le Gall [71].

In the early 2000s, the family of stable Lévy trees $(\mathcal{T}_\alpha, \alpha \in (1,2)]$—where by convention $\mathcal{T}_2$ is $\sqrt{2} \cdot \mathcal{T}_{\mathrm{Br}}$—was introduced by Duquesne and Le Gall [46, 47] in the more general framework of *Lévy trees*, building on earlier work of Le Gall and Le Jan [72]. These trees can be constructed in a way similar as above from continuous functions built from the stable Lévy processes. This construction is complex and we will not detail it here. Others constructions are possible, see e.g. [50, 56]. The stable trees are important objects of the theory of random trees. They are intimately related to continuous state branching processes, fragmentation and coalescence processes. They appear as scaling limits of various models of trees and graphs, starting with the Galton–Watson examples above and some other examples discussed in Sect. 5. In particular, it is noted that it was only proved recently that Galton–Watson trees conditioned by their number of leaves or more general arbitrary degree restrictions also converge in the scaling limit to stable trees, see Sect. 5.1.2 and the references therein.

In the last few years, the geometric and fractal aspects of stable trees have been studied in great detail: Hausdorff and packing dimensions and measures [45, 47, 48, 57]; spectral dimension [34]; spinal decompositions and invariance under uniform re-rooting [49, 64]; fragmentation into subtrees [75, 76]; and embeddings of stable trees into each other [35]. We simply point out it here that the Brownian tree is *binary*, in the sense that all its points have their degree in $\{1, 2, 3\}$ almost surely, whereas the stable trees $\mathcal{T}_\alpha, \alpha \in (1,2)$ have only points with degree in $\{1, 2, \infty\}$ almost surely (every branch point has an infinite number of "children").

Applications to Combinatorial Trees Using the connections between some families of combinatorial trees and Galton–Watson trees mentioned in Sect. 2.1, we obtain the following scaling limits (in all cases, μ_n denotes the uniform probability on the vertices of the tree T_n):

- If T_n is uniform amongst the set of rooted ordered trees with n vertices,

$$\left(\frac{T_n}{\sqrt{n}}, \mu_n\right) \xrightarrow[\mathrm{GHP}]{(\mathrm{d})} \left(\mathcal{T}_{B_r}, \mu_{\mathrm{Br}}\right).$$

- If T_n is uniform amongst the set of rooted trees with n labelled vertices,

$$\left(\frac{T_n}{\sqrt{n}}, \mu_n\right) \xrightarrow[\mathrm{GHP}]{(\mathrm{d})} \left(2\mathcal{T}_{B_r}, \mu_{\mathrm{Br}}\right).$$

- If T_n is uniform amongst the set of rooted binary ordered trees with n vertices,

$$\left(\frac{T_n}{\sqrt{n}}, \mu_n\right) \xrightarrow[\mathrm{GHP}]{(\mathrm{d})} \left(2\mathcal{T}_{B_r}, \mu_{\mathrm{Br}}\right).$$

As a consequence, this provides the behavior of several statistics of the trees, that first interested combinatorists.

We will not present the original proofs by Aldous [10] and Duquesne [44], but will rather focus on the fact that they may be recovered by using the MB-property. This is the goal of the next two sections, where we will present in a general setting some results on the scaling limits for MB-sequences of trees. As already mentioned, the main idea of the proofs of Aldous [10] and Duquesne [44] is rather based on the study of the so-called *contour functions* of the trees. We refer to Aldous and Duquesne papers, as well as Le Gall's survey [71] for details. See also Duquesne and Le Gall [46] and Kortchemski [69, 70] for further related results.

4 Scaling Limits of Markov-Branching Trees

Our goal is to set up an asymptotic criterion on the splitting probabilities (q_n) of an MB-sequence of trees so that this sequence, suitably normalized, converges to a non-trivial continuous limit. We follow here the approach of the paper [59] that found its roots in the previous work [63] were similar results where proved under stronger assumptions. A remark on these previous results is made at the end of this section.

The splitting probability q_n corresponds to a "discrete" fragmentation of the integer n into smaller integers. To set up the desired criterion, we first need to introduce a continuous counterpart for these partitions of integers, namely

$$\mathcal{S}^{\downarrow} = \left\{ \mathbf{s} = (s_1, s_2, \ldots) : s_1 \geq s_2 \geq \ldots \geq 0 \text{ and } \sum_{i \geq 1} s_i = 1 \right\}$$

which is endowed with the distance $d_{\mathcal{S}^{\downarrow}}(\mathbf{s}, \mathbf{s}') = \sup_{i \geq 1} |s_i - s'_i|$. Our main hypothesis on (q_n) then reads:

Hypothesis (H) There exist $\gamma > 0$ and ν a non-trivial σ-finite measure on $\mathcal{S}^{\downarrow}$ satisfying $\int_{\mathcal{S}^{\downarrow}} (1 - s_1)\nu(\mathrm{d}\mathbf{s}) < \infty$ and $\nu(1, 0, \ldots) = 0$, such that

$$n^{\gamma} \sum_{\lambda \in \mathcal{P}_n} q_n(\lambda) \left(1 - \frac{\lambda_1}{n}\right) f\left(\frac{\lambda_1}{n}, \ldots, \frac{\lambda_p}{n}, 0, \ldots\right) \underset{n \to \infty}{\longrightarrow} \int_{\mathcal{S}^{\downarrow}} (1 - s_1) f(\mathbf{s}) \nu(\mathrm{d}\mathbf{s}).$$

for all continuous $f : \mathcal{S}^{\downarrow} \to \mathbb{R}$.

We will see in Sect. 5 that most of the examples of splitting probabilities introduced in Sect. 2.3 satisfy this hypothesis. As a first, easy, example, consider the "basic example" introduced there (Example 2): $q_n((n)) = 1 - n^{-\alpha}$ and $q_n(\lceil n/2 \rceil, \lfloor n/2 \rfloor) = n^{-\alpha}$, $\alpha > 0$. Then, clearly, (H) is satisfied with

$$\gamma = \alpha \quad \text{and} \quad \nu(\mathrm{d}\mathbf{s}) = \delta_{\left(\frac{1}{2}, \frac{1}{2}, 0, \ldots\right)}.$$

The interpretation of the hypothesis (H) is that macroscopic branchings are rare, in the sense that the macroscopic splitting events $n \mapsto n\mathbf{s}$, $\mathbf{s} \in \mathcal{S}^{\downarrow}$ with $s_1 < 1 - \varepsilon$ occur with a probability asymptotically proportional to $n^{-\gamma} \mathbb{1}_{\{s_1 < 1-\varepsilon\}} \nu(\mathrm{d}\mathbf{s})$, for a.e. fixed $\varepsilon \in (0, 1)$.

The main result on the scaling limits of MB-trees indexed by the leaves is the following.

Theorem 4.1 ([59]) *Let $(T_n, n \geq 1)$ be a* MB*-sequence indexed by the leaves and assume that its splitting probabilities satisfy* (H). *Then there exists a compact, measured real tree $(\mathcal{T}_{\gamma,\nu}, \mu_{\gamma,\nu})$ such that*

$$\left(\frac{T_n}{n^{\gamma}}, \mu_n\right) \xrightarrow[\mathrm{GHP}]{(\mathrm{d})} \left(\mathcal{T}_{\gamma,\nu}, \mu_{\gamma,\nu}\right),$$

where μ_n is the uniform probability on the leaves of T_n.

The goal of this section is to detail the main steps of the proof of this result and to discuss some properties of the limiting measured tree, which belongs to the so-called family of *self-similar fragmentation trees* (the distribution of such a tree is entirely characterized by the parameters γ and ν). In that aim we will first study how the height of a leaf chosen uniformly at random in T_n grows (Sects. 4.1 and 4.2). Then we will review some results on self-similar fragmentation trees (Sect. 4.3). Last we will explain how one can use the scaling limit of the height of a leaf chosen at random to obtain, by induction, the scaling limit of the subtree spanned by k leaves chosen independently, for all k, and then finish the proof of Theorem 4.1 with a tightness criterion (Sect. 4.4).

There is a similar result for MB-sequences indexed by the vertices.

Theorem 4.2 ([59]) *Let $(T_n, n \geq 1)$ be a* MB*-sequence indexed by the vertices and assume that its splitting probabilities satisfy* (H) *for some $0 < \gamma < 1$. Then there exists a compact, measured real tree $(\mathcal{T}_{\gamma,\nu}, \mu_{\gamma,\nu})$ such that*

$$\left(\frac{T_n}{n^{\gamma}}, \mu_n\right) \xrightarrow[\mathrm{GHP}]{(\mathrm{d})} \left(\mathcal{T}_{\gamma,\nu}, \mu_{\gamma,\nu}\right),$$

where μ_n is the uniform probability on the vertices of T_n.

Theorem 4.2 is actually a direct corollary of Theorem 4.1, for the following reason. Consider an MB-sequence indexed by the vertices with splitting probabilities (p_n) and for all n, branch on each internal vertex of the tree T_n an edge with a leaf. This gives a tree $\overline{T}_n$ with n leaves. It is then obvious that $(\overline{T}_n, n \geq 1)$ is an MB-sequence indexed by the *leaves*, with splitting probabilities (q_n) defined by

$$q_n(\lambda_1, \ldots, \lambda_p, 1) = p_{n-1}(\lambda_1, \ldots, \lambda_p), \quad \text{for all } (\lambda_1, \ldots, \lambda_p) \in \mathcal{P}_{n-1}$$

(and $q_n(\lambda) = 0$ for all other $\lambda \in \mathcal{P}_n$). It is moreover easy to see that (q_n) satisfies (H) with parameters (γ, ν), $0 < \gamma < 1$, if and only if (p_n) does. Hence Theorem 4.1

implies Theorem 4.2, since $\overline{T}_n$, endowed with the uniform probability, is at distance less than one from (T_n, μ_n) for the GHP-distance.

We will present in Sect. 5 several applications of these two theorems. Let us just consider here the "basic example" of Sect. 2.3 (Example 2). We have already noticed that its splitting probabilities satisfy Hypothesis (H), with parameters α and $\delta_{(1/2,1/2,0,\dots)}$. Hence in this case, the corresponding sequence of MB-trees T_n divided by n^α and endowed with the uniform probability measure on its leaves converges for the GHP-topology towards a $(\alpha, \delta_{(1/2,1/2,0,\dots)})$-self-similar fragmentation tree.

Remark These two statements are also valid when replacing in (H) and in the theorems the power sequence n^γ by any regularly varying sequence with index $\gamma > 0$. We recall that a sequence (a_n) is said to vary regularly with index $\gamma > 0$ if for all $c > 0$,

$$\frac{a_{\lfloor cn \rfloor}}{a_n} \underset{n\to\infty}{\longrightarrow} c^\gamma.$$

We refer to [24] for backgrounds on that topic. For simplicity, in the following we will only works with power sequences, but the reader should have in mind that everything holds similarly for regularly varying sequences.

Convergence in Probability In [63], scaling limits are established for some MB-sequences that moreover satisfy a property of *sampling consistency*, namely that for all n, T_n is distributed as the tree with n leaves obtained by removing a leaf picked uniformly at random in T_{n+1}, as well as the adjacent edge. This consistency property is demanding and the approach developed in [59] allows to do without it. However we note that if the MB-sequence is *strongly sampling consistent*, one can actually establish under suitable conditions a *convergence in probability* of the rescaled trees, which is of course an improvement. By strongly sampling consistent, we mean that versions of the trees can be built on a same probability space so that if T_n° denotes the tree with n leaves obtained by removing an edge-leaf picked uniformly in T_{n+1}, then (T_n, T_{n+1}) is distributed as (T_n°, T_{n+1}). We refer to [63] for details.

4.1 A Markov Chain in the Markov-Branching Sequence of Trees

Consider $(T_n, n \geq 1)$ an MB-sequence of trees indexed by the leaves, with splitting distribution $(q_n, n \geq 1)$. Before studying the scaling limit of the trees in their whole, we start by studying the scaling limit of a *typical leaf*. For example, in each T_n, we mark one of the n leaves uniformly at random and we want to determine how the height of the marked leaf behaves as $n \to \infty$. In that aim, let $\star_n$ denote this marked leaf and let $\star_n(k)$ denote its ancestor at generation k, $0 \leq k \leq \mathrm{ht}(\star_n)$ (so that $\star_n(0)$ is the root of T_n and $\star_n(\mathrm{ht}(\star_n)) = \star_n$). Let also $T_n^\star(k)$ be the subtree composed of

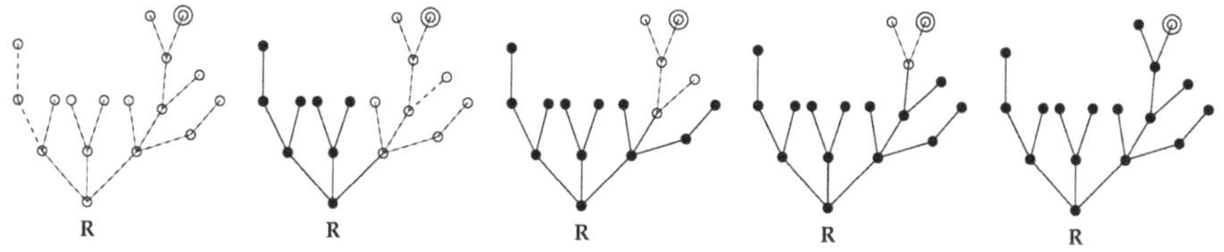

Fig. 3 A Markov chain in the Markov-Branching trees. Here $n = 9$, the marked leaf is circled and the subtrees of descendants of $\star_9(k)$ for $0 \le k \le 4$ are dotted. Moreover $X_9(0) = 9$, $X_9(1) = 5$, $X_9(2) = 3$, $X_9(3) = 2$, $X_9(4) = 1$ and $X_9(i) = 0, \forall i \geq 5$

the descendants of $\star_n(k)$ in T_n, formally,

$$T_n^\star(k) := \{v \in T_n : \star_n(k) \in [[\rho, v]]\}, \quad k \le \mathrm{ht}(\star_n)$$

and $T_n^\star(k) := \emptyset$ if $k > \mathrm{ht}(\star_n)$. We then set

$$X_n(k) := \#\left\{\text{leaves of } T_n^\star(k)\right\}, \quad \forall k \in \mathbb{Z}_+ \tag{5}$$

with the convention that $X_n(k) = 0$ for $k > \mathrm{ht}(\star_n)$ (Fig. 3).

Proposition 4.3 *The process $(X_n(k), k \geq 0)$ is a $\mathbb{Z}_+$-valued non-increasing Markov chain starting from $X_n(0) = n$, with transition probabilities*

$$p(i,j) = \sum_{\lambda \in \mathcal{P}_i} q_i(\lambda) m_j(\lambda) \frac{j}{i} \quad \textit{for all } 1 \le j \le i, \textit{ with } i \geq 1 \tag{6}$$

and $p(1,0) = q_1(\emptyset) = 1 - p(1,1)$.

Proof The Markov property is a direct consequence of the Markov branching property. Indeed, given $X_n(1) = i_1, \ldots, X_n(k-1) = i_{k-1}$, the tree $T_n^\star(k-1)$ is distributed as $T_{i_{k-1}}$ if $i_{k-1} \geq 1$ and is the emptyset otherwise. In particular, when $i_{k-1} = 0$, the conditional distribution of $X_n(k)$ is the Dirac mass at 0. When $i_{k-1} \geq 1$, we use the fact that $\star_n$ is in $T_n^\star(k-1)$, hence, still conditioning on the same event, we have that $\star_n$ is uniformly distributed amongst the i_{k-1} leaves of $T_{i_{k-1}}$. Otherwise said, given $X_n(1) = i_1, \ldots, X_n(k-1) = i_{k-1}$ with $i_{k-1} \geq 1$, $(T_n^\star(k-1), \star_n)$ is distributed as $(T_{i_{k-1}}, \star_{i_{k-1}})$ and consequently $X_n(k)$ is distributed as $X_{i_{k-1}}(1)$. Hence the Markov property of the chain $(X_n(k), k \geq 0)$. It remains to compute the transition probabilities:

$$p(n,k) = \mathbb{P}(X_n(1) = k) = \sum_{\lambda \in \mathcal{P}_n} q_n(\lambda) \mathbb{P}\left(X_n(1) = k | \Lambda_n = \lambda\right)$$

where Λ_n denotes the partition of n corresponding to the distribution of the leaves in the subtrees of T_n above the root. Since $\star_n$ is chosen uniformly amongst the set of leaves, we clearly have that

$$\mathbb{P}\left(X_n(1)=k|\Lambda_n=\lambda\right)=\frac{k}{n}\times\#\{j:\lambda_j=k\},\quad\forall k\geq 1.$$

□

Hence studying the scaling limit of the height of the marked leaf in the tree T_n reduces to studying the scaling limit of the absorption time A_n of the Markov chain $(X_n(k), k \geq 0)$ at 0:

$$A_n := \inf\left\{k \geq 0 : X_n(k) = 0\right\}$$

(to be precise, this absorption time is equal to the height of the marked leaf $+1$). The study of the scaling limit of $((X_n(k), k \geq 1), A_n)$ as $n \to \infty$ is the goal of the next section. Before getting in there, let us notice that the Hypothesis (H) on the splitting probabilities $(q_n, n \geq 1)$ of $(T_n, n \geq 1)$, together with (6), implies the following behavior of the transition probabilities $(p(n,k), k \leq n)$:

$$n^{\gamma}\sum_{k=0}^{n} p(n,k)\left(1-\frac{k}{n}\right)g\left(\frac{k}{n}\right)\underset{n\to\infty}{\longrightarrow}\int_{[0,1]} g(x)\mu(\mathrm{d}x) \tag{7}$$

for all continuous functions $g : [0,1] \to \mathbb{R}$, where the measure μ in the limit is a finite, non-zero measure on $[0,1]$ defined by

$$\int_{[0,1]} g(x)\mu(\mathrm{d}x) = \int_{\mathcal{S}^{\downarrow}} \sum_{i\geq 1} s_i(1-s_i)g(s_i)\nu(\mathrm{d}\mathbf{s}). \tag{8}$$

To see this, apply (H) to the continuous function defined by

$$f(\mathbf{s}) = \frac{\sum_{i\geq 1} s_i(1-s_i)g(s_i)}{1-s_1} \quad \text{for } \mathbf{s} \neq (1,0,\ldots)$$

and $f(1,0,\ldots) = g(1) + g(0)$.

4.2 Scaling Limits of Non-increasing Markov Chains

As discussed in the previous section, studying the height of a typical leaf in MB-trees amounts to studying the absorption time at 0 of a $\mathbb{Z}_+$-valued non-increasing Markov chain. In this section, we study in a general framework the scaling limits of $\mathbb{Z}_+$-valued non-increasing Markov chains, under appropriate assumptions on the transition probabilities. At the end of the section we will see how this applies to the height of a typical leaf in an MB-sequence. In the following,

$$(X_n(k), k \geq 0)$$

denotes a non-increasing $\mathbb{Z}_+$-valued Markov chain starting from n ($X_n(0) = n$), with transition probabilities $(p(i, j), 0 \leq j \leq i)$ such that

Hypothesis (H') $\exists\, \gamma > 0$ and μ a non-trivial finite measure on $[0, 1]$ such that

$$n^\gamma \sum_{k=0}^{n} p(n, k) \left(1 - \frac{k}{n}\right) f\left(\frac{k}{n}\right) \underset{n \to \infty}{\longrightarrow} \int_{[0,1]} f(x)\mu(\mathrm{d}x)$$

for all continuous functions $f : [0, 1] \to \mathbb{R}$.

This hypothesis implies that starting from n, *macroscopic jumps* (i.e. with size proportional to n) are rare, since for a.e. $0 < \varepsilon \leq 1$, the probability to do a jump larger than εn is of order $c_\varepsilon n^{-\gamma}$ where $c_\varepsilon = \int_{[0,1-\varepsilon]}(1-x)^{-1}\mu(\mathrm{d}x)$ (note that this may tend to ∞ when ε tends to 0).

Now, let

$$A_n := \inf\left\{k \geq 0 : X_n(i) = X_n(k), \quad \forall i \geq k\right\}$$

be the first time at which the chain enters an absorption state (note that $A_n < \infty$ a.s. since the chain is non-increasing and $\mathbb{Z}_+$-valued). In the next theorem, $\mathbb{D}([0, \infty), [0, \infty))$ denotes the set of non-negative càdlàg processes, endowed with the Skorokhod topology.

Theorem 4.4 ([58]) *Assume* (H').

(i) *Then, in* $\mathbb{D}([0, \infty), [0, \infty))$,

$$\left(\frac{X_n\left(\lfloor n^\gamma t \rfloor\right)}{n}, t \geq 0\right) \underset{n \to \infty}{\overset{(\mathrm{d})}{\longrightarrow}} \left(\exp(-\xi_{\tau(t)}), t \geq 0\right),$$

where ξ is a subordinator, i.e. a non-decreasing Lévy process, and τ is the time-change (acceleration of time)

$$\tau(t) := \inf\left\{u \geq 0 : \int_0^u \exp(-\gamma \xi_r)\mathrm{d}r \geq t\right\}, t \geq 0.$$

The distribution of ξ is characterized by its Laplace transform $\mathbb{E}[\exp(-\lambda \xi_t)] = \exp(-t\phi(\lambda))$, *with*

$$\phi(\lambda) = \mu(\{0\}) + \mu(\{1\})\lambda + \int_{(0,1)} (1 - x^\lambda)\frac{\mu(\mathrm{d}x)}{1-x}, \quad \lambda \geq 0.$$

(ii) *Moreover, jointly with the above convergence,*

$$\frac{A_n}{n^\gamma} \xrightarrow[n\to\infty]{(\mathrm{d})} \int_0^\infty \exp(-\gamma\xi_r)\mathrm{d}r = \inf\left\{t \ge 0 : \exp(-\xi_{\tau(t)}) = 0\right\}.$$

Comments For background on Lévy processes, we refer to [12]. Let us simply recall here that the law of a subordinator is characterized by three parameters: a measure on $(0,\infty)$ that codes its jumps (which here is the push-forward of $\mu(\mathrm{d}x)(1-x)\mathbb{1}_{\{x\in(0,1)\}}$ by the application $x \mapsto -\ln(x)$), a linear drift (here $\mu(\{1\})$) and a killing rate at which the process jumps to $+\infty$ (here $\mu(\{0\})$).

Main Ideas of the Proof of Theorem 4.4

(i) Let $Y_n(t) := n^{-1}X_n(\lfloor n^\gamma t\rfloor)$, for $t \ge 0, n \in \mathbb{N}$. First, using Aldous' tightness criterion [23, Theorem 16.10] and (H′), one can check that the sequence $(Y_n, n \ge 1)$ is tight. It is then sufficient to prove that every possible limit in distribution of subsequences of (Y_n) are distributed as $\exp(-\xi_\tau)$. Let Y' be such a limit and $(n_k, k \ge 1)$ a sequence such that Y_{n_k} converges to Y' in distribution. In the limit, we actually prefer to deal with ξ than with ξ_τ, and for this reason we start by changing time in Y_n by setting

$$\tau_{Y_n}(t) := \inf\left\{u \ge 0 : \int_0^u Y_n^{-\gamma}(r)\mathrm{d}r > t\right\} \quad \text{and } Z_n(t) := Y_n\left(\tau_{Y_n}(t)\right), t \ge 0.$$

One can then easily check that (Z_{n_k}) converges in distribution to Z' where $Z' = Y' \circ \tau_{Y'}$, with $\tau_{Y'}(t) := \inf\left\{u \ge 0 : \int_0^u (Y'(r))^{-\gamma}\mathrm{d}r > t\right\}$. It is also easy to reverse the time-change and get that

$$Y'(t) = Z'\left(\tau_{Y'}^{-1}(t)\right) = Z'\left(\inf\left\{u \ge 0 : \int_0^u Z'^\gamma(r)\mathrm{d}r > t\right\}\right), \quad t \ge 0.$$

With this last equality, we see that it just remains to prove that Z' is distributed as $\exp(-\xi)$. This can be done in three steps:

(a) Observe the following (easy!) fact: if P is the transition function of a Markov chain M with countable state space $\subset \mathbb{R}$, then for any positive function f such that $f^{-1}(\{0\})$ is absorbing,

$$f(M(k))\prod_{i=0}^{k-1}\frac{f(M(i))}{Pf(M(i))}, \quad k \ge 0$$

is a martingale. As a consequence: for all $\lambda \geq 0$ and $n \geq 1$, if we let $G_n(\lambda) := \mathbb{E}\left[(X_n(1)/n)^\lambda\right]$, then,

$$M_n^{(\lambda)}(t) := Z_n^\lambda(t) \left(\prod_{i=0}^{\lfloor n^\gamma \tau_{Y_n}(t) \rfloor - 1} G_{X_n(i)}(\lambda) \right)^{-1}, \quad t \geq 0$$

is a martingale.

(b) Under (H′), $1 - G_n(\lambda) \underset{n\to\infty}{\sim} n^{-\gamma}\phi(\lambda)$. Together with the convergence in distribution of (Z_{n_k}) to Z' and the definition of $M_n^{(\lambda)}$, this leads to the convergence (this is the most technical part)

$$M_{n_k}^{(\lambda)} \underset{k\to\infty}{\overset{(d)}{\longrightarrow}} (Z')^\lambda \exp(\phi(\lambda)\cdot),$$

and the martingale property passes to the limit.

(c) Hence $(Z')^\lambda \exp(\phi(\lambda)\cdot)$ is a martingale for all $\lambda \geq 0$. Using Laplace transforms, it is then easy to see that this implies in turn that $-\ln Z'$ is a non-decreasing process with independent and stationary increments (hence a subordinator), with Laplace exponent ϕ.

Hence $Z' \overset{(d)}{=} \exp(-\xi)$.

(ii) We do not detail this part and refer to [58, Section 4.3]. Let us simply point out that it is *not* a direct consequence of the convergence of (Y_n) to $\exp(-\xi_\tau)$ since convergence of functions in $\mathbb{D}([0,\infty),[0,\infty))$ does not lead, in general, to the convergence of their absorption times (when they exist). □

This result leads to the following corollary.

Corollary 4.5 *Let $(T_n, n \geq 1)$ be a* MB*-sequence indexed by the leaves, with splitting probabilities satisfying* (H) *with parameters (γ, ν). For each n, let $\star_n$ be a leaf chosen uniformly amongst the n leaves of T_n. Then,*

$$\frac{\mathsf{ht}(\star_n)}{n^\gamma} \underset{n\to\infty}{\overset{(d)}{\longrightarrow}} \int_0^\infty \exp(-\gamma \xi_r) \mathrm{d}r$$

where ξ is a subordinator with Laplace exponent $\phi(\lambda) = \int_{\mathcal{S}^\downarrow} \sum_{i\geq 1}\left(1 - s_i^\lambda\right) s_i \nu(\mathrm{d}s)$, $\lambda \geq 0$.

Proof As seen at the end of the previous section, under (H) the transition probabilities of the Markov chain (5) satisfy assumption (H′) with parameters γ and μ, with μ defined by (8). The conclusion follows with Theorem 4.4 (ii). □

Further Reading Apart from applications to Markov-Branching trees, Theorem 4.4 can be used to describe the scaling limits of various stochastic processes, e.g. random walks with a barrier or the number of collisions in Λ-coalescent processes, see [58]. Recently, Bertoin and Kortchemski [18] set up results similar to Theorem 4.4 for *non-monotone* Markov chains and develop several applications, to random walks conditioned to stay positive, to the number of particles in some coagulation-fragmentations processes, to random planar maps (see [20] for this last point). Also in [61] similar convergences for bivariate Markov chains towards time-changed Markov additive processes are studied. This will have applications to dynamical models of tree growth in a broader context than the one presented in Sect. 5.3, and more generally to multi-type MB-trees.

4.3 Self-Similar Fragmentation Trees

Self-similar fragmentation trees are random compact measured real trees that describe the genealogical structure of self-similar fragmentation processes with a negative index. It turns out that this set of trees is closely related to the set of trees arising as scaling limits of MB-trees. We start by introducing the self-similar fragmentation processes, following Bertoin [14], and then turn to the description of their genealogical trees, which were first introduced in [57] and then in [91] in a broader context.

4.3.1 Self-Similar Fragmentation Processes

Fragmentation processes are continuous-time processes that describe the evolution of an object that splits repeatedly and randomly as time passes. In the models we are interested in, the fragments are characterized by their mass alone, other characteristics, such as their shape, do not come into account. Many researchers have been working on such models. From a historical perspective, it seems that Kolmogorov [68] was the first in 1941. Since the early 2000s, there has been a full treatment of fragmentation processes satisfying a self-similarity property. We refer to Bertoin's book [14] for an overview of work in this area and a deepening of the results presented here.

We will work on the space of masses

$$\mathcal{S}^{\downarrow}_{-} = \left\{ \mathbf{s} = (s_1, s_2, \ldots) : s_1 \geq s_2 \geq \ldots \geq 0 \text{ and } \sum_{i \geq 1} s_i \leq 1 \right\},$$

which contains the set $\mathcal{S}^{\downarrow}$, and which is equipped with the same metric $d_{\mathcal{S}^{\downarrow}}$.

Definition 4.6 Let $\alpha \in \mathbb{R}$. An α-self-similar fragmentation process is an $\mathcal{S}^{\downarrow}_{-}$-valued Markov process $(F(t), t \geq 0)$ which is continuous in probability and such that, for all $t_0 \geq 0$, given that $F(t_0) = (s_1, s_2, \ldots)$, the process $(F(t_0 + t), t \geq 0)$ is distributed as the process G obtained by considering a sequence $(F^{(i)}, i \geq 1)$ of i.i.d. copies of F and then defining $G(t)$ to be the decreasing rearrangement of the sequences $s_i F^{(i)}(s_i^\alpha t)$, $i \geq 1$, for all $t \geq 0$.

In the following, we will always consider processes starting from a unique mass equal to 1, i.e. $F(0) = (1, 0, \ldots)$. At time t, the sequence $F(t)$ should be understood as the decreasing sequence of the masses of fragments present at that time.

It turns out that such processes indeed exist and that their distributions are characterized by three parameters: the *index of self-similarity* $\alpha \in \mathbb{R}$, an *erosion coefficient* $c \geq 0$ that codes a continuous melt of the fragment (when $c = 0$ there is no erosion) and a *dislocation measure* ν, which is a measure ν on $\mathcal{S}_-^\downarrow$ such that $\int_{\mathcal{S}_-^\downarrow} (1 - s_1)\nu(\mathrm{d}\mathbf{s}) < \infty$. The role of the parameters α and ν can be specified as follows when $c = 0$ and ν is finite: then, each fragment with mass m waits a random time with exponential distribution with parameter $\nu(\mathcal{S}_-^\downarrow)$ and then splits in fragments with masses $m\mathbf{S}$, where $\mathbf{S}$ is distributed according to $\nu/\nu(\mathcal{S}_-^\downarrow)$, independently of the splitting time. When ν is infinite, the fragments split immediately, see [14, Chapter 3] for further details.

The index α has an enormous influence on the behavior of the process: when $\alpha = 0$, all fragments split at the same rate, whereas when $\alpha > 0$ fragments with small masses split slower and when $\alpha < 0$ fragments with small masses split faster. In this last case the fragments split so quickly that the whole initial object is reduced to "dust" in finite time, almost surely, i.e. $\inf\{t \geq 0 : F(t) = (0, \ldots)\} < \infty$ a.s.

The Tagged Fragment Process We turn to a connection with the results seen in the previous section. Pick a point uniformly at random in the initial object (this object can be seen as an interval of length 1, for example), independently of the evolution of the process and let $F_*(t)$ be the mass of the fragment containing this marked point at time t. The process $F_*(t)$ is non-increasing and more precisely,

Theorem 4.7 (Bertoin [14], Theorem 3.2 and Corollary 3.1) *The process F_* can be written as*

$$F_*(t) = \exp(-\xi_{\tau(t)}), \quad \forall t \geq 0,$$

where ξ is a subordinator with Laplace exponent

$$\phi(\lambda) = c + \int_{\mathcal{S}_-^\downarrow} \Big(1 - \sum_{i \geq 1} s_i\Big)\nu(\mathrm{d}\mathbf{s}) + c\lambda + \int_{\mathcal{S}_-^\downarrow} \sum_{i \geq 1} (1 - s_i^\lambda) s_i \nu(\mathrm{d}\mathbf{s}), \quad \lambda \geq 0$$

and τ is a time-change depending on the parameter α, $\tau(t) = \inf\{u \geq 0 : \int_0^u \exp(\alpha\xi_r)\mathrm{d}r > t\}$.

4.3.2 Self-Similar Fragmentation Trees

It was shown in [57] that to every self-similar fragmentation process with a negative index $\alpha = -\gamma < 0$, no erosion ($c = 0$) and a dislocation measure ν satisfying $\nu(\sum_{i\geq 1} s_i < 1) = 0$ (we say that ν is *conservative*), there is an associated compact rooted measured tree that describes its genealogy. We denote such a tree by $(\mathcal{T}_{\gamma,\nu}, \mu_{\gamma,\nu})$ and precise that the measure $\mu_{\gamma,\nu}$ is fully supported by the set of leaves of $\mathcal{T}_{\gamma,\nu}$ and non-atomic. In [91], Stephenson more generally constructed and studied compact rooted measured trees that describe the genealogy of any self-similar fragmentation process with a negative index. However in this survey, we restrict ourselves to the family of trees $(\mathcal{T}_{\gamma,\nu}, \mu_{\gamma,\nu})$, with $\gamma > 0$ and ν conservative.

The connection between a tree $(\mathcal{T}_{\gamma,\nu}, \mu_{\gamma,\nu})$ and the fragmentation process it is related to can be summarized as follows: for all $t \geq 0$, consider the connected components of $\{v \in \mathcal{T}_{\gamma,\nu} : \mathrm{ht}(v) > t\}$, the set of points in $\mathcal{T}_{\gamma,\nu}$ that have a height strictly larger than t, and let $F(t)$ denote the decreasing rearrangement of the $\mu_{\gamma,\nu}$-masses of these components. Then F is a fragmentation process, with index of self-similarly $-\gamma$, dislocation measure ν and no erosion. Besides, we note that $\mathcal{T}_{\gamma,\nu}$ possesses a fractal property, in the sense that if we fix a $t \geq 0$ (deterministic) and consider a point x at height t, then any subtree of $\mathcal{T}_{\gamma,\nu}$ descending from this point x (i.e. any connected component of $\{v \in \mathcal{T}_{\gamma,\nu} : x \in [[\rho, v]]\}$), having, say, a $\mu_{\gamma,\nu}$-mass m, is distributed as $m^{\gamma}\mathcal{T}_{\gamma,\nu}$.

First Examples The Brownian tree and the α-stable trees that arise as scaling limits of Galton–Watson trees all belong to the family of self-similar fragmentation trees. More precisely,

- Bertoin [13] notices that the Brownian tree $(\mathcal{T}_{\mathrm{Br}}, \mu_{\mathrm{Br}})$ is a self-similar fragmentation tree and calculates its characteristics: $\gamma = 1/2$ and $\nu_{\mathrm{Br}}(s_1 + s_2 < 1) = 0$ and

$$\nu_{\mathrm{Br}}(s_1 \in \mathrm{d}x) = \frac{\sqrt{2}}{\sqrt{\pi} x^{3/2}(1-x)^{3/2}}, \qquad 1/2 < x < 1.$$

The fact that $\nu_{\mathrm{Br}}(s_1+s_2 < 1) = 0$ corresponds to the fact the tree is binary: every branch point has two descendants trees, and in the corresponding fragmentation, every splitting events gives two fragments.
- Miermont [75] proves that each stable tree $\mathcal{T}_{\alpha}$ is self-similar and calculates its characteristics when $\alpha \in (1,2)$: $\gamma = 1 - 1/\alpha$ and

$$\int_{\mathcal{S}^{\downarrow}} f(\mathbf{s})\nu_{\alpha}(\mathrm{d}\mathbf{s}) = C_{\alpha}\mathbb{E}\left[T_1 f\left(\frac{\Xi_i}{T_1}, i \geq 1\right)\right],$$

where

$$C_{\alpha} = \frac{\alpha(\alpha-1)\Gamma(1-1/\alpha)}{\Gamma(2-\alpha)}$$

and $(\Xi_i, i \geq 1)$ is the sequence of lengths, ranked in decreasing order, of intervals between successive atoms of a Poisson measure on $\mathbb{R}_+$ with intensity $(\alpha\Gamma(1-1/\alpha))^{-1}\mathrm{d}r/r^{1+1/\alpha}$, and $T_1 = \sum_i \Xi_i$.

Hausdorff Dimension We first quickly recall the definition of Hausdorff dimension, which is a quantity that measures the "size" of metric spaces. We refer to the book of Falconer [53] for more details on that topic and for an introduction to fractal geometry in general. For all $r > 0$, the r-dimensional Hausdorff measure of a metric space (Z, d_Z) is defined by

$$\mathcal{M}^r(Z) := \lim_{\varepsilon \to 0} \inf_{\{(C_i)_{i\in\mathbb{N}}:\mathrm{diam}C_i \leq \varepsilon\}} \left\{ \sum_{i\geq 1} \mathrm{diam}(C_i)^r : Z \subset \cup_{i\geq 1} C_i \right\}.$$

where the infimum is taken over all coverings of Z by countable families of subsets $C_i \subset Z, i \in \mathbb{N}$, all with a diameter smaller than ε. The function $r > 0 \mapsto \mathcal{M}^r(Z) \in [0, \infty]$ is finite, non-zero at most one point. The Hausdorff dimension of Z is then given by

$$\dim_{\mathrm{H}}(Z) = \inf\left\{r > 0 : \mathcal{M}^r(Z) = 0\right\} = \sup\left\{r > 0 : \mathcal{M}^r(Z) = \infty\right\}.$$

The Hausdorff dimension of a fragmentation tree depends mainly on its index of self-similarity. Let $\mathcal{L}(\mathcal{T}_{\gamma,\nu})$ denote the set of leaves of $\mathcal{T}_{\gamma,\nu}$. Then we know that,

Theorem 4.8 ([57]) *If* $\int_{\mathcal{S}^{\downarrow}} (s_1^{-1} - 1)\nu(\mathrm{ds}) < \infty$, *then almost surely*

$$\dim_{\mathrm{H}}(\mathcal{L}(\mathcal{T}_{\gamma,\nu})) = \frac{1}{\gamma} \quad \textit{and} \quad \dim_{\mathrm{H}}(\mathcal{T}_{\gamma,\nu}) = \max\left(\frac{1}{\gamma}, 1\right).$$

In particular, the Hausdorff dimension of the Brownian tree is 2, and more generally the Hausdorff dimension of the α-stable tree, $\alpha \in (1, 2)$, is $\alpha/(\alpha - 1)$. This recovers a result of Duquesne and Le Gall [47] proved in the framework of Lévy trees (we note that the intersection between the set of Lévy trees and that of self-similar fragmentation trees is exactly the set of stable Lévy trees).

Height of a Typical Leaf The measured tree $(\mathcal{T}_{\gamma,\nu}, \mu_{\gamma,\nu})$ has been constructed in such a way that if we pick a leaf L at random in $\mathcal{T}_{\gamma,\nu}$ according to $\mu_{\gamma,\nu}$ and we consider for each $t \geq 0$ the $\mu_{\gamma,\nu}$-mass of the connected component of $\{v \in \mathcal{T}_{\gamma,\nu} : \mathrm{ht}(v) > t\}$ that contains this marked leaf (with the convention that this mass is 0 if $\mathrm{ht}(L) \leq t$), then we obtain a process which is distributed as the tagged fragment of the corresponding fragmentation process, as defined in the previous section. In particular, the height of L is distributed as the absorption time of the process $\exp(-\xi_\tau)$ introduced in Theorem 4.7, i.e.

$$\mathrm{ht}(L) \overset{(\mathrm{d})}{=} \int_0^\infty \exp(-\gamma\xi_r)\mathrm{d}r \tag{9}$$

where ξ is a subordinator with Laplace exponent $\phi(\lambda) = \int_{\mathcal{S}^{\downarrow}} \sum_{i\geq 1}(1 - s_i^{\lambda})s_i \nu(\mathrm{d}\mathbf{s}), \lambda \geq 0$. This is exactly the distribution of the limit appearing in Corollary 4.5.

4.4 Scaling Limits of Markov-Branching Trees

We can now explain the main steps of the proof of Theorem 4.1. In that aim, let $(T_n, n \geq 1)$ denote an MB-sequence indexed by the leaves with splitting probabilities satisfying (H), with parameters (γ, ν) in the limit. We actually only give here a hint of the proof of the convergence of the rescaled trees and refer to [59, Section 4.4] to see how to incorporate the measures. The proof of the convergence of the rescaled trees consists in three main steps:

First Step: Convergence of the Height of a Typical Leaf Keeping the notations previously introduced, $\mathrm{ht}(\star_n)$ for the height of a typical leaf in T_n and $\mathrm{ht}(L)$ for the height of a typical leaf in a fragmentation tree $(\mathcal{T}_{\gamma,\nu}, \mu_{\gamma,\nu})$, we get by (9) and Corollary 4.5 that

$$\frac{\mathrm{ht}(\star_n)}{n^{\gamma}} \xrightarrow[n\to\infty]{(\mathrm{d})} \mathrm{ht}(L).$$

Second Step: Convergence of Finite-Dimensional Marginals For all integers $k \geq 2$, let $T_n(k)$ be the subtree of T_n spanned by the root and k (different) leaves picked independently, uniformly at random. Similarly, let $\mathcal{T}_{\gamma,\nu}(k)$ be the subtree of $\mathcal{T}_{\gamma,\nu}$ spanned by the root and k leaves picked independently at random according to the measure $\mu_{\gamma,\nu}$. Then (under (H)),

$$\frac{T_n(k)}{n^{\gamma}} \xrightarrow[n\to\infty]{(\mathrm{d})} \mathcal{T}_{\gamma,\nu}(k). \tag{10}$$

This can be proved by induction on k. For $k = 1$, this is Step 1 above. For $k \geq 2$, we use the induction hypothesis and the MB-property. Here is the main idea. Consider the decomposition of T_n into subtrees above its first branch point in $T_n(k)$ and take only into account the subtrees having marked leaves. We obtain $m \geq 2$ subtrees with, say, $n_1, \ldots, n_m$ leaves respectively ($\sum_{i=1}^m n_i \leq n$), and each of these trees have $k_1 \geq 1, \ldots k_m \geq 1$ marked leaves ($\sum_{i=1}^m k_i = k$). Given m, $n_1, \ldots, n_m, k_1, \ldots k_m$, the MB-property ensures that the m subtrees are independent with respective distributions that of $T_{n_1}(k_1), \ldots, T_{n_m}(k_m)$. An application of the induction hypothesis to these subtrees leads to the expected result. We refer to [59, Section 4.2] for details.

Third Step: A Tightness Criterion To get the convergence for the GH-topology, the previous result must be completed with a tightness criterion. The idea is to use the following well known result.

Theorem 4.9 ([23], Theorem 3.2) *If $X_n, X, X_n(k), X(k)$ are r.v. in a metric space (E, d) such that $X_n(k) \xrightarrow[n\to\infty]{(d)} X(k)$, $\forall k$ and $X(k) \xrightarrow[k\to\infty]{(d)} X$ and for all $\varepsilon > 0$,*

$$\lim_{k\to\infty} \limsup_{n\to\infty} \mathbb{P}\left(d(X_n, X_n(k)) > \varepsilon\right) = 0 \tag{11}$$

then $X_n \xrightarrow[n\to\infty]{(d)} X$.

In our context, the finite-dimensional convergence (10) has already been checked. Moreover, since $\mu_{\gamma,\nu}$ is fully supported on the set of leaves of $\mathcal{T}_{\gamma,\nu}$, we see by picking an infinite sequence of i.i.d. leaves according to $\mu_{\gamma,\nu}$, that there exist versions of the $\mathcal{T}_{\gamma,\nu}(k), k \geq 1$ that converge almost surely to $\mathcal{T}_{\gamma,\nu}$ as $k \to \infty$. It remains to establish the tightness criterion (11) for $T_n, T_n(k)$, with respect to the distance d_{GH}. The main tool is the following bounds:

Proposition 4.10 *Under* (H), *for all $p > 0$, there exists a finite constant C_p such that*

$$\mathbb{P}\left(\frac{\mathsf{ht}(T_n)}{n^\gamma} \geq x\right) \leq \frac{C_p}{x^p}, \quad \forall x > 0, \forall n \geq 1.$$

The proof holds by induction on n, using (H) and the MB-property. We refer to [59, Section 4.3] for details and to see how, using again the MB-property, this helps to control the distance between T_n and $T_n(k)$, to get that for $\varepsilon > 0$:

$$\lim_{k\to\infty} \limsup_{n\to\infty} \mathbb{P}\left(d_{\mathrm{GH}}\left(\frac{T_n(k)}{n^\gamma}, \frac{T_n}{n^\gamma}\right) \geq \varepsilon\right) = 0$$

as required.

5 Applications

We now turn to the description of the scaling limits of various models of random trees that are closely linked to the MB-property.

5.1 *Galton–Watson Trees*

5.1.1 Galton–Watson Trees with *n* Vertices

A first application of Theorem 4.2 is that it permits to recover the classical results of Aldous and Duquesne (grouped together in Theorem 3.1) on the scaling limits of Galton–Watson trees conditioned to have n vertices. To see this, one just has to check the two following lemmas, for η a critical offspring distribution, $\eta(1) \neq 1$, and $(p_n^{\mathrm{GW},\eta})$ the associated splitting distributions defined in (3).

Lemma 5.1 *If η has a finite variance σ^2, then $(p_n^{\mathrm{GW},\eta})$ satisfies* (H) *with*

$$\gamma = 1/2 \quad \text{and} \quad \nu = \frac{\sigma}{2}\nu_{Br}.$$

Lemma 5.2 *If $\eta(k) \sim \kappa k^{-\alpha-1}$ for some $\alpha \in (1,2)$, then $(p_n^{\mathrm{GW},\eta})$ satisfies* (H) *with*

$$\gamma = 1 - 1/\alpha \quad \text{and} \quad \nu = \left(\kappa\Gamma(2-\alpha)\alpha^{-1}(\alpha-1)^{-1}\right)^{1/\alpha}\nu_\alpha.$$

The measures ν_{Br} and ν_α are the dislocation measures of the Brownian and α-stable tree, respectively, and are defined in Sect. 4.3.2. Together with the scaling limit results on MB-trees, this gives Theorem 3.1 (i) and (ii) respectively. The proofs of these lemmas are not completely obvious. We give here a rough idea of the main steps of the proof of Lemma 5.1, and refer to [59, Section 5] for more details and for the proof of Lemma 5.2.

Sketch of the Main Steps of the Proof of Lemma 5.1 Recall that T^η denotes a Galton–Watson tree with offspring distribution η. To simplify, we assume that the support of η generates $\mathbb{Z}$, so that $\mathbb{P}(\#_{\text{vertices}}T^\eta = n) > 0$ for all n large enough. The Otter-Dwass formula (or cyclic lemma) [86, Chapter 6] then implies that

$$\mathbb{P}(\#_{\text{vertices}}T^\eta = n) = n^{-1}\mathbb{P}(S_n = -1)$$

where S_n is a random walk with i.i.d. increments of law $(\eta(i+1), i \geq -1)$. Together with the local limit Theorem, which ensures that $\mathbb{P}(S_n = -1) \underset{n\to\infty}{\sim} (2\pi\sigma^2 n)^{-1/2}$, this leads to

$$\mathbb{P}(\#_{\text{vertices}}T^\eta = n) \underset{n\to\infty}{\sim} (2\pi\sigma^2)^{-1/2}n^{-3/2}.$$

(We note that this argument is also fundamental in the study of large Galton–Watson trees via their contour functions). Then the idea is to use this approximation in the definition of $p_n^{\mathrm{GW},\eta}$ to show that the two sums

$$\sqrt{n}\sum_{\lambda\in\mathcal{P}_n} p_n^{\mathrm{GW},\eta}(\lambda)\left(1-\frac{\lambda_1}{n}\right) f\left(\frac{\lambda}{n}\right) \quad \text{and}$$

$$\frac{\sigma}{\sqrt{2\pi}}\frac{1}{n}\sum_{\lambda_1=\lceil n/2\rceil}^{n} f\left(\frac{\lambda_1}{n},\frac{n-\lambda_1}{n},\ldots\right)\left(\frac{\lambda_1}{n}\right)^{-3/2}\left(\frac{n-\lambda_1}{n}\right)^{-3/2}$$

are asymptotically equivalent (this is the technical part), for all continuous functions $f : \mathcal{S}^{\downarrow} \to \mathbb{R}$. The conclusion follows, since the second sum is a Riemann sum that converges to $\sigma(\sqrt{2\pi})^{-1} \int_{1/2}^{1} f(x, 1-x, \ldots) x^{-3/2}(1-x)^{-3/2} \mathrm{d}x$.

5.1.2 Galton–Watson Trees with Arbitrary Degree Constraints

One may then naturally wonder if Theorem 4.1 could also be used to get the scaling limits of Galton–Watson trees conditioned to have n leaves. The answer is yes, and moreover this can be done in a larger context, using a simple generalization of Theorems 4.1 and 4.2 to MB-trees with arbitrary degree constraints. This generalization was done by Rizzolo [89] using an idea similar to the one presented below Theorem 4.2 to get this theorem from Theorem 4.1. It is quite heavy to state neatly, so we let the reader see the paper [89] and focus here on the applications developed in this paper to Galton–Watson trees.

The classical theorems of Aldous and Duquesne on conditioned Galton–Watson trees can be extended to Galton–Watson trees conditioned to have a number of vertices with out-degree in a given set. To be more precise, fix $A \subset \mathbb{Z}_+$ and consider an offspring distribution η with mean 1 and variance $0 < \sigma^2 < \infty$. For integers n for which such a conditioning is possible, let $T_n^{\eta,A}$ denote a version of a η-Galton–Watson tree conditioned to have exactly n vertices with out-degree in A. For example, if $A = \mathbb{Z}_+$, this is the model of the previous section, whereas if $A = \{0\}$, $T_n^{\eta,A}$ is a η-Galton–Watson tree conditioned to have n leaves.

Theorem 5.3 (Kortchemski [69] and Rizzolo [89]) *As $n \to \infty$,*

$$\left(\frac{T_n^{\eta,A}}{\sqrt{n}}, \mu_n^{\eta,A}\right) \xrightarrow[\mathrm{GHP}]{(\mathrm{d})} \left(\frac{2}{\sigma\sqrt{\eta(A)}}\mathcal{T}_{\mathrm{Br}}, \mu_{\mathrm{Br}}\right).$$

The proof of Rizzolo [89] relies on his theorem on scaling limits of MB-trees with arbitrary degree constraints. The most technical part is to evaluate the splitting probabilities, which is done by generalizing the Otter-Dwass formula. The proof of Kortchemski [69] is more in the spirit of the proofs of Aldous and Duquesne and consists in studying the contour functions of the conditioned trees. We note that [69] also includes cases where η has an infinite variance, and is in the domain of attraction of a stable distribution (the limit is then a multiple of a stable tree). It should be possible to recover this more general case via the approach of Rizzolo.

An Example of Application to Combinatorial Trees Indexed by the Number of Leaves Let T_n be a tree uniformly distributed amongst the set of rooted ordered trees with n leaves with no vertex with out-degree 1. One checks, using (2), that T_n is distributed as an η-Galton–Watson tree conditioned to have n leaves, with η defined by $\eta(i) = (1 - 2^{-1/2})^{i-1}, i \geq 2$, $\eta(1) = 0$ and $\eta(0) = 2 - 2^{1/2}$. The variance of η is $4(\sqrt{2}-1)$, so that finally,

$$\left(\frac{T_n}{\sqrt{n}}, \mu_n\right) \xrightarrow[\text{GHP}]{(\mathrm{d})} \left(\frac{1}{2^{1/4}(\sqrt{2}-1)}\mathcal{T}_{\mathrm{Br}}, \mu_{\mathrm{Br}}\right)$$

where μ_n is the uniform probability on the leaves of T_n.

5.2 *Pólya Trees*

The above results on conditioned Galton–Watson trees give the scaling limits of several sequences of combinatorial trees, as already mentioned. There is however a significant case which does not fall within the Galton–Watson framework, that of *uniform Pólya trees*. By Pólya trees we simply mean rooted finite trees (non-ordered, non-labelled). They are named after Pólya [87] who developed an analytical treatment of this family of trees, based on generating functions. In this section, we let $T_n(\mathrm{P})$ be uniformly distributed amongst the set of Pólya trees with n vertices.

These trees are more complicated to study than uniform rooted trees with labelled vertices, or uniform rooted, ordered trees, because of their lack of symmetry. In this direction, Drmota and Gittenberger [43] showed that the shape of $T_n(\mathrm{P})$ *is not* a conditioned Galton–Watson tree. However Aldous [9] conjectured in 1991 that the scaling limit of $(T_n(\mathrm{P}))$ should nevertheless be the Brownian tree, up to a multiplicative constant. Quite recently, several papers studied the scaling limits of Pólya trees, with different points of view. Using techniques of analytic combinatorics, Broutin and Flajolet [26] studied the scaling limit of the height of a uniform *binary* Pólya tree with n vertices, whereas Drmota and Gittenberger [43] studied the profile of $T_n(\mathrm{P})$ (the profile is the sequence of the sizes of each generation of the tree) and showed that it converges after an appropriate rescaling to the local time of a Brownian excursion. Marckert and Miermont [74] obtained a full scaling limit picture of uniform *binary* Pólya trees: by appropriate trimming procedures, they showed that rescaled by $\sqrt{n}$, they converge in distribution towards a multiple of the Brownian tree.

More recently, with different methods, the following result was proved.

Theorem 5.4 (Haas–Miermont [59] and Panagiotou–Stufler [81]) *As $n \to \infty$,*

$$\left(\frac{T_n(\mathrm{P})}{\sqrt{n}}, \mu_n(\mathrm{P})\right) \xrightarrow[\text{GHP}]{(\mathrm{d})} \left(c_{\mathrm{P}}\mathcal{T}_{\mathrm{Br}}, \mu_{\mathrm{Br}}\right), \quad c_{\mathrm{P}} \sim 1.491$$

where $\mu_n(\mathrm{P})$ denotes the uniform probability on the vertices of $T_n(\mathrm{P})$.

The proof of [59] uses connections with MB-trees, whereas that of [81] uses, still, connections with Galton–Watson trees. Let us first quickly discuss the MB point of view. It is easy to check that the sequence $(T_n(\mathrm{P}))$ is not Markov-Branching (this is left as an exercise!), however it is not far from being so. It is actually possible to couple this sequence with a Markov-Branching sequence $(T'_n(\mathrm{P}))$ such that

$$\mathbb{E}\left[d_{\mathrm{GHP}}(n^{-\varepsilon}T_n(\mathrm{P}), n^{-\varepsilon}T'_n(\mathrm{P}))\right] \underset{n\to\infty}{\longrightarrow} 0, \quad \forall \varepsilon > 0$$

and $(T_n(\mathrm{P}))$ and $(T'_n(\mathrm{P}))$ have the same splitting probabilities (p_n) (by splitting probabilities for trees that are not MB, we mean the distribution of the sizes of the subtrees above the root). These splitting probabilities are given here by

$$p_{n-1}(\lambda) = \frac{\prod_{j=1}^{n-1} \#F_j(m_j(\lambda))}{\#\mathbb{T}_n}, \quad \text{for } \lambda \in \mathcal{P}_{n-1}$$

where $m_j(\lambda) = \{i : \lambda_i = j\}$, $\#\mathbb{T}_n$ is the number of rooted trees with n vertices and $F_j(k)$ denotes the set of multisets with k elements in $\mathbb{T}_j$ (with the convention $F_j(0) := \{\emptyset\}$). It remains to check that these splitting probabilities satisfy (H) with appropriate parameters and to do this, we use the result of Otter [79]:

$$\#\mathbb{T}_n \underset{n\to\infty}{\sim} \mathrm{c}\frac{\kappa^n}{n^{3/2}}, \quad \text{for some } \mathrm{c} > 0, \kappa > 1.$$

Very roughly, this allows to conclude that the two following sums

$$\sqrt{n} \sum_{\lambda\in\mathcal{P}_n} p_n(\lambda)\left(1 - \frac{\lambda_1}{n}\right) f\left(\frac{\lambda}{n}\right) \quad \text{and}$$

$$\frac{\mathrm{c}}{n} \sum_{\lambda_1=\lceil (n-1)/2\rceil}^{n-1} f\left(\frac{\lambda_1}{n}, \frac{n-\lambda_1}{n}, 0, \ldots\right)\left(\frac{\lambda_1}{n}\right)^{-3/2}\left(\frac{n-\lambda_1}{n}\right)^{-3/2}$$

are asymptotically equivalent, so that finally, using that the second sum is a Riemann sum, (H) holds with parameters $\gamma = 1/2$ and $\nu = \nu_{\mathrm{Br}}/c_P$, with $c_{\mathrm{P}} = \sqrt{2}/(\mathrm{c}\sqrt{\pi})$. The method of [81] is different. It consists in showing that asymptotically $T_n(\mathrm{P})$ can be seen as a large finite-variance critical Galton–Watson tree of random size concentrated around a constant times n on which small subtrees of size $O(\log(n))$ are attached. The conclusion then follows from the classical result by Aldous on scaling limits of Galton–Watson trees.

Both methods can be adapted to Pólya trees with other degree constraints. In [59] uniform Pólya trees with n vertices having out-degree in $\{0, m\}$ for some fixed integer m, or out-degree at most m, are considered. More generally, in [81], uniform Pólya trees with n vertices having out-degree in a fixed set A (containing at least 0 and an integer larger than 2) are studied. In all cases, the trees rescaled by $\sqrt{n}$ converge in distribution towards a multiple of the Brownian tree.

Further Result To complete the picture on combinatorial trees asymptotics, we mention a recent result by Stufler on *unrooted* trees, that was conjectured by Aldous, but remained open for a while.

Theorem 5.5 (Stufler [92]) *Let $T_n^*(\mathrm{P})$ be uniform amongst the set of unrooted trees with n vertices (unordered, unlabelled). Then,*

$$\frac{T_n^*(\mathrm{P})}{\sqrt{n}} \xrightarrow[\mathrm{GH}]{(\mathrm{d})} c_{\mathrm{P}}\mathcal{T}_{\mathrm{Br}}$$

(with the same c_{P} as in Theorem 5.4).

The main idea to prove this scaling limit of unrooted uniform trees consists in using a decomposition due to Bodirsky et al. [25] to approximate $T_n^*(\mathrm{P})$ by uniform *rooted* Pólya trees and then use Theorem 5.4. This result more generally holds for unrooted trees with very general degree constraints.

5.3 Dynamical Models of Tree Growth

As mentioned in Sect. 2.2, the prototype example of Rémy's algorithm $(T_n(\mathrm{R}), n \geq 1)$ is strongly connected to Galton–Watson trees since the shape of $T_n(\mathrm{R})$ (to which has been subtracted the edge between the root and the first branch point) is distributed as the shape of a binary critical Galton–Watson tree conditioned to have $2n - 1$ vertices. This implies that

$$\left(\frac{T_n(\mathrm{R})}{\sqrt{n}}, \mu_n(\mathrm{R})\right) \xrightarrow[\mathrm{GHP}]{(\mathrm{d})} \left(2\sqrt{2}\mathcal{T}_{\mathrm{Br}}, \mu_{\mathrm{Br}}\right).$$

Similar scaling limits results actually extends to most of the tree-growth models seen in Sect. 2.2. To see this, it suffices to check that their splitting probabilities satisfy Hypothesis (H). Technically, this mainly relies on Stirling's formula and/or balls in urns schemes. Note however that the convergence in distribution is not fully satisfactory in these cases, since the trees are recursively built on a same probability space, and we may hope to have convergence in a stronger sense. We will see below that this is indeed the case.

5.3.1 Ford's Alpha Model

For Ford's α-model, with $\alpha \in (0, 1)$, it is easy to check (see [63]) that the splitting probabilities $q_n^{\mathrm{Ford},\alpha}$ satisfy hypothesis (H) with $\gamma = \alpha$ and $\nu = \nu_{\mathrm{Ford},\alpha}$, where $\nu_{\mathrm{Ford},\alpha}$ is a binary measure on $\mathcal{S}^{\downarrow}$ $(\nu_{\mathrm{Ford},\alpha}(s_1 + s_2 < 1) = 0)$ defined by

$$\nu_{\mathrm{Ford},\alpha}(s_1 \in \mathrm{d}x) = \frac{\mathbb{1}_{\{1/2 \leq x \leq 1\}}}{\Gamma(1-\alpha)} \left(\alpha(x(1-x))^{-\alpha-1} + (2-4\alpha)(x(1-x))^{-\alpha}\right) \mathrm{d}x.$$

This, together with Theorem 4.2 leads for $\alpha \in (0, 1)$ to the convergence:

Theorem 5.6 ([63] and [59]) *For all $\alpha \in (0, 1)$,*

$$\left(\frac{T_n(\alpha)}{n^\alpha}, \mu_n(\alpha)\right) \xrightarrow[\mathrm{GHP}]{(\mathrm{d})} \left(\mathcal{T}_{\alpha,\nu_{\mathrm{Ford},\alpha}}, \mu_{\alpha,\nu_{\mathrm{Ford},\alpha}}\right).$$

This result was actually first proved in [63], using the fact that the sequence $(T_n(\alpha))$ is Markov-Branching and consistent. Chen and Winkel [32] then improved this result by showing that the convergence holds in probability.

For $\alpha = 1/2$ (Rémy's algorithm), note that we recover the result obtained via the Galton–Watson approach. Note also that the case $\alpha = 1$ is not included in the hypotheses of the above theorem, however the trees $T_n(\alpha)$ are then deterministic (comb trees) and it is clear that they converge after rescaling by n to a segment of length 1, equipped with the Lebesgue measure. This tree is a *general fragmentation tree* as introduced by Stephenson [91], with pure erosion (and no dislocation). When $\alpha = 0$, we observe a different regime, the height of a typical leaf in the tree growth logarithmically, and there is no convergence in the GH-sense of the whole tree.

5.3.2 *k*-Ary Growing Trees

Observing the asymptotic behavior of the sequence of trees constructed via Rémy's algorithm, it is natural to wonder how this may change when deciding to branch at each step $k - 1$ branches on the pre-existing tree, instead of one. For this k-ary model, it was shown in [60] that q_n^k satisfies (H) with $\gamma = 1/k$ and $\nu = \nu_k$ where

$$\nu_k(\mathrm{ds}) = \frac{(k-1)!}{k(\Gamma(\frac{1}{k}))^{k-1}} \prod_{i=1}^{k} s_i^{-(1-1/k)} \left(\sum_{i=1}^{k} \frac{1}{1-s_i}\right) \mathbb{1}_{\{s_1 \geq s_2 \geq \ldots \geq s_k\}} \mathrm{ds},$$

is supported on the simplex of dimension $k - 1$. Together with Theorem 4.1 this gives the limit in distribution of the sequence $(T_n(k), n \geq 1)$. Besides, using some connections with the Chinese Restaurant Processes of Dubins and Pitman (see [86, Chapter 3] for a definition) and more general urns schemes, it was shown that these models converge in probability (however this second approach did not give the distribution of the limiting tree.) Together, these two methods lead to:

Theorem 5.7 ([60]) *Let $\mu_n(k)$ be the uniform measure on the leaves of $T_n(k)$. Then,*

$$\left(\frac{T_n(k)}{n^{1/k}}, \mu_n(k)\right) \xrightarrow[\mathrm{GHP}]{\mathbb{P}} (\mathcal{T}_k, \mu_k)$$

where $(\mathcal{T}_k, \mu_k)$ is a self-similar fragmentation tree, with index of self-similarity $1/k$ and dislocation measure ν_k.

Interestingly, using the approximation by discrete trees, it is possible to show that randomized versions of the limiting trees $\mathcal{T}_k, k \geq 2$—note that $\mathcal{T}_2$ is the Brownian tree up to a scaling factor—can be embedded into each other so as to form an increasing (in k) sequence of trees [60, Section 5].

In Sect. 6.1 we will discuss a generalization of this model. Here, we glue at each step star-trees with $k-1$ branches. However more general results are available when deciding to glue more general tree structures, with possibly a random number of leaves.

5.3.3 Marginals of Stable Trees

For $\beta \in (1, 2]$, the sequence $(T_n(\beta), n \geq 1)$ built by Marchal's algorithm provides, for each n, a tree that is distributed as the shape of the subtree of the stable tree $\mathcal{T}_\beta$ spanned by n leaves taken independently according to μ_β. Duquesne and Le Gall [46] showed that $T_n(\beta)$ is distributed as a Galton–Watson tree whose offspring distribution has probability generating function $z + \beta^{-1}(1-z)^\beta$, conditioned to have n leaves. As so, it is not surprising that appropriately rescaled it should converge to the β-stable tree. Marchal [73] proved an almost-sure *finite-dimensional convergence*, whereas the results of [63] give the convergence in probability for the GHP-topology. Additional manipulations even lead to an almost-sure convergence for the GHP-topology:

Theorem 5.8 ([35]) *Let $\mu_n(\beta)$ be the uniform measure on the leaves of $T_n(\beta)$. Then*

$$\left(\frac{T_n(\beta)}{n^\beta}, \mu_n(\beta)\right) \underset{\text{GHP}}{\overset{\text{a.s.}}{\longrightarrow}} (\beta \mathcal{T}_\beta, \mu_\beta)$$

Using this convergence, it was shown in [35] that randomized versions of the stable trees $\mathcal{T}_\beta, 1 < \beta \leq 2$ can be embedded into each other so as to form a decreasing (in β) sequence of trees.

To complete these results, we mention that Chen et al. [33] propose a model that interpolate between the α-model of Ford and Marchal's recursive construction of the marginals of stable trees, and determine there scaling limits, relying on the results of [63].

5.4 *Cut-Trees*

The notion of the cut-trees was introduced in Example 5 of Sect. 2.3.

Cut-Tree of a Uniform Cayley Tree We use the notation of Example 5, Sect. 2.3 and let C_n be a uniform Cayley tree and T_n^{cut} its cut-tree. Relying essentially on Stirling's formula, one gets that $q_n^{\text{Cut,Cayley}}$ satisfies (H) with $\gamma = 1/2$ and $\nu =$

$\nu_{\mathrm{Br}}/2$, which shows that the rescaled cut-tree $T_n^{\mathrm{cut}}/\sqrt{n}$ endowed with the uniform measure on its leaves converges in distribution to $(2\mathcal{T}_{\mathrm{Br}}, \mu_{\mathrm{Br}})$. This was noticed in [15] and used to determine the scaling limits of the number of steps needed to isolated by edges delation a fixed number of vertices in C_n. Actually, Bertoin and Miermont [19] improve this result by showing the joint convergence

Theorem 5.9 (Bertoin–Miermont [19])

$$\left(\frac{C_n}{\sqrt{n}}, \frac{T_n^{\mathrm{cut}}}{\sqrt{n}}\right) \xrightarrow[\mathrm{GHP}]{(\mathrm{d})} \left(2\mathcal{T}_{\mathrm{Br}}, 2\overline{\mathcal{T}_{\mathrm{Br}}}\right)$$

where $\overline{\mathcal{T}_{\mathrm{Br}}}$ is a tree constructed from $\mathcal{T}_{\mathrm{Br}}$, that can be interpreted as its cut-tree, and that is distributed as $\mathcal{T}_{\mathrm{Br}}$.

Bertoin and Miermont [19] actually more generally extend this result to cut-trees of Galton–Watson trees with a critical offspring distribution with finite variance. This in turn was generalized by Dieuleveut [41] to Galton–Watson trees with a critical offspring distribution in the domain of attraction of a stable law. See also [5, 28, 29] for related results.

Cut-Tree of a Uniform Recursive Tree On the other hand, note that $q_n^{\mathrm{Cut,Recursive}}$ does not satisfy (H). However, Bertoin showed in [17] that in this case, the cut-tree T_n rescaled by $n/\ln(n)$ converges for the GHP-topology to a segment of length 1, equipped with the Lebesgue measure.

6 Further Perspectives

6.1 *Multi-Type Markov-Branching Trees and Applications*

It is possible to enrich trees with *types*, by deciding that each vertex of a tree carries a type, which is an element of a finite or countable set. This multi-type setting is often used in the context of branching processes, where individuals with different types may evolve differently, and had been widely studied. For the trees point of view, scaling limits of multi-type Galton–Watson trees conditioned to have a given number of vertices have been studied by Miermont [77] when both the set of types and the covariance matrix of the offspring distributions are finite, by Berzunza [21] when the set of types is finite with offspring distributions in the domain of attraction of a stable distribution and by de Raphélis [40] when the number of types is infinite, under a finite variance-type assumption. The Brownian and stable trees appear in the scaling limits.

One may more generally be interested in multi-type Markov-Branching trees, which are sequences of trees with vertices carrying types, where, roughly, the subtrees above the root are independent and with distributions that only depend on their size and on the type of their root. In a work in progress [62], results similar to Theorems 4.1 and 4.2 are set up for multi-type MB-trees, when the set of types is finite. Interestingly, different regimes appear in the scaling limits (multi-type or standard fragmentation trees), according to whether the rate of type change is faster than or equal to or slower than the rate of macroscopic branchings.

This should lead to new proofs of the results obtained in [21, 77]. This should also lead to other interesting applications, in particular to dynamical models of tree growth. In these growing models one starts from a finite alphabet of trees and then glues recursively trees by choosing at each step one tree at random in the alphabet and grafting it uniformly on an edge of the pre-existing tree. This generalizes the k-ary construction studied in Sects. 2.2 and 5.3.2, and is connected to multi-type MB-trees. In this general setting multi-type fragmentation trees will appear in the scaling limits.

6.2 Local Limits

This survey deals with scaling limits of random trees. There is another classical way to consider limits of sequences of trees (or graphs), that of *local limits*. This approach is quite different and provides other information on the asymptotics of the trees (e.g. on the limiting behavior of the degrees of vertices). Roughly, a sequence of finite rooted trees (t_n) is said to converge locally to a limit t if for all $R > 0$, the restriction of t_n to a ball of radius R centered at the root converges to the restriction of t to a ball of radius R centered at the root. The trees are therefore not rescaled and the limit is still a discrete object.

For results on the local limits of random models related to the ones considered here, we refer to: Abraham and Delmas [1, 2] and the references therein for Galton–Watson trees, Stefánsson [90] for Ford's α-model and Pagnard [80] for general MB-sequences and the study of the volume growth of their local limits. We also mention the related work by Broutin and Mailler [27] that uses local limits of some models of MB-trees to study asymptotics of And/Or trees, that code boolean functions.

6.3 Related Random Geometric Structures

The discrete trees form a subclass of graphs and are generally simpler to study. There exist however several models of graphs (that are not trees) whose asymptotic study can be conducted by using trees. Different approaches are possible and it is not our purpose to present them here. However we still give some references that are related to some models of trees presented here (in particular Galton–Watson trees) to the interested reader (the list is not exhaustive):

- on random graphs converging to the Brownian tree: [7, 22, 31, 39, 67, 82, 93]
- on random graphs converging to tree-like structures: [36–38]

- on the Erdős–Rényi random graph in the critical window and application to the minimum spanning tree of the complete graph: [4, 6]
- on random maps (which are strongly connected to labeled trees): [78] and all the references therein.

In most of these works the Brownian tree intervenes in the construction of the continuous limit.

Acknowledgements I warmly thank the organizers of the *XII Simposio de Probabilidad y Procesos Estocásticos*, held in Mérida, as well as my collaborators on the topic of Markov-Branching trees, Grégory Miermont, Jim Pitman, Robin Stephenson and Matthias Winkel. Many thanks also to the referee for a very careful reading of a first version of this manuscript that led to many improvements in the presentation.

This work was partially supported by the ANR GRAAL ANR–14–CE25–0014.

References

1. R. Abraham, J.-F. Delmas, Local limits of conditioned Galton-Watson trees: the condensation case. Electron. J. Probab. **19**, 1–29 (2014)
2. R. Abraham, J.-F. Delmas, Local limits of conditioned Galton-Watson trees: the infinite spine case. Electron. J. Probab. **19**, 1–19 (2014)
3. R. Abraham, J.-F. Delmas, P. Hoscheit, A note on the Gromov-Hausdorff-Prokhorov distance between (locally) compact metric measure spaces. Electron. J. Probab. **18**(14), 1–21 (2013)
4. L. Addario-Berry, N. Broutin, C. Goldschmidt, The continuum limit of critical random graphs. Probab. Theory Relat. Fields **152**, 367–406 (2012)
5. L. Addario-Berry, N. Broutin, C. Holmgren, Cutting down trees with a Markov chainsaw. Ann. Appl. Probab. **24**, 2297–2339 (2014)
6. L. Addario-Berry, N. Broutin, C. Goldschmidt, G. Miermont, The scaling limit of the minimum spanning tree of the complete graph. Ann. Probab. **45**(5), 3075–3144 (2017)
7. M. Albenque, J.-F. Marckert, Some families of increasing planar maps. Electron. J. Probab. **13**(56), 1624–1671 (2008)
8. D. Aldous, The continuum random tree. I. Ann. Probab. **19**, 1–28 (1991)
9. D. Aldous, The continuum random tree. II. An overview, in *Stochastic Analysis (Durham, 1990)*. London Mathematical Society Lecture Note Series, vol. 167 (Cambridge University Press, Cambridge, 1991), pp. 23–70
10. D. Aldous, The continuum random tree III. Ann. Probab. **21**, 248–289 (1993)
11. D. Aldous, Probability distributions on cladograms, in *Random Discrete Structures (Minneapolis, MN, 1993)*. The IMA Volumes in Mathematics and its Applications, vol. 76 (Springer, New York, 1996), pp. 1–18
12. J. Bertoin, *Lévy Processes*. Cambridge Tracts in Mathematics, vol. 121 (Cambridge University Press, Cambridge, 1996)
13. J. Bertoin, Self-similar fragmentations. Ann. Inst. Henri Poincaré Probab. Stat. **38**, 319–340 (2002)
14. J. Bertoin, *Random Fragmentation and Coagulation Processes*. Cambridge Studies in Advanced Mathematics, vol. 102 (Cambridge University Press, Cambridge, 2006)
15. J. Bertoin, Fires on trees. Ann. Inst. Henri Poincaré Probab. Stat. **48**, 909–921 (2012)
16. J. Bertoin, Sizes of the largest clusters for supercritical percolation on random recursive trees. Random Struct. Algorithm **44**, 29–44 (2014)
17. J. Bertoin, The cut-tree of large recursive trees. Ann. Inst. Henri Poincaré Probab. Stat. **51**, 478–488 (2015)

18. J. Bertoin, I. Kortchemski, Self-similar scaling limits of Markov chains on the positive integers. Ann. Appl. Probab. **26**, 2556–2595 (2016)
19. J. Bertoin, G. Miermont, The cut-tree of large Galton-Watson trees and the Brownian CRT. Ann. Appl. Probab. **23**, 1469–1493 (2013)
20. J. Bertoin, N. Curien, I. Kortchemski, Random planar maps & growth-fragmentations. Ann. Probab. **46**(1), 207–260 (2018)
21. G. Berzunza, On scaling limits of multitype Galton-Watson trees with possibly infinite variance. Preprint (2016). arXiv:1605.04810
22. J. Bettinelli, Scaling limit of random planar quadrangulations with a boundary. Ann. Inst. Henri Poincaré Probab. Stat. **51**, 432–477 (2015)
23. P. Billingsley, *Convergence of Probability Measures*. Wiley Series in Probability and Statistics: Probability and Statistics, 2nd edn. (Wiley, New York, 1999). A Wiley-Interscience Publication
24. N.H. Bingham, C.M. Goldie, J.L. Teugels, *Regular Variation*. Encyclopedia of Mathematics and its Applications, vol. 27 (Cambridge University Press, Cambridge, 1989)
25. M. Bodirsky, É. Fusy, M. Kang, S. Vigerske, Boltzmann samplers, Pólya theory, and cycle pointing. SIAM J. Comput. **40**, 721–769 (2011)
26. N. Broutin, P. Flajolet, The distribution of height and diameter in random non-plane binary trees. Random Struct. Algorithm **41**, 215–252 (2012)
27. N. Broutin, C. Mailler, And/or trees: a local limit point of view. Random Struct. Algorithm (to appear)
28. N. Broutin, M. Wang, Reversing the cut tree of the Brownian continuum random tree. Electron. J. Probab. **22**(80), 1–23 (2017)
29. N. Broutin, M. Wang, Cutting down p-trees and inhomogeneous continuum random trees. Bernoulli, 23(4A), 2380–2433 (2017)
30. N. Broutin, L. Devroye, E. McLeish, M. de la Salle, The height of increasing trees. Random Struct. Algorithm **32**, 494–518 (2008)
31. A. Caraceni, The scaling limit of random outerplanar maps. Ann. Inst. Henri Poincaré Probab. Stat. **52**, 1667–1686 (2016)
32. B. Chen, M. Winkel, Restricted exchangeable partitions and embedding of associated hierarchies in continuum random trees. Ann. Inst. Henri Poincaré Probab. Stat. **49**, 839–872 (2013)
33. B. Chen, D. Ford, M. Winkel, A new family of Markov branching trees: the alpha-gamma model. Electron. J. Probab. **14**(15), 400–430 (2009)
34. D. Croydon, B. Hambly, Spectral asymptotics for stable trees. Electron. J. Probab. **15**(57), 1772–1801 (2010)
35. N. Curien, B. Haas, The stable trees are nested. Probab. Theory Relat. Fields **157**(1), 847–883 (2013)
36. N. Curien, I. Kortchemski, Random stable looptrees. Electron. J. Probab. **19**(108), 1–35 (2014)
37. N. Curien, I. Kortchemski, Percolation on random triangulations and stable looptrees. Probab. Theory Relat. Fields **163**, 303–337 (2015)
38. N. Curien, T. Duquesne, I. Kortchemski, I. Manolescu, Scaling limits and influence of the seed graph in preferential attachment trees. J. Éc. Polytech. Math. **2**, 1–34 (2015)
39. N. Curien, B. Haas, I. Kortchemski, The CRT is the scaling limit of random dissections. Random Struct. Algorithms **47**, 304–327 (2015)
40. L. de Raphélis, Scaling limit of multitype Galton-Watson trees with infinitely many types. Ann. Inst. Henri Poincaré Probab. Stat. **53**, 200–225 (2017)
41. D. Dieuleveut, The vertex-cut-tree of Galton-Watson trees converging to a stable tree. Ann. Appl. Probab. **25**, 2215–2262 (2015)
42. M. Drmota, *Random Trees: An Interplay Between Combinatorics and Probability* (Springer, Vienna, 2009)
43. M. Drmota, B. Gittenberger, The shape of unlabeled rooted random trees. Eur. J. Comb. **31**, 2028–2063 (2010)
44. T. Duquesne, A limit theorem for the contour process of conditioned Galton-Watson trees. Ann. Probab. **31**, 996–1027 (2003)
45. T. Duquesne, The exact packing measure of Lévy trees. Stoch. Process. Appl. **122**, 968–1002 (2012)

46. T. Duquesne, J.-F. Le Gall, *Random Trees, Lévy Processes and Spatial Branching Processes.* Astérisque (Société Mathématique de France, Paris, 2002), pp. vi+147
47. T. Duquesne, J.-F. Le Gall, Probabilistic and fractal aspects of Lévy trees. Probab. Theory Relat. Fields **131**, 553–603 (2005)
48. T. Duquesne, J.-F. Le Gall, The Hausdorff measure of stable trees. ALEA Lat. Am. J. Probab. Math. Stat. **1**, 393–415 (2006)
49. T. Duquesne, J.-F. Le Gall, On the re-rooting invariance property of Lévy trees. Electron. Commun. Probab. **14**, 317–326 (2009)
50. T. Duquesne, M. Winkel, Growth of Lévy trees. Probab. Theory Relat. Fields **139**, 313–371 (2007)
51. S.N. Evans, *Probability and Real Trees. Lectures from the 35th Summer School on Probability Theory Held in Saint-Flour, July 6–23, 2005.* Lecture Notes in Mathematics, vol. 1920 (Springer, Berlin, 2008)
52. S. Evans, J. Pitman, A. Winter, Rayleigh processes, real trees, and root growth with re-grafting. Probab. Theory Relat. Fields **134**(1), 918–961 (2006)
53. K. Falconer, *Fractal Geometry: Mathematical Foundations and Applications*, 3rd ed. (John Wiley & Sons, Chichester, 2014)
54. P. Flajolet, R. Sedgewick, *Analytic Combinatorics* (Cambridge University Press, Cambridge, 2009)
55. D. Ford, Probabilities on cladograms: introduction to the alpha model. Preprint (2005). arXiv:math/0511246
56. C. Goldschmidt, B. Haas, A line-breaking construction of the stable trees. Electron. J. Probab. **20**, 1–24 (2015)
57. B. Haas, G. Miermont, The genealogy of self-similar fragmentations with negative index as a continuum random tree. Electron. J. Probab. **9**(4), 57–97 (2004)
58. B. Haas, G. Miermont, Self-similar scaling limits of non-increasing Markov chains. Bernoulli J. **17**, 1217–1247 (2011)
59. B. Haas, G. Miermont, Scaling limits of Markov branching trees with applications to Galton-Watson and random unordered trees. Ann. Probab. **40**, 2589–2666 (2012)
60. B. Haas, R. Stephenson, Scaling limits of k-ary growing trees. Ann. Inst. Henri Poincaré Probab. Stat. **51**, 1314–1341 (2015)
61. B. Haas, R. Stephenson, Bivariate Markov chains converging to Lamperti transform Markov Additive Processes. Stoch. Process. Appl. (to appear)
62. B. Haas, R. Stephenson, Multitype Markov Branching trees (In preparation)
63. B. Haas, G. Miermont, J. Pitman, M. Winkel, Continuum tree asymptotics of discrete fragmentations and applications to phylogenetic models. Ann. Probab. **36**, 1790–1837 (2008)
64. B. Haas, J. Pitman, M. Winkel, Spinal partitions and invariance under re-rooting of continuum random trees. Ann. Probab. **37**, 1381–1411 (2009)
65. S. Janson, Random cutting and records in deterministic and random trees. Random Struct. Algorithms **29**, 139–179 (2006)
66. S. Janson, Simply generated trees, conditioned Galton-Watson trees, random allocations and condensation: extended abstract, in *23rd International Meeting on Probabilistic, Combinatorial, and Asymptotic Methods for the Analysis of Algorithms (AofA'12). Discrete Mathematics and Theoretical Computer Science Proceedings, AQ* (Association of Discrete Mathematics and Theoretical Computer Science, Nancy, 2012), pp. 479–490
67. S. Janson, S.Ö. Stefánsson, Scaling limits of random planar maps with a unique large face. Ann. Probab. **43**, 1045–1081 (2015)
68. A. Kolmogorov, Uber das logarithmisch normale Verteilungsgesetz der Dimensionen der Teilchen bei Zerstuckelung. C.R. Acad. Sci. U.R.S.S. **31**, 99–101 (1941)
69. I. Kortchemski, Invariance principles for Galton-Watson trees conditioned on the number of leaves. Stoch. Process. Appl. **122**, 3126–3172 (2012)
70. I. Kortchemski, A simple proof of Duquesne's theorem on contour processes of conditioned Galton-Watson trees, in *Séminaire de Probabilités XLV.* Lecture Notes in Mathematics, vol. 2078 (Springer, Cham, 2013), pp. 537–558

71. J.-F. Le Gall, Random real trees. Ann. Fac. Sci. Toulouse Math. (6) **15**, 35–62 (2006)
72. J.-F. Le Gall, Y. Le Jan, Branching processes in Lévy processes: the exploration process. Ann. Probab. **26**, 213–252 (1998)
73. P. Marchal, A note on the fragmentation of a stable tree, in *Fifth Colloquium on Mathematics and Computer Science. Discrete Mathematics and Theoretical Computer Science Proceedings, AI* (Association of Discrete Mathematics and Theoretical Computer Science, Nancy, 2008), pp. 489–499
74. J.-F. Marckert, G. Miermont, The CRT is the scaling limit of unordered binary trees. Random Struct. Algorithms **38**, 467–501 (2011)
75. G. Miermont, Self-similar fragmentations derived from the stable tree. I. Splitting at heights. Probab. Theory Relat. Fields **127**, 423–454 (2003)
76. G. Miermont, Self-similar fragmentations derived from the stable tree. II. Splitting at nodes. Probab. Theory Relat. Fields **131**, 341–375 (2005)
77. G. Miermont, Invariance principles for spatial multitype Galton-Watson trees. Ann. Inst. Henri Poincaré Probab. Stat. **44**, 1128–1161 (2008)
78. G. Miermont, *Aspects of Random Maps, Lecture Notes of the 2014 Saint–Flour Probability Summer School*. Preliminary draft: http://perso.ens-lyon.fr/gregory.miermont/coursSaint-Flour.pdf
79. R. Otter, The number of trees. Ann. Math. (2) **49**, 583–599 (1948)
80. C. Pagnard, Local limits of Markov-Branching trees and their volume growth. Electron. J. Probab. **22**(95), 1–53 (2017)
81. K. Panagiotou, B. Stufler, Scaling limits of random Pólya trees. Probab. Theory Relat. Fields **170**, 801–820 (2018)
82. K. Panagiotou, B. Stufler, K. Weller, Scaling limits of random graphs from subcritical classes. Ann. Probab. **44**, 3291–3334 (2016)
83. A. Panholzer, Cutting down very simple trees. Quaest. Math. **29**, 211–227 (2006)
84. J.L. Pavlov, The asymptotic distribution of the maximum size of trees in a random forest. Teor. Verojatnost. i Primenen. **22**, 523–533 (1977)
85. J. Pitman, Coalescent random forests. J. Combin. Theory Ser. A **85**, 165–193 (1999)
86. J. Pitman, *Combinatorial Stochastic Processes. Lectures from the 32nd Summer School on Probability Theory Held in Saint-Flour, July 7–24, 2002*. Lecture Notes in Mathematics, vol. 1875 (Springer, Berlin, 2006)
87. G. Pólya, Kombinatorische Anzahlbestimmungen für Gruppen, Graphen und chemische Verbindungen. Acta Math. **68**, 145–254 (1937)
88. J.-L. Rémy, *Un procédé itératif de dénombrement d'arbres binaires et son application à leur génération aléatoire*. RAIRO Inform. Théor. **19**, 179–195 (1985)
89. D. Rizzolo, Scaling limits of Markov branching trees and Galton-Watson trees conditioned on the number of vertices with out-degree in a given set. Ann. Inst. Henri Poincaré Probab. Stat. **51**, 512–532 (2015)
90. S. Stefánsson, The infinite volume limit of Ford's alpha model. Acta Phys. Pol. B Proc. Suppl. **2**(3), 555–560 (2009)
91. R. Stephenson, General fragmentation trees. Electron. J. Probab. **18**(101), 1–45 (2013)
92. B. Stufler, The continuum random tree is the scaling limit of unlabelled unrooted trees. Preprint (2016). arXiv:1412.6333
93. B. Stufler, Random enriched trees with applications to random graphs. Preprint (2016). arXiv:1504.02006

Optimality of Two-Parameter Strategies in Stochastic Control

Kazutoshi Yamazaki

Abstract In this note, we study a class of stochastic control problems where the optimal strategies are described by two parameters. These include a subset of singular control, impulse control, and two-player stochastic games. The parameters are first chosen by the two continuous/smooth fit conditions, and then the optimality of the corresponding strategy is shown by verification arguments. Under the setting driven by a spectrally one-sided Lévy process, these procedures can be efficiently performed owing to the recent developments of scale functions. In this note, we illustrate these techniques using several examples where the optimal strategy and the value function can be concisely expressed via scale functions.

Keywords Singular control · Impulse control · Zero-sum games · Optimal stopping · Spectrally one-sided Lévy processes · Scale functions

AMS 2010 Subject Classifications 60G51, 93E20, 49J40

1 Introduction

In stochastic control, the objective is to optimally control a stochastic process to minimize or maximize the expected value of a given payoff, which is determined by the paths of the control and/or controlled processes. In other words, we want to identify an *optimal strategy* that attains the minimal or maximal expected value, which is referred to the *(optimal) value function*. Essentially, all real-life phenomena contain uncertainty. Consequently the problem of stochastic control

K. Yamazaki (✉)
Department of Mathematics, Faculty of Engineering Science, Kansai University, Suita-shi, Osaka, Japan
e-mail: kyamazak@kansai-u.ac.jp

© Springer International Publishing AG, part of Springer Nature 2018

D. Hernández-Hernández et al. (eds.), *XII Symposium of Probability and Stochastic Processes*, Progress in Probability 73,
https://doi.org/10.1007/978-3-319-77643-9_2

arises everywhere. It is well studied in, among other fields, finance (e.g., portfolio optimization, asset pricing, and risk management), economics (e.g., search, real options, and games), insurance, inventory management, and queues.

Because stochastic control has a wide range of applications and is studied in a variety of fields, there are many different possible modeling approaches. A model can be categorized based on (1) a discrete/continuous time, (2) a discrete/continuous state, and (3) a finite/infinite horizon.

In this note, we focus on a relatively simple class of stochastic control problems where analytical solutions can be obtained. We assume the continuous-time, infinite-horizon case with the state space given by $\mathbb{R}$ or its subset. In addition, we assume randomness to be modeled by a one-dimensional *spectrally one-sided Lévy process*, or a Lévy process with only one-sided jumps that does not have a monotone path almost surely (a.s.). As the title of this note suggests, we are particularly interested in cases where *two parameters* are sufficient to describe the optimal strategy. While one-parameter optimal strategies are ubiquitous, to the best of our knowledge the study of two-parameter strategies is rather rare.

1.1 One-Parameter Strategies

In most stochastic control problems that admit analytical solutions, an optimal strategy can typically be described by one parameter.

In the continuous-time, infinite-horizon *optimal stopping* driven by a one-dimensional Markov process, the stopping and waiting regions are separated by *free boundaries*, and, in many cases, the boundary is a single point. In the American/Russian perpetual (vanilla) options driven by a Lévy process, it is known, as in [3] and [37], that it is optimal to exercise when the process itself or its reflected process goes above or below a certain barrier for the first time. In the quickest detection of a Wiener process [45] where we want to promptly detect the unobservable sudden change of the drift of the process, it is optimal to stop when the posterior probability process exceeds some level for the first time. There are a number of other examples for which the first crossing time of a boundary is optimal. See [17, 31, 33] and also the book by Peskir and Shiryaev [41].

In singular control, again, the controlling and waiting regions are typically separated by a single point. Well-studied examples include de Finetti's dividend problem, in which we want to maximize the total expected dividends accumulated until ruin [or the first time the (controlled) surplus process goes below zero]. A majority of the existing literature focuses on the optimality of the barrier strategy that pays dividends so that the surplus process is reflected at the barrier. In the spectrally negative Lévy model, it has been shown by Loeffen [34] that a barrier strategy is optimal on the condition that the Lévy measure has a completely monotone density. On the other hand, for the spectrally positive Lévy case, optimality is guaranteed as shown in [8]. Recently, these results have been extended to cases in which a strategy is assumed to be *absolutely continuous* with respect to the Lebesgue measure: the optimal strategy can again be described by a single threshold, and the so-called *refraction strategy* is optimal; see [32] and [50].

In the continuous-time inventory model (with the assumption that backorders are allowed), one wants to find an optimal replenishment strategy that minimizes the sum of the inventory and controlling costs. In the spectrally negative Lévy case, under e.g. the convexity assumption regarding the inventory cost and with the absence of a fixed cost, it has been shown to be optimal to replenish the item so that the inventory does not drop below a certain level (see Section 7 of [49]). The absolutely continuous case has been studied by Hernández-Hernández et al. [27], in which they showed the optimality of a refraction strategy.

1.2 Two-Parameter Strategies

In view of the above examples of one-parameter strategies, it is not difficult to see that by a simple modification to the problem setting, more parameters are needed to describe the optimal strategy. Here, we list several examples where one additional parameter is also needed.

1.2.1 Two-Sided Singular Control

In the above examples of singular control, we assumed control to be one-sided: we can only decrease or increase the underlying process. However, there are versions in which it is two-sided and we can both decrease and increase the process.

In the extension of de Finetti's problem with *capital injections*, the surplus process can also be increased by injecting capital. Typically, the problem requires that capital be injected so that the surplus process never goes below zero. In inventory control, we can think of a version in which the item can be replenished and also sold so as to avoid a shortage or excess of an inventory, respectively.

1.2.2 Impulse Control

We can consider another extension from singular control by adding a fixed cost. Namely, in addition to the cost (or reward) that is proportional to the amount of modification, a fixed cost is incurred each time it is modified. In this case, it is clear that one parameter is no longer sufficient to describe the optimal strategy. Instead, we can expect that the (s, S)-strategy (more commonly called the (s, S)-policy) is a reasonable candidate. In other words, given two threshold levels s and S, whenever the process goes above (or below) s, it is pushed down (or up) to S. The optimality of an (s, S)-strategy is often a primary objective in the impulse control literature.

1.2.3 Zero-Sum Games Between Two Players

In a (stochastic) game, multiple players aim to maximize their own expected payoffs. However, the payoff depends not only on the actions of one player but also on those of the others. The primary objective of game theory is to identify, if any, a *Nash equilibrium (saddle point)*, which is a set of strategies such that no player can increase her expected payoff by solely changing her strategy, unless the other players also change their strategies.

Consider a case with two players in which a common payoff is maximized by one player and is minimized by the other. Under settings similar to those described in Sect. 1.1 above, each player's strategy is described by one parameter. Consequently, the equilibrium is described by two parameters.

1.3 Fluctuation Theory of Spectrally One-Sided Lévy Processes

In this note, we assume throughout that the underlying (uncontrolled) process is a spectrally negative Lévy process. The spectrally positive Lévy process is its dual and hence the case driven by this process is also covered. While spectrally one-sided Lévy processes are not necessarily desirable for realistic models, at least analytically, it has a great advantage to work with these sets of processes.

Over the last decade, significant developments in the fluctuation theory of spectrally one-sided Lévy processes have been presented (see, e.g., the textbooks by Bertoin [13], Doney [16], and Kyprianou [30]). Various fluctuation identities are known to be written using the so-called *scale functions*, and these include essentially all the expectations needed to compute the net present values (NPVs) of the payoffs under the one- and two-parameter strategies described above.

The scale function is defined by its Laplace transform written in terms of the Laplace exponent of the process. We see in this note that, despite its concise characterization, it still contains sufficient information to solve the problem.

1.4 Solution Procedures

Using the expected NPVs of payoffs under each two-parameter strategy, written explicitly in terms of the scale function, the classical "*guess and verify*" approach can be performed in a straightforward manner. Here, we briefly illustrate each step below.

1.4.1 Selection of the Two Parameters

As the form of the candidate strategy is already conjectured, the guessing part essentially is to decide on the values of the two parameters. Because we need to identify two values, naturally, we need two equations.

Before discussing the two-parameter case, we start with the one-parameter case to gain some intuition. As reviewed above in Sect. 1.1, the parameter usually corresponds to the value of a barrier. Here, we temporarily use $u_a(x)$ for the expected NPV when the parameter/barrier is a and the starting value of the process is x.

In this case, the most intuitive and straightforward approach is to use the first-order condition. Namely, we first obtain the parameter, say a^*, that minimizes or maximizes $a \mapsto u_a(x)$. Naturally, we expect (given that the barrier is in the interior of the state space) the derivative $\partial u_a(x)/\partial a|_{a=a^*}$ to vanish. This can be easily accomplished because $u_a(x)$ is written using the scale function, whose smoothness has been well studied (see Remark 2.1 below).

Alternatively, we can apply what is known as *continuous/smooth fit,* which basically chooses the barrier a^* so that *the degree of smoothness of* $u_a(\cdot)$ *at* a *increases by one by setting* $a = a^*$. The smoothness at the barrier is in general dependent on the regularity (see Sect. 2.1 below for its definition). In optimal stopping and impulse control, we expect the value function to be continuous (resp. continuously differentiable) at the barrier when it is irregular (resp. regular) for the controlling/stopping region. On the other hand, for singular control, we expect it to be continuously differentiable (resp. twice continuously differentiable) at the barrier when it is irregular (resp. regular).

At least for the Lévy case, these two methods tend to lead to the same condition, i.e., some function, say $a \mapsto g(a)$, of the barrier level a (and not x) vanishes; see Fig. 1. In addition, under a suitable assumption, it is typically a strictly monotone function. Hence, the candidate barrier can be defined as its unique root. Detailed discussion of the equivalence of these two methods for optimal stopping problems is presented in [18].

We now discuss the two-parameter case. Let us temporarily use $v_{a,b}(x)$ for the expected NPV under the strategy parametrized by (a, b) when the starting value of the process is x.

Again, the first approach is to use the first-order condition. This time, we apply it with respect to the two parameters (a, b), or equivalently, we compute the partial derivatives $\partial v_{a,b}(x)/\partial a$ and $\partial v_{a,b}(x)/\partial b$ and choose parameters so that both vanish simultaneously. The second approach is to use continuous/smooth fit at the barriers (with an additional condition for the case of impulse control). Again, we end up having the same two equations, e.g., $\Lambda(a, b) = 0$ and $\lambda(a, b) = 0$.

The difficulty here is that this time we need to show the existence of solutions to the two equations, which are typically nonlinear functions. However, the two equations tend to be related in that one is the partial derivative of the other, i.e., $\lambda(a, b) = \partial \Lambda(a, b)/\partial b$. In other words, we want to obtain the curve $b \mapsto \Lambda(a^*, b)$ that touches and becomes tangent to the x-axis at b^* (see Fig. 2).

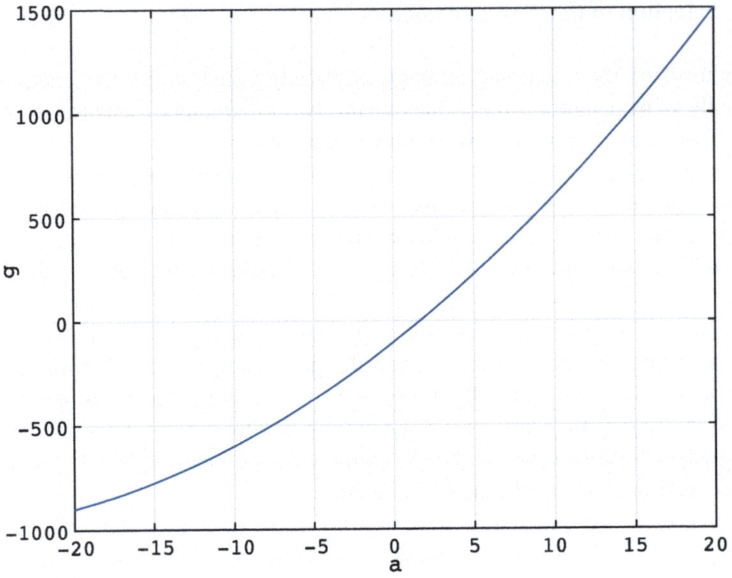

Fig. 1 (One-parameter case) Typical function $a \mapsto g(a)$ obtained when the first-order or continuous/smooth fit condition is applied. The desired parameter becomes its unique root

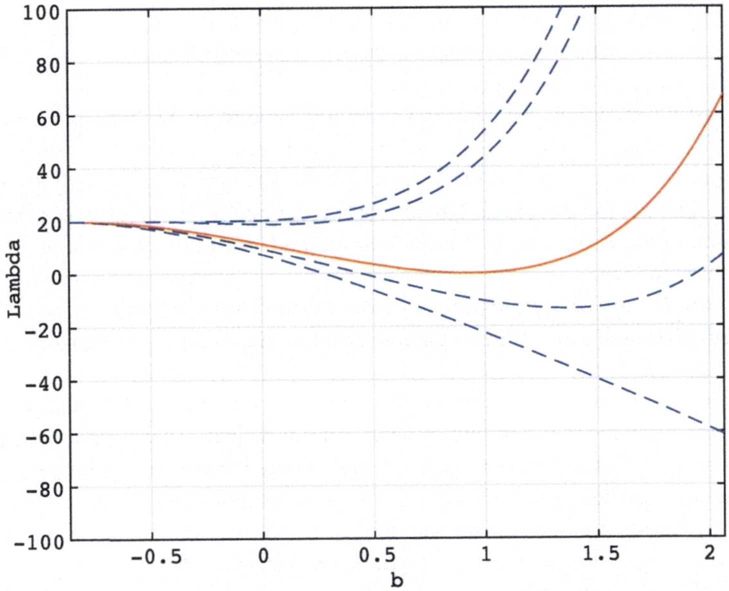

Fig. 2 (Two-parameter case) Typical function obtained when the first-order or continuous/smooth fit condition is applied. The plot is the curve $b \mapsto \Lambda(a, b)$ on $[a, \infty)$ for different values of a. Typically the desired values (a^*, b^*) become those for which $\lambda(a^*, b^*) = \partial\Lambda(a^*, b)/\partial b|_{b=b^*} = 0$. In other words, we must determine the starting point a^* such that the curve becomes tangent to the x-axis at b^*, as in the solid curve in the plot

1.4.2 Verification of Optimality

After we select the values of the two parameters, say (a^*, b^*), we must verify the optimality of the corresponding strategy. The so-called *verification lemma* gives a sufficient condition for optimality that commonly requires:

(1) the smoothness of v_{a^*,b^*},
(2) that v_{a^*,b^*} solves the variational inequalities.

The imposed conditions must be sufficient so that the discounted process of $v_{a^*,b^*}(\cdot)$ (killed upon exiting the state space), driven by any controlled process, is a local sub/super-martingale. In general, the forms of the variational inequalities are well known (see e.g. [39]). However, its technical details must be customized, and, in particular, we need to take care of the tails of v_{a^*,b^*} and the Lévy measure. Because of the localizing arguments needed to apply Itô's formula, at the end, we must take a limit and interchange it over integrals.

Regarding condition (1), we choose the values of (a^*, b^*) at the guessing step so that v_{a^*,b^*} is "sufficiently smooth," although the smoothness at the boundary may not be sufficient to apply the usual version of Itô's formula (so we may need the Meyer-Itô version). For stochastic calculus for Lévy processes, see [44] and [1].

Showing (2) is usually the hardest part and sometimes it fails. The variational inequalities must hold at each point in the state space, which is separated into the waiting and controlling regions. In our examples when the state space is $\mathbb{R}$, except for the impulse control case, the waiting region is given by (a^*, b^*), whereas the controlling region is $(-\infty, a^*) \cup (b^*, \infty)$. At a point in the waiting region (a^*, b^*), the proof is normally simple because the discounted process of $v_{a^*,b^*}(\cdot)$ driven by the underlying process is a martingale (see Sect. 2.7.3). On the other hand, the proof for the point in (b^*, ∞) (resp. $(-\infty, a^*)$) tends to be difficult for the spectrally negative (resp. positive) Lévy case. Intuitively, this is because the process can jump from one region to the other, where the form of v_{a^*,b^*} changes.

1.5 Comparison with Other Approaches

The classical approach to the stochastic control of Lévy processes involves integro-differential equations (IDEs).

First, we identify the candidate value function as the solution to an IDE with its boundary conditions given by the desired continuity/smoothness at the barriers. Except for special cases, this cannot be solved analytically, and hence verification arguments must be made using this implicit representation of the candidate value function. This is especially difficult when the Lévy measure is infinite.

A clear advantage of using the fluctuation theory approach described above is that, if the function v_{a^*,b^*} can be computed using the scale function, computation is much more direct and simple. While the scale function in general does not admit analytically closed expression, solution methods do not require details of its form.

Typically, the selection of parameters can be performed by its asymptotic property at zero (see Sect. 2.3 below) and, for verification, we can use some general properties of the scale function.

Another advantage is that it can deal with cases with jumps of infinite activity/variation without any additional work. The IDE approach must often assume that the jump part of the underlying process is a compound Poisson process. However, there are a number of important examples with infinite Lévy measures, such as variance gamma, CGMY, and normal inverse Gaussian processes, as well as classical ones, such as the gamma process and a subset of stable processes.

1.6 Computation

Using these approaches, the value function and the selected parameters are written in terms of the scale function. Hence, their computation is essentially equivalent to that of the scale function. Because the scale function is defined by its Laplace transform written in terms of the Laplace exponent, it must be inverted either analytically or numerically.

Some classes of Lévy processes have rational forms of Laplace exponents. For these processes, analytical forms of scale functions can be easily obtained by partial fraction decomposition. Among them, the case with i.i.d. phase-type jumps [2] is particularly important, because at least in principle it can approximate any Lévy process. This means that any scale function can be approximated by the scale function of this process. Egami and Yamazaki [19] conducted a sequence of numerical experiments to confirm the accuracy of this approximation.

Alternatively, the scale function can always be directly computed via numerical Laplace inversion. As discussed in [29], the scale function can be written as the difference between an exponential function (whose parameter is defined by $\Phi(q)$ in the current note) and the resolvent (potential) term (see the third equation in (2.8) below). Hence, the computation is reduced to that of the resolvent term. This is a bounded function that asymptotically converges to zero, and hence, numerical Laplace inversion can be quickly and accurately conducted. For more details, we refer readers to Section 5 of [29].

In this note, we review these techniques, using several examples of two-sided singular control, impulse control and games, as reviewed in Sect. 1.2 above. Our aim is not to offer rigorous arguments. Instead, we present a guide on how we can apply the existing results in the fluctuation theory and scale function to solve stochastic control problems. For more technical details, we refer readers to the original works we cite throughout the note.

The rest of this note is organized as follows:

In Sect. 2, we review the spectrally negative Lévy process and the scale function. In particular, we review the fluctuation identities as well as some important properties of the scale function that we use later in the note.

In Sect. 3, we examine two-sided singular control, which we introduced in Sect. 1.2.1. First, we give the formulation and review several examples. Then, we discuss how to choose the two parameters via continuous/smooth fit and demonstrate its optimality via verification arguments. In particular, we focus on the problems considered in Bayraktar et al. [8] and Baurdoux and Yamazaki [6] and illustrate how to follow these solution procedures.

In Sect. 4, we consider impulse control, as addressed in Sect. 1.2.2. While the techniques used are similar to those used for singular control, there are several major differences and new challenges in the solution. In particular, we use the case in Yamazaki [49] to illustrate the steps necessary to solve the problem.

In Sect. 5, we discuss two-player optimal stopping games, as introduced in Sect. 1.2.3, with a special focus on the problem studied by Egami et al. [20]. We also make some remarks regarding other forms of two-player zero-sum games.

Throughout this study, we use $f(x+) := \lim_{y \downarrow x} f(y)$ and $f(x-) := \lim_{y \uparrow x} f(y)$ to indicate the right- and left-hand limits, respectively, for any function f whenever they exist. We let $\Delta \xi_t := \xi_t - \xi_{t-}$, for any process with left limits ξ. Finally, for any interval $\mathcal{I} \subset \mathbb{R}$, let $\overline{\mathcal{I}} := \sup \mathcal{I}$, $\underline{\mathcal{I}} := \inf \mathcal{I}$, and $\mathcal{I}^o$ be the interior of $\mathcal{I}$.

2 Spectrally Negative Lévy Processes and Scale Functions

In this section, we review the spectrally negative Lévy process and its fluctuation theory. We shall also review the scale function and list the fluctuation identities as well as some important properties that are frequently used in stochastic control. Note that the spectrally positive Lévy process is its dual, and the results introduced here can be directly applied as well.

Defined on a probability space $(\Omega, \mathcal{F}, \mathbb{P})$, let X be a spectrally negative Lévy process with its Laplace exponent X given by

$$\psi(s) := \log \mathbb{E}\left[e^{sX_1}\right] = \gamma s + \frac{1}{2}\sigma^2 s^2 + \int_{(-\infty,0)} (e^{sz} - 1 - sz1_{\{z>-1\}})\nu(dz), \quad s \geq 0, \tag{2.1}$$

where ν is a Lévy measure with the support $(-\infty, 0)$ that satisfies the integrability condition $\int_{(-\infty,0)} (1 \wedge |z|^2)\nu(dz) < \infty$. For every $x \in \mathbb{R}$, let $\mathbb{P}_x$ be the conditional probability under which $X_0 = x$ (in particular, we let $\mathbb{P} \equiv \mathbb{P}_0$), and $\mathbb{E}_x$ and $\mathbb{E}$ be the corresponding expectation operators. Let $\mathbb{F}$ be the filtration generated by X.

The path variation of the process is particularly important in stochastic control, especially when we apply continuous/smooth fit as we shall see in later sections. For the case of a Lévy process, it has paths of *bounded variation* a.s. or otherwise it has paths of *unbounded variation* a.s. The former holds if and only if $\sigma = 0$ and

$\int_{(-1,0)} |z| \, \nu(\mathrm{d}z) < \infty$; in this case, the expression (2.1) can be simplified to

$$\psi(s) = \delta s + \int_{(-\infty,0)} (e^{sz} - 1)\nu(\mathrm{d}z), \quad s \geq 0,$$

with $\delta := \gamma - \int_{(-1,0)} z \, \nu(\mathrm{d}z)$.

Throughout the note, we exclude the case in which X is the negative of a subordinator (i.e., X is monotonically decreasing a.s.). This assumption implies that $\delta > 0$ when X is of bounded variation.

2.1 *Path Variations and Regularity*

As defined in Definition 6.4 of [30], we call a point x *regular* for an open or closed set B if $\mathbb{P}_x\{T_B = 0\} = 1$ where

$$T_B := \inf\{t > 0 : X_t \in B\},$$

and *irregular* if $\mathbb{P}_x\{T_B = 0\} = 0$; here and throughout the note, let $\inf \varnothing = \infty$. By Blumenthal's zero-one law, the probability $\mathbb{P}_x\{T_B = 0\}$ is either 0 or 1, and hence any point is either regular or irregular.

As summarized in Section 8 of [30], for any spectrally negative Lévy process X, the point 0 is regular for $(0, \infty)$, meaning that, if the process starts at 0, it enters $(0, \infty)$ immediately. On the other hand, 0 is regular for $(-\infty, 0)$ if and only if the process has paths of unbounded variation.

We shall see in later sections that the smoothness of the value function at (free) boundaries depends on their regularity.

2.2 *Scale Functions*

Fix $q \geq 0$. For any spectrally negative Lévy process X, its q-scale function

$$W^{(q)} : \mathbb{R} \to [0, \infty)$$

is a function that is zero on $(-\infty, 0)$, continuous and strictly increasing on $[0, \infty)$, and is characterized by the Laplace transform:

$$\int_0^\infty e^{-sx} W^{(q)}(x) \mathrm{d}x = \frac{1}{\psi(s) - q}, \qquad s > \Phi(q), \tag{2.2}$$

where

$$\Phi(q) := \sup\{\lambda \geq 0 : \psi(\lambda) = q\}.$$

Here, the Laplace exponent ψ in (2.1) is known to be zero at the origin and convex on $[0, \infty)$. We also define, for $x \in \mathbb{R}$,

$$\begin{aligned}
\overline{W}^{(q)}(x) &:= \int_0^x W^{(q)}(y)\mathrm{d}y, \\
Z^{(q)}(x) &:= 1 + q\overline{W}^{(q)}(x), \\
\overline{Z}^{(q)}(x) &:= \int_0^x Z^{(q)}(z)\mathrm{d}z = x + q\int_0^x \int_0^z W^{(q)}(w)\mathrm{d}w\mathrm{d}z.
\end{aligned}$$

Because $W^{(q)}(x) = 0$ for $-\infty < x < 0$, we have

$$\overline{W}^{(q)}(x) = 0, \quad Z^{(q)}(x) = 1 \quad \text{and} \quad \overline{Z}^{(q)}(x) = x, \quad x \leq 0. \tag{2.3}$$

We shall also define, when $\psi'(0+) > -\infty$,

$$R^{(q)}(x) := \overline{Z}^{(q)}(x) + \frac{\psi'(0+)}{q}, \quad x \in \mathbb{R}.$$

In Fig. 3, we show sample plots of the scale function $W^{(q)}$ on $[0, \infty)$ for the cases of bounded and unbounded variation. Its behaviors as $x \downarrow 0$ and $x \uparrow \infty$ are reviewed later in this section.

2.3 Smoothness of Scale Functions

A particularly important property of the scale function, which is helpful in applying continuous/smooth fit, is its behaviors around zero: as in Lemmas 3.1 and 3.2 of [29],

$$W^{(q)}(0) = \begin{cases} 0, & \text{if } X \text{ is of unbounded variation,} \\ \frac{1}{\delta}, & \text{if } X \text{ is of bounded variation,} \end{cases} \tag{2.4}$$

$$W^{(q)\prime}(0+) := \lim_{x \downarrow 0} W^{(q)\prime}(x) = \begin{cases} \frac{2}{\sigma^2}, & \text{if } \sigma > 0, \\ \infty, & \text{if } \sigma = 0 \text{ and } \nu(-\infty, 0) = \infty, \\ \frac{q+\nu(-\infty,0)}{\delta^2}, & \text{if } \sigma = 0 \text{ and } \nu(-\infty, 0) < \infty. \end{cases} \tag{2.5}$$

Note that these can be confirmed in Fig. 3.

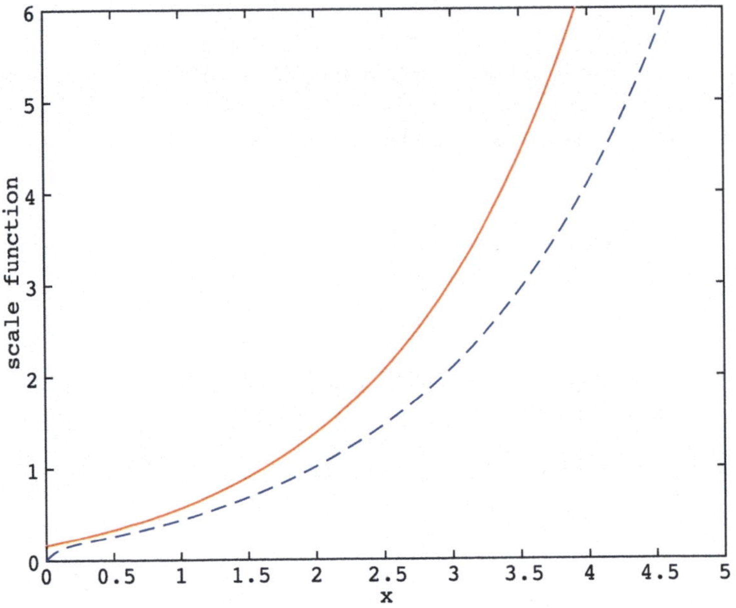

Fig. 3 Plots of the scale function $W^{(q)}$ on $[0, \infty)$. The solid red curve is for the case of bounded variation; the dotted blue curve is for the case of unbounded variation (with $\sigma > 0$). As reviewed in (2.4), its behaviors around zero depend on the path variation of the process. In addition, as in (2.13), it increases exponentially as $x \to \infty$

As we shall see in later sections, when considering continuity/smoothness at the lower barrier, the difference between the right-hand and left-hand limits often becomes the product of $W^{(q)}(0)$ and some function, say $\Lambda(a, b)$, of the two parameters (barriers) (a, b) to be selected: for these to match, the parameters (a, b) must be chosen so that $\Lambda(a, b)$ vanishes if $W^{(a)}(0) > 0$.

When $W^{(q)}(0) = 0$ (or equivalently X is of unbounded variation), then the value function is expected to be smoother. Repeating the same procedure for its derivative, one gets that the difference between the right-hand and left-hand limits becomes the product of $W^{(q)\prime}(0+)$ and $\Lambda(a, b)$; in this case, (a, b) must be chosen so that $\Lambda(a, b) = 0$.

At the upper boundary, the smoothness tends to be the same for both bounded and unbounded variation cases: this gives another equation $\lambda(a, b) = 0$ where $\lambda(a, b)$ is the partial derivative of $\Lambda(a, b)$ with respect to b.

Regarding the smoothness of the scale function on $\mathbb{R}\backslash\{0\}$, we have the following; see [15] for more comprehensive results. These smoothness results are important in order to apply Itô's formula where the (candidate) value function must be C^2 (resp. C^1) for the case of unbounded (resp. bounded) variation.

Remark 2.1 If X is of unbounded variation or the Lévy measure does not have an atom, then it is known that $W^{(q)}$ is $C^1(\mathbb{R}\backslash\{0\})$. Hence,

(1) $Z^{(q)}$ is $C^1(\mathbb{R}\backslash\{0\})$ and $C^0(\mathbb{R})$ for the bounded variation case, while it is $C^2(\mathbb{R}\backslash\{0\})$ and $C^1(\mathbb{R})$ for the unbounded variation case,

(2) $\overline{Z}^{(q)}$ is $C^2(\mathbb{R}\backslash\{0\})$ and $C^1(\mathbb{R})$ for the bounded variation case, while it is $C^3(\mathbb{R}\backslash\{0\})$ and $C^2(\mathbb{R})$ for the unbounded variation case.

In addition, if $\sigma > 0$, then $W^{(q)}$ is $C^2(\mathbb{R}\backslash\{0\})$.

2.4 *Fluctuation Identities for Spectrally Negative Lévy Processes*

Here we shall list some fluctuation identities for the spectrally negative Lévy process X.

2.4.1 Two-Sided Exit

The most well-known application of the scale function is as follows. Let us define the first down- and up-crossing times, respectively, of X by

$$T_b^- := \inf\{t > 0 : X_t < b\} \quad \text{and} \quad T_b^+ := \inf\{t > 0 : X_t > b\}, \quad b \in \mathbb{R}. \tag{2.6}$$

Then, for any $b > 0$ and $x \le b$,

$$\begin{aligned} \mathbb{E}_x\left[e^{-qT_b^+} 1_{\{T_b^+ < T_0^-\}}\right] &= \frac{W^{(q)}(x)}{W^{(q)}(b)}, \\ \mathbb{E}_x\left[e^{-qT_0^-} 1_{\{T_b^+ > T_0^-\}}\right] &= Z^{(q)}(x) - Z^{(q)}(b)\frac{W^{(q)}(x)}{W^{(q)}(b)}, \\ \mathbb{E}_x\left[e^{-qT_0^-}\right] &= Z^{(q)}(x) - \frac{q}{\Phi(q)} W^{(q)}(x). \end{aligned} \tag{2.7}$$

2.4.2 Resolvent Measures

The scale function can express concisely the q-resolvent (potential) measure. As summarized in Theorem 8.7 and Corollaries 8.8 and 8.9 of [30] (see also Bertoin

[14], Emery [22], and Suprun [46]), we have

$$\mathbb{E}_x\Big[\int_0^{T_0^- \wedge T_b^+} e^{-qt} 1_{\{X_t \in \mathrm{d}y\}} \mathrm{d}t\Big] = \Big[\frac{W^{(q)}(x) W^{(q)}(b-y)}{W^{(q)}(b)} - W^{(q)}(x-y)\Big]\mathrm{d}y, \quad b > 0,\ x \le b,$$
$$\mathbb{E}_x\Big[\int_0^{T_0^-} e^{-qt} 1_{\{X_t \in \mathrm{d}y\}} \mathrm{d}t\Big] = \Big[e^{-\Phi(q)y} W^{(q)}(x) - W^{(q)}(x-y)\Big]\mathrm{d}y, \tag{2.8}$$
$$\mathbb{E}_x\Big[\int_0^{\infty} e^{-qt} 1_{\{X_t \in \mathrm{d}y\}} \mathrm{d}t\Big] = \Big[\frac{e^{\Phi(q)(x-y)}}{\psi'(\Phi(q))} - W^{(q)}(x-y)\Big]\mathrm{d}y.$$

Now define, for any measurable function h and $s \in \mathbb{R}$,

$$\Psi(s; h) := \int_0^\infty e^{-\Phi(q)y} h(y+s)\mathrm{d}y = \int_s^\infty e^{-\Phi(q)(y-s)} h(y)\mathrm{d}y,$$
$$\varphi_s(x; h) := \int_s^x W^{(q)}(x-y) h(y)\mathrm{d}y, \quad x \in \mathbb{R}.$$

Here $\varphi_s(x; h) = 0$ for any $x \le s$ because $W^{(q)}$ is uniformly zero on $(-\infty, 0)$. Then it is clear that

$$\mathbb{E}_x\Big[\int_0^{T_a^- \wedge T_b^+} e^{-qt} h(X_t)\mathrm{d}t\Big] = \frac{W^{(q)}(x-a)}{W^{(q)}(b-a)}\varphi_a(b; h) - \varphi_a(x; h), \quad b > a,\ x \le b,$$
$$\mathbb{E}_x\Big[\int_0^{T_a^-} e^{-qt} h(X_t)\mathrm{d}t\Big] = \Psi(a; h) W^{(q)}(x-a) - \varphi_a(x; h), \quad x, a \in \mathbb{R},$$

where we assume for the latter that $\Psi(a; h)$ is well-defined and finite.

2.5 Fluctuation Identities for the Infimum and Reflected Processes

Let us define the *running infimum and supremum processes*

$$\underline{X}_t := \inf_{0 \le t' \le t} X_{t'} \quad \text{and} \quad \overline{X}_t := \sup_{0 \le t' \le t} X_{t'}, \quad t \ge 0.$$

Then, the processes reflected from above at b and below at a are given, respectively, by

$$\bar{Y}_t^b := X_t - D_t^b \quad \text{and} \quad \underline{Y}_t^a := X_t + U_t^a, \quad t \ge 0,$$

where

$$D_t^b := (\overline{X}_t - b) \vee 0 \quad \text{and} \quad U_t^a := (a - \underline{X}_t) \vee 0, \quad t \geq 0,$$

are the cumulative amounts of reflections that push the processes downward and upward, respectively.

2.5.1 Fluctuation Identities for the Infimum Process

By Corollary 2.2 of [29],

$$\begin{aligned}\mathbb{E}\Big[\int_0^\infty e^{-qt} 1_{\{-\underline{X}_t \in \mathrm{d}y\}} \mathrm{d}t\Big] &= \frac{1}{\Phi(q)} W^{(q)}(\mathrm{d}y) - W^{(q)}(y)\mathrm{d}y \\ &= \frac{1}{\Phi(q)} [\Theta^{(q)}(y)\mathrm{d}y + W^{(q)}(0)\delta_0(\mathrm{d}y)],\end{aligned}$$

where $W^{(q)}(\mathrm{d}y)$ is the measure such that $W^{(q)}(y) = \int_{[0,y]} W^{(q)}(\mathrm{d}z)$ (see [30, (8.20)]) and δ_0 is the Dirac measure at zero. Here, for all $y > 0$,

$$\Theta^{(q)}(y) := W^{(q)\prime}(y+) - \Phi(q) W^{(q)}(y) > 0. \tag{2.9}$$

See another probabilistic interpretation of this function in Section 3.3 in [47]. This function often appears in stochastic control. See in particular Sects. 4 and 5.1.1 below and also [47].

2.5.2 Fluctuation Identities for $\bar{Y}_t^b$

Fix $a < b$. Define the first down-crossing time of $\bar{Y}_t^b$ as:

$$\overline{\tau}_{a,b} := \inf\{t > 0 : \bar{Y}_t^b < a\}.$$

First, the Laplace transform of $\overline{\tau}_{a,b}$ is given, as in Proposition 2(ii) of [43], by

$$\mathbb{E}_x[e^{-q\overline{\tau}_{a,b}}] = Z^{(q)}(x-a) - qW^{(q)}(b-a)\frac{W^{(q)}(x-a)}{W^{(q)\prime}((b-a)+)}, \quad x \leq b.$$

Second, using its resolvent given in Theorem 1(ii) of [43], we have, for $x \leq b$,

$$\begin{aligned}&\mathbb{E}_x\Big[\int_0^{\overline{\tau}_{a,b}} e^{-qt} h(\bar{Y}_t^b)\mathrm{d}t\Big] \\ &= \frac{W^{(q)}(x-a)}{W^{(q)\prime}((b-a)+)} \Big[W^{(q)}(0)h(b) + \int_a^b h(y) W^{(q)\prime}(b-y)\mathrm{d}y\Big] - \varphi_a(x;h).\end{aligned}$$

Finally, as in Proposition 1 of [4], the discounted cumulative amount of reflection from above is given by

$$\mathbb{E}_x\Big[\int_{[0,\overline{\tau}_{a,b}]} e^{-qt}\mathrm{d}D_t^b\Big] = \frac{W^{(q)}(x-a)}{W^{(q)\prime}((b-a)+)}, \quad x \le b.$$

2.5.3 Fluctuation Identities for $\underline{Y}_t^a$

Fix $a < b$. Define the first up-crossing time of $\underline{Y}_t^a$ as:

$$\underline{\tau}_{a,b} := \inf\{t > 0 : \underline{Y}_t^a > b\}.$$

First, as in page 228 of [30], its Laplace transform is concisely given by

$$\mathbb{E}_x[e^{-q\underline{\tau}_{a,b}}] = \frac{Z^{(q)}(x-a)}{Z^{(q)}(b-a)}, \quad x \le b.$$

Second, by Theorem 1(i) of [43], for any $x \le b$,

$$\mathbb{E}_x\Big[\int_0^{\underline{\tau}_{a,b}} e^{-qt}h(\underline{Y}_t^a)\mathrm{d}t\Big] = \frac{Z^{(q)}(x-a)}{Z^{(q)}(b-a)}\varphi_a(b;h) - \varphi_a(x;h).$$

Finally, as in the proof of Theorem 1 of [4], the discounted cumulative amount of reflection from below, given $\psi'(0+) > -\infty$, is

$$\mathbb{E}_x\Big[\int_0^{\underline{\tau}_{a,b}} e^{-qt}\mathrm{d}U_t^a\Big] = -R^{(q)}(x-a) + Z^{(q)}(x-a)\frac{R^{(q)}(b-a)}{Z^{(q)}(b-a)}, \quad x \le b.$$

2.6 *Fluctuation Identities for Doubly Reflected Lévy Processes*

Fix $a < b$. As a variant of the reflected processes addressed above, the *doubly reflected Lévy process* is given by

$$Y_t^{a,b} := X_t + U_t^{a,b} - D_t^{a,b}, \quad t \ge 0. \tag{2.10}$$

This process is reflected at the two barriers a and b so as to stay on the interval $[a, b]$; see page 165 of [4] for the construction of the processes $U^{a,b}$, $D^{a,b}$, and $Y^{a,b}$. To put it simply, $U^{a,b}$ is activated whenever $Y^{a,b}$ attempts to downcross a so that $Y^{a,b}$ stays at or above a; similarly, $D^{a,b}$ is activated so that $Y^{a,b}$ stays at or below b.

First, as in Theorem 1 of [4], for $x \le b$,

$$\begin{aligned}
\mathbb{E}_x \left[\int_{[0,\infty)} e^{-qt} \mathrm{d}D_t^{a,b} \right] &= \frac{Z^{(q)}(x-a)}{qW^{(q)}(b-a)}, \\
\mathbb{E}_x \left[\int_{[0,\infty)} e^{-qt} \mathrm{d}U_t^{a,b} \right] &= -R^{(q)}(x-a) + \frac{Z^{(q)}(b-a)}{qW^{(q)}(b-a)} Z^{(q)}(x-a),
\end{aligned} \tag{2.11}$$

where we assume $\psi'(0+) > -\infty$ for the latter.

Second, using the q-resolvent density of $Y^{a,b}$ given in Theorem 1 of [42], we have, for $x \le b$,

$$\begin{aligned}
\mathbb{E}_x \left[\int_{[0,\infty)} e^{-qt} h(Y_t^{a,b}) \mathrm{d}t \right] &= \int_a^b h(y) \left[\frac{Z^{(q)}(x-a) W^{(q)\prime}(b-y)}{qW^{(q)}(b-a)} - W^{(q)}(x-y) \right] \mathrm{d}y \\
&\quad + h(b) \Big[Z^{(q)}(x-a) \frac{W^{(q)}(0)}{qW^{(q)}(b-a)} \Big].
\end{aligned} \tag{2.12}$$

2.7 *Other Properties of the Scale Function*

Here we list some other properties of the scale function that are often useful in solving stochastic control problems.

2.7.1 Asymptotics as $x \to \infty$

Suppose $q > 0$. It is known that the scale function $W^{(q)}$ increases exponentially: we have

$$W^{(q)}(x)/e^{\Phi(q)x} \xrightarrow{x \to \infty} \psi'(\Phi(q))^{-1}. \tag{2.13}$$

By this, the following limits are also immediate:

$$\lim_{x \to \infty} \frac{W^{(q)\prime}(x+)}{W^{(q)}(x)} = \Phi(q), \quad \lim_{x \to \infty} \frac{Z^{(q)}(x)}{W^{(q)}(x)} = \frac{q}{\Phi(q)} \quad \text{and} \quad \lim_{x \to \infty} \frac{\overline{Z}^{(q)}(x)}{W^{(q)}(x)} = \frac{q}{\Phi^2(q)}.$$

Note also that, for $s \in \mathbb{R}$ and any measurable function h such that $\Psi(s; h)$ is well-defined,

$$\lim_{x \to \infty} \frac{\varphi_s(x; h)}{W^{(q)}(x-s)} = \Psi(s; h). \tag{2.14}$$

2.7.2 Log-Concavity

The scale function $W^{(q)}$ is known to be log-concave: as in (8.18) and Lemma 8.2 of [30],

$$\frac{W^{(q)\prime}(y+)}{W^{(q)}(y)} \leq \frac{W^{(q)\prime}(x+)}{W^{(q)}(x)}, \quad y > x > 0.$$

In addition, $W^{(q)\prime}(x-) \geq W^{(q)\prime}(x+)$ for all $x > 0$. These properties are sometimes needed for the monotonicity of related functions; see Sects. 4.3.2 and 5.2.1 below.

2.7.3 Martingale Properties

Let $\mathcal{L}$ be the infinitesimal generator associated with the process X applied to a *sufficiently smooth* function h (i.e. C^1 [resp. C^2] for the case X is of bounded [resp. unbounded] variation): for $x \in \mathbb{R}$,

$$\begin{aligned}\mathcal{L}h(x) &:= \gamma h'(x) + \frac{1}{2}\sigma^2 h''(x) \\ &+ \int_{(-\infty,0)} \left[h(x+z) - h(x) - h'(x)z 1_{\{-1<z<0\}}\right] \nu(\mathrm{d}z), \qquad (2.15)\end{aligned}$$

$$(\text{resp. } \mathcal{L}h(x) := \delta h'(x) + \int_{(-\infty,0)} [h(x+z) - h(x)]\, \nu(\mathrm{d}z)).$$

The variational inequalities are written using this generator with h replaced with the candidate value function. Typically, it makes sense (except at the selected boundaries), thanks to its smoothness that can be confirmed by that of the scale function as in Remark 2.1. At the boundaries, for optimal stopping and impulse control, the function may not be smooth enough and hence (2.15) is not well-defined, although its right and left limits normally exist and are finite. In such cases, the Meyer-Itô formula (see, e.g., Theorem 71 of Protter [44]) is used in the proof of verification lemma.

One useful known fact regarding the generator (2.15) is as follows. By Proposition 2 of [4] and as in the proof of Theorem 8.10 of [30], the processes

$$e^{-q(t\wedge T_0^- \wedge T_B^+)} Z^{(q)}(X_{t\wedge T_0^- \wedge T_B^+}) \quad \text{and} \quad e^{-q(t\wedge T_0^- \wedge T_B^+)} R^{(q)}(X_{t\wedge T_0^- \wedge T_B^+}), \quad t \geq 0,$$

for any $B > 0$ are martingales, where we assume $\psi'(0+) > -\infty$ for the latter. Thanks to the smoothness of $Z^{(q)}$ and $\overline{Z}^{(q)}$ on $(0, \infty)$ as in Remark 2.1, we obtain

$$(\mathcal{L} - q)Z^{(q)}(y) = (\mathcal{L} - q)R^{(q)}(y) = 0, \quad y > 0. \qquad (2.16)$$

The same result holds for $W^{(q)}$ and

$$(\mathcal{L} - q)W^{(q)}(y) = 0, \quad y > 0, \tag{2.17}$$

on condition that it is sufficiently smooth.

Another useful known fact is that, as in the proof of Lemma 4.5 of [17], if h is continuous,

$$(\mathcal{L} - q)\varphi_s(x; h) = h(x), \quad x > s. \tag{2.18}$$

These properties are often sufficient to prove that the candidate value function is harmonic in the waiting (non-controlling) region.

2.8 Some Further Notations

Before closing this section, we shall define, if they exist, the following threshold levels.

Definition 2.1 Given a closed interval $\mathcal{I} \subset \mathbb{R}$ and a measurable function h, let $\overline{a} = \overline{a}(h) \in \mathcal{I}$ be such that $h(x) < 0$ for $x \in (-\infty, \overline{a}) \cap \mathcal{I}$, and $h(x) > 0$ for $x \in (\overline{a}, \infty) \cap \mathcal{I}$, if such a value exists. If $h(x) < 0$ for all $x \in \mathcal{I}$, then we set $\overline{a} = \overline{a}(h) = \overline{\mathcal{I}}$. If $h(x) > 0$ for $x \in \mathcal{I}$, then we set $\overline{a} = \overline{a}(h) = \underline{\mathcal{I}}$.

Definition 2.2 Given a closed interval $\mathcal{I} \subset \mathbb{R}$ and a measurable function h such that $\Psi(x; h)$ is well-defined and finite for all $x \in \mathcal{I}$, let $\underline{a} = \underline{a}(h) \in \mathcal{I}$ be such that $\Psi(x; h) < 0$ for $x \in (-\infty, \underline{a}) \cap \mathcal{I}$, and $\Psi(x; h) > 0$ for $x \in (\underline{a}, \infty) \cap \mathcal{I}$, if such a value exists. If $\Psi(x; h) < 0$ for all $x \in \mathcal{I}$, then we set $\underline{a} = \underline{a}(h) = \overline{\mathcal{I}}$. If $\Psi(x; h) > 0$ for $x \in \mathcal{I}$, then we set $\underline{a} = \underline{a}(h) = \underline{\mathcal{I}}$.

These values for a suitably chosen (often monotone) function h give us particularly important information. Typically, as in the examples shown in later sections, the values of $\underline{a}$ and $\overline{a}$ can act as upper or lower bounds of the two parameters (a^*, b^*) to be chosen. See, in particular, Sects. 3.3.3, 4.2.1 and 5.1.1 and also Tables 1, 2, and 3.

In addition, the value $\underline{a}$ can be understood as the optimal parameter a^* when the other parameter is $b^* = \infty$. We will also see that the value $\overline{a}$ is important in the verification step; see Lemmas 3.1(2), 4.1(2), and 5.2(2).

3 Two-Sided Singular Control

In this section, we consider the singular control problem where one can increase and also decrease the underlying process. An admissible strategy $\pi := \left\{(U_t^\pi, D_t^\pi); t \geq 0\right\}$ is given by a pair of nondecreasing, right-continuous, and $\mathbb{F}$-adapted processes with $U^{\pi}_{0-} = D^{\pi}_{0-} = 0$ such that the controlled process

$$Y^{\pi}_t := X_t + U^{\pi}_t - D^{\pi}_t, \quad t \geq 0,$$

stays in some given closed interval $\mathcal{I}$ uniformly in time. Let Π be the set of all admissible strategies.

We consider the sum of the running and controlling costs; its expected NPV is given by

$$v^{\pi}(x) := \mathbb{E}_x\Big[\int_0^{\infty} e^{-qt} f(Y^{\pi}_t)\mathrm{d}t + \int_{[0,\infty)} e^{-qt}\left(C_U \mathrm{d}U^{\pi}_t + C_D \mathrm{d}D^{\pi}_t\right)\Big], \quad x \in \mathbb{R},$$

for $q > 0$, some continuous and piecewise continuously differentiable function f on $\mathcal{I}$ and fixed constants $C_U, C_D \in \mathbb{R}$ satisfying

$$C_U + C_D > 0. \tag{3.1}$$

Here, if $x < \underline{\mathcal{I}}$ (resp. $x > \overline{\mathcal{I}}$), then $U^{\pi}_0 = \Delta U^{\pi}_0 = \underline{\mathcal{I}} - x$ (resp. $D^{\pi}_0 = \Delta D^{\pi}_0 = x - \overline{\mathcal{I}}$) so that $Y^{\pi}_0 \in \mathcal{I}$.

The problem is to compute the value function given by

$$v(x) := \inf_{\pi \in \Pi} v^{\pi}(x), \quad x \in \mathbb{R},$$

and the optimal strategy that attains it, if such a strategy exists.

Throughout this and next sections, let us also use the slope-changed version of f given by

$$\tilde{f}(x) := f(x) + C_U q x, \quad x \in \mathbb{R}. \tag{3.2}$$

The roles and significance of this function will be clear shortly. We also assume the following so that the expected NPV associated with U^{π}_t is finite.

Assumption 3.1 *We assume* $\mathbb{E}X_1 = \psi'(0+) > -\infty$.

Example 3.1 In the optimal dividend problem with capital injections driven by a spectrally negative Lévy process, it is required that the controlled risk process stay nonnegative uniformly in time (i.e. $\mathcal{I} = [0, \infty)$). One wants to maximize the expected NPV of dividends minus that for capital injections. This is a maximization problem with U^{π}_t and D^{π}_t being, respectively, the cumulative amounts of capital injections and dividends until $t \geq 0$. We can formulate this as a minimization problem as above by setting $C_D = -1$ and $C_U = \beta$ where $\beta > 1$ is the unit cost of capital injection. Here f is assumed to be uniformly zero. This problem has been solved by Avram et al. [4] for a general spectrally negative Lévy process.

Example 3.2 In the dual model of Example 3.1, it is assumed that the underlying process is a spectrally positive Lévy process. By flipping the processes with respect to the origin, it is easy to see that the problem is equivalent to the above formulation driven by a spectrally negative Lévy process with $\mathcal{I} = (-\infty, 0]$, $C_D = \beta$ and $C_U = -1$. This problem has been solved by Bayraktar et al. [8] for a general spectrally positive Lévy process.

Example 3.3 A version of continuous-time inventory control considers the case where inventory can be increased (replenished) and decreased (sold). With the absence of fixed costs and if backorders are allowed, the problem can be formulated as above with $\mathcal{I} = \mathbb{R}$. Currency rate control (see, e.g., [28, 38]), where a central bank controls the currency rate so as to prevent it from going too high or too low, can also be modeled in the same way. The classical Brownian motion and continuous diffusion models have been solved by Harrison and Taksar [24] and Matomäki [36], respectively. In Baurdoux and Yamazaki [6], it has been solved for a general spectrally negative Lévy process. In this note, we assume that f is convex for this example.

3.1 The Double Reflection Strategy

In all the examples above, the optimal strategy is shown to be a *double barrier strategy* $\pi_{a,b} := \{U^{a,b}, D^{a,b}\}$ with the resulting controlled process being the doubly reflected Lévy process given in (2.10).

By (2.11) and (2.12), we can directly compute, for $a < b$,

$$v_{a,b}(x) := \mathbb{E}_x\Big[\int_0^\infty e^{-qt} f(Y_t^{a,b})\mathrm{d}t + \int_{[0,\infty)} e^{-qt}(C_U \mathrm{d}U_t^{a,b} + C_D \mathrm{d}D_t^{a,b})\Big], \quad x \in \mathbb{R}.$$

For $x \leq b$, it is given by

$$\begin{aligned} v_{a,b}(x) = {} & \frac{\Lambda(a,b)}{qW^{(q)}(b-a)} Z^{(q)}(x-a) - C_U R^{(q)}(x-a) \\ & + \frac{f(a)}{q} Z^{(q)}(x-a) - \varphi_a(x; f) \end{aligned} \tag{3.3}$$

where

$$\Lambda(a,b) := C_D + C_U + \varphi_a(b; \tilde{f}'), \quad b \geq a. \tag{3.4}$$

For $x > b$, we have $v_{a,b}(x) = v_{a,b}(b) + C_D(x-b)$.

Remark 3.1 In particular, when $f \equiv 0$ (as in Examples 3.1 and 3.2 above), for $a < b$,

$$\Lambda(a,b) = C_D + C_U Z^{(q)}(b-a),$$

$$v_{a,b}(x) = \frac{C_D + C_U Z^{(q)}(b-a)}{qW^{(q)}(b-a)} Z^{(q)}(x-a) - C_U R^{(q)}(x-a), \quad x \le b;$$

see [4] and [8].

3.2 Smoothness of the Value Function

Focusing on the set of double barrier strategies, the first step is to narrow down to a candidate optimal strategy by deciding on the threshold values, say a^* and b^*. Because the spectrally negative Lévy process can reach any point with positive probability, we must have that $[a^*, b^*] \subset \mathcal{I}$.

As we have discussed in Sect. 1.4.1, the two parameters can be identified by the first-order condition or the smooth fit condition. The first approach uses the first-order conditions at a^* and b^*; because a^* and b^* must minimize $v_{a,b}$ over a and b, partial derivatives $\partial v_{a,b}(x)/\partial a|_{a=a^*,b=b^*}$ and $\partial v_{a,b}(x)/\partial b|_{a=a^*,b=b^*}$ must vanish, at least when the minimizers are in the interior of $\mathcal{I}$. The second approach uses the condition that the value function is smooth. Here, we focus on the second smoothness approach because the computation is slightly easier, and we need to confirm the smoothness of v_{a^*,b^*} after all when we verify its optimality.

In singular control, the value function normally admits twice continuous differentiability (resp. continuous differentiability) at each interior point in $\mathcal{I}$ when it is regular (resp. irregular). Thanks to the smoothness of the scale function as in Remark 2.1, the only points of v_{a^*,b^*} we need to pay attention are a^* and b^* where the functions are pasted together. Due to the asymmetry of the spectrally negative Lévy process, what we observe at these two points will be different. Here, recall the definition of regularity and its relation with the path variation of the process as reviewed in Sect. 2.1.

Regarding the smoothness of the value function at the lower barrier a^*,

(1) if a^* is regular for $(-\infty, a^*)$ (or equivalently X is of unbounded variation), then the twice continuous differentiability at a^* is expected;
(2) if a^* is irregular for $(-\infty, a^*)$ (or equivalently X is of bounded variation), then the continuous differentiability at a^* is expected.

Regarding the smoothness at the upper barrier b^*, because it is always regular for (b^*, ∞), twice-differentiability is expected at b^* regardless of the path variation of X.

These procedures can be carried out in a straightforward fashion by using the expression (3.3) in terms of the scale function. By taking derivatives in (3.3) and using (3.2),

$$\begin{aligned} v'_{a,b}(x) &= \frac{\Lambda(a,b)}{W^{(q)}(b-a)} W^{(q)}(x-a) - C_U - \varphi_a(x; \tilde{f}'), \quad a < x < b, \\ v''_{a,b}(x+) &= \frac{\Lambda(a,b)}{W^{(q)}(b-a)} W^{(q)\prime}((x-a)+) \\ &\quad - \int_a^x W^{(q)\prime}(x-y)\tilde{f}'(y)\mathrm{d}y - \tilde{f}'(x+)W^{(q)}(0), \quad a < x < b. \end{aligned} \tag{3.5}$$

In view of the former of (3.5), by (3.4),

$$\begin{aligned} v'_{a,b}(b-) &= C_D = v'_{a,b}(b+), \\ v'_{a,b}(a+) &= \frac{\Lambda(a,b)}{W^{(q)}(b-a)} W^{(q)}(0) - C_U = \frac{\Lambda(a,b)}{W^{(q)}(b-a)} W^{(q)}(0) + v'_{a,b}(a-). \end{aligned} \tag{3.6}$$

In other words, the continuous differentiability of $v_{a,b}$ holds at b regardless of the path variation. On the other hand, in view of (2.5), while the differentiability at a holds for the case of unbounded variation, it only holds if

$$\mathfrak{C}_a : \frac{\Lambda(a,b)}{W^{(q)}(b-a)} = 0 \tag{3.7}$$

for the case of bounded variation. Here, the case $b = \infty$ is understood as $\lim_{b\to\infty} \Lambda(a,b)/W^{(q)}(b-a) = 0$ where by (2.14) we can show that

$$\lim_{b\to\infty} \frac{\Lambda(a,b)}{W^{(q)}(b-a)} = \Psi(a; \tilde{f}'). \tag{3.8}$$

In view of the latter of (3.5),

$$\begin{aligned} v''_{a,b}(b-) &= \frac{\Lambda(a,b)}{W^{(q)}(b-a)} W^{(q)\prime}((b-a)-) - \lambda(a,b), \\ v''_{a,b}(a+) &= \frac{\Lambda(a,b)}{W^{(q)}(b-a)} W^{(q)\prime}(0+) - \tilde{f}'(a+)W^{(q)}(0), \end{aligned}$$

where

$$\lambda(a,b) := \frac{\partial}{\partial b}\Lambda(a,b-) = \int_a^b W^{(q)\prime}(b-y)\tilde{f}'(y)\mathrm{d}y + \tilde{f}'(b-)W^{(q)}(0), \quad b > a. \tag{3.9}$$

For the unbounded variation case where the continuous differentiability at a automatically holds, again by (2.5), its twice continuous differentiability holds on condition that $\mathfrak{C}_a$ holds. Now, for both the bounded and unbounded variation cases, the twice continuous differentiability at b holds if

$$\mathfrak{C}_b : \frac{\Lambda(a,b)}{W^{(q)}(b-a)} W^{(q)\prime}((b-a)-) - \lambda(a,b) = 0. \tag{3.10}$$

In particular, *on condition that* $\mathfrak{C}_a$ *holds*, the condition $\mathfrak{C}_b$ can be simplified to

$$\mathfrak{C}_b' : \lambda(a,b) = 0. \tag{3.11}$$

Remark 3.2 When $f \equiv 0$, the conditions $\mathfrak{C}_a$ and $\mathfrak{C}_b$, respectively, are simplified to

$$\mathfrak{C}_a^0 : \frac{C_D + C_U Z^{(q)}(b-a)}{W^{(q)}(b-a)} = 0, \tag{3.12}$$

$$\mathfrak{C}_b^0 : \frac{C_D + C_U Z^{(q)}(b-a)}{W^{(q)}(b-a)} W^{(q)\prime}((b-a)-) - qC_U W^{(q)}(b-a) = 0. \tag{3.13}$$

These conditions on a and b can be used to identify the pairs (a^*, b^*). However, these do not necessarily hold unless $a^*, b^* \in \mathcal{I}^o$. Here, we give examples where a^* and/or b^* become boundaries of $\mathcal{I}$.

Remark 3.3

(1) In Example 3.1, it is expected, because $\beta > 1$ (the unit cost of capital injection is higher than the unit reward of dividend), that capital is injected only when it is necessary to make the company alive, and hence $a^* = 0$.
(2) Similarly, under the formulation with the underlying spectrally negative Lévy process described in Example 3.2, it is expected that $b^* = 0$.
(3) In Example 3.3, if the increment of f as $|x| \to \infty$ is at most linear and small in comparison to the unit controlling costs C_U and C_D, it may not be desirable to activate at all the processes U^π and/or D^π. Hence, $a^* = -\infty$ and/or $b^* = \infty$.

3.3 Existence of (a^*, b^*)

The first challenge is to show the existence of such (a^*, b^*). Here, we assume the following.

Assumption 3.2 *We assume that* $\overline{a} \equiv \overline{a}(f')$ *(see Definition 2.1) exists and is finite, where* f' *is understood as its right-hand derivative if not differentiable.*

We shall see that $\overline{a}$ is a point such that a^* lies on the left of $\overline{a}$ and b^* lies on its right; see Table 1.

Table 1 Summary of the key functions and parameters in Examples 3.2 and 3.3

Example 3.2	
$\Lambda(a, b)$	$:= C_D + C_U Z^{(q)}(b-a)$
$\tilde{f}'(b)$	$:= C_U q$
a^*	$:= a$ of $(a, 0)$ such that $\mathfrak{C}_a^0$ holds
$< \overline{a}$	$:= 0 = \overline{a}(\tilde{f}')$
$= b^*$	$:= 0 = \overline{\mathcal{I}}$
Example 3.3	
$\Lambda(a, b)$	$:= C_D + C_U + \varphi_a(b; \tilde{f}')$
$\tilde{f}'(b)$	$:= f'(b) + C_U q$
$\underline{a}$	$:= \underline{a}(\tilde{f}')$
$\leq a^*$	$:= a$ of (a, b) such that $\mathfrak{C}_a$ and $\mathfrak{C}_b$ hold simultaneously
$< \overline{a}$	$:= \overline{a}(\tilde{f}')$
$< b^*$	$:= b$ of (a, b) such that $\mathfrak{C}_a$ and $\mathfrak{C}_b$ hold simultaneously

For Example 3.3, when $b^* = \infty$, $a^* = \underline{a}$

3.3.1 The Case of Example 3.1

It is clear that Assumption 3.2 is satisfied with $\overline{a} = 0$. As in Remark 3.3(1), $a^* = 0 = \overline{a} = \underline{\mathcal{I}}$. Therefore, the condition $\mathfrak{C}_a^0$ has no effect and we only require $\mathfrak{C}_b^0$ which reduces to

$$\frac{C_D + C_U Z^{(q)}(b)}{W^{(q)}(b)} W^{(q)\prime}(b-) - q C_U W^{(q)}(b) = 0. \tag{3.14}$$

Hence, $b^* > 0 = \overline{a} = \overline{\mathcal{I}}$ can be chosen as the smallest value of b such that (3.14) holds. This matches the condition given in (5.6) of [4].

3.3.2 The Case of Example 3.2

Again, Assumption 3.2 is satisfied with $\overline{a} = 0$. Because $C_D = \beta$ and $C_U = -1$, there is a unique $a^* < 0 = \overline{a}$ that satisfies $\mathfrak{C}_a^0$ or equivalently that

$$C_D + C_U Z^{(q)}(-a^*) = 0. \tag{3.15}$$

Hence, the candidate optimal strategy is given by $a^* = -(Z^{(q)})^{-1}(-C_D/C_U) = -(Z^{(q)})^{-1}(\beta)$ and $b^* = 0$. This matches the result in [8].

3.3.3 The Case of Example 3.3

For Example 3.3, we want a pair (a^*, b^*) such that (3.7) and (3.10) hold simultaneously. Equivalently, we want (a^*, b^*) such that the function $b \mapsto \Lambda(a^*, b)$ attains a (local) minimum 0 at b^* (if $b^* < \infty$). Note that, for any $a \in \mathbb{R}$, $b \mapsto \Lambda(a, b)$ starts at $\Lambda(a, a) = C_D + C_U > 0$.

In this case, $\overline{a}$ always exists by the assumption that f is convex. In addition, Assumption 3.2 requires that it is finite. Recall now Definition 2.2. The convexity assumption and Assumption 3.2 guarantee that $\underline{a} = \underline{a}(\tilde{f}')$ also exists and is finite (with the understanding that $\tilde{f}'$ is the right-hand derivative if it is not differentiable). Note that necessarily $\underline{a} < \overline{a}$.

Figure 4 shows some sample plots of $b \mapsto \Lambda(a, b)$ and $b \mapsto \lambda(a, b)$. As observed in these plots, we shall show that a^* must lie on $[\underline{a}, \overline{a})$.

To see this, when $a \geq \overline{a}$, then $\Lambda(a, \cdot)$ is uniformly positive because $\lambda(a, b) \geq 0$ for $b > a$ in view of (3.9). In addition, by the convergence (3.8) and how $\underline{a}$ is chosen, $\lim_{b \to \infty} \Lambda(a, b) = \infty$ if $a > \underline{a}$, $\lim_{b \to \infty} \Lambda(a, b) = -\infty$ if $a < \underline{a}$, and (3.8) becomes zero if $a = \underline{a}$. On the other hand, for any $a < \overline{a}$ and $a < b$,

$$\frac{\partial}{\partial a} \Lambda(a+, b) = -\tilde{f}'(a+) W^{(q)}(b - a) > 0. \tag{3.16}$$

This implies that the infimum $a \mapsto \inf_{b > a} \Lambda(a, b)$ is monotonically increasing. Hence, the desired a^* such that $\Lambda(a^*, \cdot)$ touches the x-axis, if it exists, must lie on $(\underline{a}, \overline{a})$.

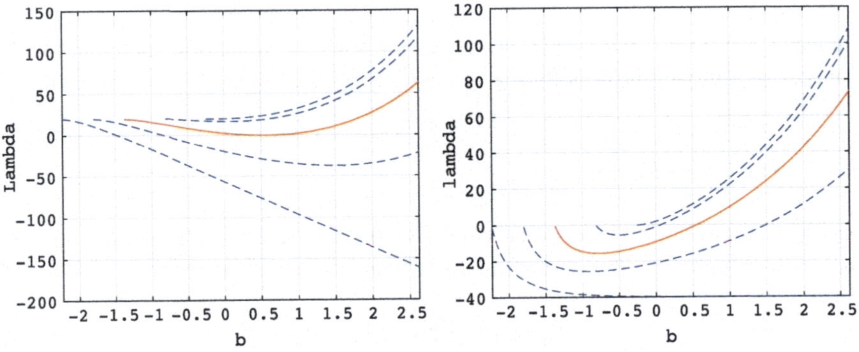

$b \mapsto \Lambda(a, b)$ $b \mapsto \lambda(a, b)$

Fig. 4 Existence of (a^*, b^*) for Example 3.3. Plots of $b \mapsto \Lambda(a, b)$ on $[a, \infty)$ for the starting values $a = \underline{a}, (\underline{a} + a^*)/2, a^*, (a^* + \overline{a})/2, \overline{a}$ are shown. The solid curve in red corresponds to the one for $a = a^*$; the point at which $\Lambda(a^*, \cdot)$ is tangent to the x-axis (or $\lambda(a^*, \cdot)$ vanishes) becomes b^*. The function $\Lambda(\underline{a}, \cdot)$ is monotonically decreasing while $\Lambda(\overline{a}, \cdot)$ is monotonically increasing. Equivalently, $\lambda(\underline{a}, \cdot)$ is uniformly negative while $\lambda(\overline{a}, \cdot)$ is uniformly positive

By these observations, one can attempt to decrease the value of a starting at $\overline{a}$ until we arrive at (1) a point a^* such that $\inf_{b>a^*} \Lambda(a^*, b) = 0$ or (2) the point $\underline{a}$, whichever comes first. For each case, we set (a^*, b^*) as follows.

(1) We set (a^*, b^*) such that $0 = \inf_{b>a^*} \Lambda(a^*, b) = \Lambda(a^*, b^*)$. Hence, $\mathfrak{C}_a$ holds. If in addition, $b \mapsto \lambda(a^*, b)$ is continuous at b^*, then $\mathfrak{C}'_b$ also holds as well.
(2) We set $a^* = \underline{a}$ and $b^* = \infty$. By (3.8), $\lim_{b\to\infty} \Lambda(a^*, b)/W^{(q)}(b - a^*) = 0$, or equivalently $\mathfrak{C}_a$ holds.

Remark 3.4 In Examples 3.2 and 3.3, by construction, $\Lambda(a^*, x) \geq 0$ for $x \in [a^*, b^*]$.

3.4 Variational Inequalities and Verification

Below, we shall focus on the case $a^* \in \mathcal{I}^o$ and hence $\mathfrak{C}_a$ is satisfied (this excludes Example 3.1): the value function becomes, by (3.3), for all $x \leq b^*$,

$$\begin{aligned} v_{a^*,b^*}(x) &= -C_U R^{(q)}(x - a^*) + \frac{f(a^*)}{q} Z^{(q)}(x - a^*) - \varphi_{a^*}(x; f) \\ &= -C_U \Big(\frac{\psi'(0+)}{q} + x \Big) + \frac{\tilde{f}(a^*)}{q} Z^{(q)}(x - a^*) - \varphi_{a^*}(x; \tilde{f}). \end{aligned} \tag{3.17}$$

By (3.4) and (3.5),

$$v'_{a^*,b^*}(x) = -\Lambda(a^*, x) + C_D, \quad a^* \leq x \leq b^*. \tag{3.18}$$

The verification of optimality asks that our candidate value function v_{a^*,b^*} solves the variational inequalities:

$$\begin{aligned} &(\mathcal{L} - q)v_{a^*,b^*}(x) + f(x) \geq 0, \quad x \in \mathcal{I}^o, \\ &\min(v'_{a^*,b^*}(x) + C_U, C_D - v'_{a^*,b^*}(x)) \geq 0, \quad x \in (-\infty, \overline{\mathcal{I}}], \\ &[(\mathcal{L} - q)v_{a^*,b^*}(x) + f(x)] \min(v'_{a^*,b^*}(x) + C_U, C_D - v'_{a^*,b^*}(x)) = 0, \quad x \in \mathcal{I}^o. \end{aligned} \tag{3.19}$$

Notice that, when $\underline{\mathcal{I}} > -\infty$, the middle condition is required to hold for the extended set $(-\infty, \overline{\mathcal{I}}]$ because X can jump instantaneously to the region $(-\infty, \underline{\mathcal{I}})$ (and then immediately pushed up to $\underline{\mathcal{I}}$). Here, the generator $\mathcal{L}v_{a^*,b^*}$ makes sense due to the smoothness obtained above of v_{a^*,b^*} and because v_{a^*,b^*} is linear below a^* and Assumption 3.1 is given.

In order to show that these are sufficient conditions for optimality, in general we need additional assumptions on the tail property of f and the Lévy measure. This is necessary because verification arguments first localize in order to use Itô's

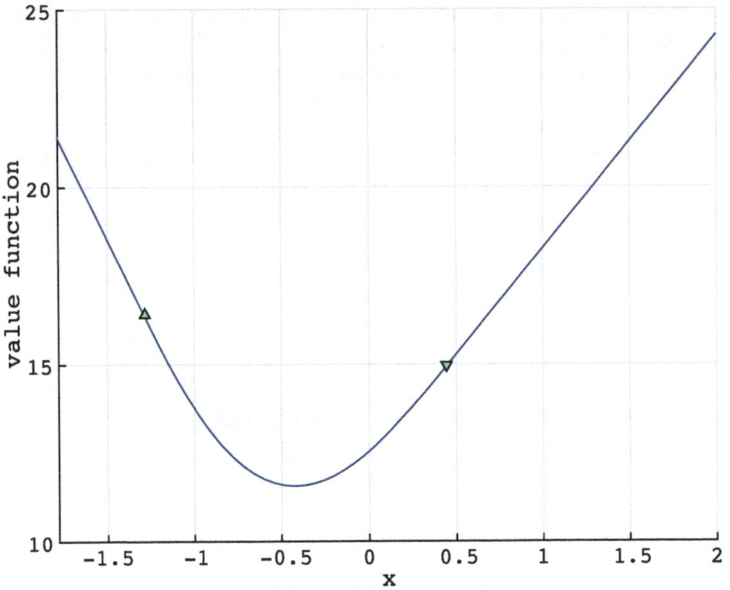

Fig. 5 A sample plot of the value function for Example 3.3 when X is of unbounded variation. The up-pointing and down-pointing triangles show the points at a^* and b^*, respectively. It can be confirmed that it is twice differentiable at a^* and b^*

formula. After the localization arguments, one needs to interchange the limits over expectations. To this end, it is typically required that $|f|$ only increases moderately and/or the Lévy measure does not have a heavy tail.

Showing (3.19) is the main challenge and the proof needs to be customized for each problem. However, some inequalities of (3.19) are easily shown without strong assumptions on the function f (Fig. 5).

Lemma 3.1 *Suppose $\mathfrak{C}_a$ holds.*

(1) *We have $(\mathcal{L} - q)v_{a^*,b^*}(x) + f(x) = 0$ for $a^* < x < b^*$.*
(2) *If Assumption 3.2 holds with $a^* \leq \overline{a}$, then $(\mathcal{L} - q)v_{a^*,b^*}(x) + f(x) \geq 0$ on $(-\infty, a^*)$.*
(3) *If $\Lambda(a^*, x) \geq 0$ for $x \in [a^*, b^*]$, then $v'_{a^*,b^*}(x) \leq C_D$ on $(-\infty, \overline{\mathcal{I}}]$.*

Proof

(1) This is immediate by the results summarized in Sect. 2.7.3 in view of the first equality of (3.17).
(2) By the second equality of (3.17), $v_{a^*,b^*}(x) = [-C_U\psi'(0+) + \tilde{f}(a^*)]/q - C_U x$, for $x < a^*$, and hence $(\mathcal{L}-q)v_{a^*,b^*}(x) + f(x) = \tilde{f}(x) - \tilde{f}(a^*)$. This is positive by $x \leq a^* < \overline{a}$ and by how $\overline{a}$ is chosen.

(3) In view of (3.18), this inequality holds for $x \in [a^*, b^*]$. For $x \in (-\infty, a^*)$, we have $v'_{a^*,b^*}(x) = -C_U$, which is smaller than C_D by (3.1). Finally, for $x \in (b^*, \infty) \cap \mathcal{I}$, we have $v'_{a^*,b^*}(x) = C_D$. □

For Examples 3.2 and 3.3, by the fact that $a^* < \overline{a}$ as discussed in Sects. 3.3.1 and 3.3.2, and also by Remark 3.4, the conditions in Lemma 3.1 hold. Hence, the only pieces left to show in (3.19) are

(1') $-C_U \le v'_{a^*,b^*}(x)$ for all $x \in (a^*, b^*)$,
(2') $(\mathcal{L} - q)v_{a^*,b^*}(x) + f(x) \ge 0$ for $x \in (b^*, \infty) \cap \mathcal{I}^o$.

These conditions unfortunately do not hold generally and must be checked individually. Here we give brief illustrations on how these hold for Examples 3.2 and 3.3.

In Example 3.2, (1') holds immediately because, with $C_U = -1 < 0$,

$$v'_{a^*,b^*}(x) = -C_U Z^{(q)}(x - a^*) \ge -C_U.$$

In addition, (2') holds trivially because $(b^*, \infty) \cap \mathcal{I}^o = \varnothing$.

In Example 3.3, thanks to the assumption that f is convex, $x \mapsto \Lambda(a^*, x)$ is first decreasing and decreasing (see Fig. 4). This together with (3.18) and the smoothness at a^* and b^*, the function v_{a^*,b^*} is convex on $\mathbb{R}$ and hence (1') holds.

The hardest part for Example 3.3 is to show (2'); the difficulty comes from the fact that the process can jump from (b^*, ∞) to the regions $(-\infty, a^*)$ and (a^*, b^*) where the form of v_{a^*,b^*} changes. In [6] under the convexity assumption, they use contradiction arguments similar to [25, 34], where they show, for $x > b^*$,

$$(\mathcal{L} - q)(v_{a^*,b^*} - v_{a(x),x})(x-) := \lim_{y \uparrow x}(\mathcal{L} - q)(v_{a^*,b^*} - v_{a(x),x})(y) \ge 0, \qquad (3.20)$$

where $a(x)$ is the unique value of a such that $\Lambda(a, x) = 0$. This implies (2') because if both (3.20) and $(\mathcal{L} - q)v_{a^*,b^*}(x) + f(x) < 0$ hold simultaneously, then

$$0 > (\mathcal{L} - q)v_{a^*,b^*}(x) + f(x) \ge (\mathcal{L} - q)v_{a(x),x}(x-) + f(x),$$

which contradicts with $(\mathcal{L} - q)v_{a(x),x}(x-) + f(x) = 0$ that can be shown similarly to Lemma 3.1(1). The proof depends heavily on the convexity of f, with which the function $y \mapsto \Lambda(x, y)$ is first decreasing and then increasing. We refer the reader to [6] for more careful analysis.

We conclude this section with a summary of the functions and parameters that played key roles in Examples 3.2 and 3.3. Some similarities and differences with the problems to be considered in later sections can be seen by comparing this with Tables 2 and 3 below.

Table 2 Summary of the key functions and parameters in Example 4.3

$\Lambda(a, b)$	$:= \Phi(q)\Psi(s; \tilde{f})\overline{W}^{(q)}(S-s) + K - \varphi_s(S; \tilde{f})$
$\tilde{f}'(b)$	$:= f'(b) + C_U q$
s^*	$:= s$ of (s, S) such that $\mathfrak{C}_s$ and $\mathfrak{C}_S$ hold simultaneously
$< \underline{a}$	$:= \underline{a}(\tilde{f}')$
$< S^*$	$:= S$ of (s, S) such that $\mathfrak{C}_s$ and $\mathfrak{C}_S$ hold simultaneously

It can be shown that $s^*, S^* \to \underline{a}$ as $K \downarrow 0$

4 Impulse Control

In impulse control, a strategy $\pi := \{U_t^\pi; t \geq 0\}$ is given by $U_t^\pi = \sum_{i:T_i^\pi \leq t} u_i^\pi$, $t \geq 0$, where $\{T_i^\pi; i \geq 1\}$ is an increasing sequence of $\mathbb{F}$-stopping times and u_i^π, for $i \geq 1$, is an $\mathcal{F}_{T_i^\pi}$-measurable random variable such that $u_i^\pi \in \mathcal{A}$, $i \geq 1$, a.s. for some $\mathcal{A} \subset \mathbb{R}$.

The corresponding controlled process is given by $Y^\pi = \{Y_t^\pi; t \geq 0\}$ where $Y_{0-}^\pi = 0$ and

$$Y_t^\pi := X_t + U_t^\pi, \quad t \geq 0.$$

The time horizon is given by $T_{\mathcal{I}^c}^\pi := \inf\{t > 0 : Y_t^\pi \notin \mathcal{I}\}$ for some given closed interval $\mathcal{I}$ and U^π must be such that

$$Y_t^\pi \in \mathcal{I}, \quad 0 \leq t \leq T_{\mathcal{I}^c}^\pi \text{ at which } \Delta U_t^\pi > 0 \quad a.s. \tag{4.1}$$

Let Π be the set of all admissible strategies.

With f, some continuous and piecewise continuously differentiable function on $\mathcal{I}$, and $q > 0$, the problem is to compute the value function

$$v(x) := \inf_{\pi \in \Pi} v^\pi(x)$$

where

$$v^\pi(x) := \mathbb{E}_x\Big[\int_0^{T_{\mathcal{I}^c}^\pi} e^{-qt} f(Y_t^\pi)\mathrm{d}t + \sum_{0 \leq t \leq T_{\mathcal{I}^c}^\pi} e^{-qt}[C_U|\Delta U_t^\pi| + K]1_{\{|\Delta U_t^\pi|>0\}}\Big], \quad x \in \mathbb{R},$$

and to obtain an admissible strategy that minimizes it, if such a strategy exists. The constant C_U is the *proportional cost*, which is not necessarily restricted to be a positive value. On the other hand, K is the *fixed cost* and must be strictly positive. Again in this section, we assume Assumption 3.1 (note that this is not necessarily needed for Example 4.1 below).

Example 4.1 In the optimal dividend problem with fixed costs driven by a spectrally negative Lévy process, each time dividend is paid, a fixed cost K is incurred. In addition, the problem is terminated at ruin (i.e. $\mathcal{I} = [0, \infty)$). The condition (4.1) means that one cannot pay more than the remaining surplus.

The objective is to maximize the total expected discounted dividends minus that for fixed costs. We can formulate this as a minimization problem as above by setting $C_U = -1$, U_t^π being the negative of the cumulative amount of dividends until $t \geq 0$, and $\mathcal{A} = (-\infty, 0)$. Here, f is assumed to be zero. This problem has been solved by Loeffen [35] for a spectrally negative Lévy process under a log-convexity assumption on the Lévy density.

Example 4.2 In the dual model of Example 4.1, it is assumed that the underlying process is a spectrally positive Lévy process. By flipping the processes with respect to the origin, it is easy to see that it is equivalent to the above formulation driven by a spectrally negative Lévy process with $\mathcal{A} = (0, \infty)$, $\mathcal{I} = (-\infty, 0]$ and $C_U = -1$. This problem has been solved by Bayraktar et al. [9] for a general spectrally positive Lévy process.

Example 4.3 Continuous-time inventory control often uses this model. Here, the function f corresponds to the cost of holding and shortage when $x > 0$ and $x < 0$, respectively. With the assumption that backorders are allowed, the problem is infinite-horizon ($\mathcal{I} = \mathbb{R}$). Bensoussan et al. [10, 12] considered the case of a spectrally negative compound Poisson process perturbed by a Brownian motion with $\mathcal{A} = (0, \infty)$. It has been generalized by Yamazaki [49] to a general spectrally negative Lévy model. As in Example 3.3, we assume that f is convex. Assume also that $\psi'(0+) > -\infty$.

4.1 The (s, S)-Strategy

With the fixed cost $K > 0$ incurred each time the control U^π is activated, it is clear that the reflection strategy is no longer feasible; instead one needs to solve the tradeoff between controlling the process and minimizing the number of activation of U^π. In this sense, the (s, S)-strategy is a natural candidate for an optimal strategy: whenever the process goes below (resp. above) a level s, it pushes the process up (resp. down) to S when $s < S$ (resp. $S < s$).

Suppose $\pi^{s,S} := \{U_t^{s,S}; t \geq 0\}$ is the (s, S)-strategy, and $Y^{s,S}$ and $T_{\mathcal{I}^c}^{s,S}$ are the corresponding controlled process and the termination time, respectively. By using the results summarized in Sect. 2.4, it is a simple exercise to compute the corresponding expected NPV of costs:

$$v_{s,S}(x) := \mathbb{E}_x \left[\int_0^{T_{\mathcal{I}^c}^{s,S}} e^{-qt} f(Y_t^{s,S}) \mathrm{d}t + \sum_{0 \leq t \leq T_{\mathcal{I}^c}^{s,S}} e^{-qt} [C_U |\Delta U_t^{s,S}| + K] 1_{\{|\Delta U_t^{s,S}| > 0\}} \right], \quad x \in \mathbb{R}. \tag{4.2}$$

To see this, for the case $s < S$, it is noted (from the construction of the process $Y^{s,S}$) that $\mathbb{P}_x$-a.s., $Y_t^{s,S} = X_t$ for $0 \le t < T_s^-$ and $\Delta U_{T_s^-}^{s,S} = S - X_{T_s^-}$ on $\{T_s^- < T_{\mathcal{I}^c}^{s,S}\}$. By these and the strong Markov property of $Y^{s,S}$, the expectation (4.2) must satisfy, for every $x > s$,

$$v_{s,S}(x) = \mathbb{E}_x\Big[\int_0^{T_s^- \wedge T_{\mathcal{I}^c}^{s,S}} e^{-qt} f(X_t)\mathrm{d}t\Big] + \mathbb{E}_x\Big[e^{-qT_s^-}(C_U(S - X_{T_s^-}) + K)1_{\{T_s^- < T_{\mathcal{I}^c}^{s,S}\}}\Big]$$
$$+ \mathbb{E}_x\Big[e^{-qT_s^-}1_{\{T_s^- < T_{\mathcal{I}^c}^{s,S}\}}\Big] v_{s,S}(S). \tag{4.3}$$

Here the expectations on the right hand side can be computed by the identities given in Sect. 2. By setting $x = S$ on both sides, we can solve for $v_{s,S}(S)$; substituting this back in, we obtain $v_{s,S}(x)$ for $x \in \mathbb{R}$. In particular, for the computation when $\mathcal{I} = \mathbb{R}$, see (4.13) below.

The case $s > S$ is even simpler because then there is no overshoot at the time it reaches s: we have, for $x < s$,

$$v_{s,S}(x) = \mathbb{E}_x\Big[\int_0^{T_s^+ \wedge T_{\mathcal{I}^c}^{s,S}} e^{-qt} f(X_t)\mathrm{d}t\Big]$$
$$+ \mathbb{E}_x\Big[e^{-qT_s^+}1_{\{T_s^+ < T_{\mathcal{I}^c}^{s,S}\}}\Big][v_{s,S}(S) + C_U(s - S) + K].$$

We can similarly obtain first $v_{s,S}(S)$ and then, by substituting this back in, $v_{s,S}(x)$, for $x \in \mathbb{R}$. See, e.g., [35] for explicit expressions when $f \equiv 0$.

Remark 4.1 The same technique can be used to compute also the two-sided extension (i.e. $\mathcal{A} = \mathbb{R}\backslash\{0\}$) of the (s, S)-strategy: in this case, the strategy is specified by four parameters, say, (d, D, U, u). The controller pushes the process up to D as soon as it goes below d and pushes down to U as soon as it goes above u, while he does not intervene whenever it is within the set (d, u). See [48] for the fluctuation identities.

4.2 Smoothness of the Value Function

Focusing on the set of (s, S)-strategies, the first step again is to narrow down to a candidate optimal strategy by deciding on the values of s and S, which we call s^* and S^*. Again, as there are two values to be identified, naturally we need two equations to identify these.

(1) As is clear from what we have seen in the previous section, the value function is expected to satisfy some continuity/smoothness at the point s^*. In comparison to the case of singular control, *the degree of smoothness is decreased by one* in the case of impulse control. This can be summarized as follows:

When $s^* < S^*$ (where v_{s^*,S^*} is linear below s^* and hence $v'_{s^*,S^*}(s^*-) = -C_U$),

(a) if s^* is regular for $(-\infty, s^*)$ (or equivalently X is of unbounded variation), then the continuous differentiability at s^* is expected;
(b) if s^* is irregular for $(-\infty, s^*)$ (or equivalently X is of bounded variation), then the continuity at s^* is expected.

When $s^* > S^*$ (where v_{s^*,S^*} is linear above s^* and hence $v'_{s^*,S^*}(s^*+) = C_U$), because s^* is regular for (s^*, ∞) for any spectrally negative Lévy process, the continuous differentiability at s^* is expected.

It is noted that alternatively one can use the first-order condition on s^* so that $\partial v_{s,S}/\partial s|_{s=s^*,S=S^*}$ vanishes: we typically arrive at the same equation.

(2) The other equation can be obtained by what we postulate at the point S^*. This is less intuitive than (1). However, if we consider the first-order condition at S^* so that $\partial v_{s,S}/\partial S|_{s=s^*,S=S^*}$ vanishes, easy computation derives that it tends to be equivalent to the condition $v'_{s^*,S^*}(S^*) = -C_U$ (resp. $v'_{s^*,S^*}(S^*) = C_U$) when $s^* < S^*$ (resp. $s^* > S^*$).

From the above discussions, when $s^* < S^*$, except for the case X is of bounded variation, we arrive at the function that satisfies

$$v'_{s^*,S^*}(s^*) = v'_{s^*,S^*}(S^*) = -C_U.$$

Due to this fact, it is often easier if we deal with a modified function

$$\tilde{v}_{s,S}(x) := v_{s,S}(x) + C_U x; \tag{4.4}$$

by this, some terms tend to disappear and computation gets simplified. When $S^* < s^*$, then the sign of the coefficient of C_U is flipped.

In impulse control, while the two equations that identify the two unknown parameters (s^*, S^*) are slightly different from the singular control case for (a^*, b^*) as in Sect. 3.2, we shall see that these two equations possess a similar relation to those obtained for (a^*, b^*). Namely, the desired pair (s^*, S^*) is such that a function of two variables and its partial derivative with respect to one of the parameters vanish simultaneously.

4.2.1 The Case of Example 4.3

For Example 4.3, we shall see that the desired (s^*, S^*) are those (s, S) such that

$$\mathfrak{C}_s : \frac{\Lambda(s, S)}{\overline{\Theta}^{(q)}(S - s)} = 0, \tag{4.5}$$

$$\mathfrak{C}_S : \frac{\Theta^{(q)}(S - s)}{\overline{\Theta}^{(q)}(S - s)} \Lambda(s, S) - \lambda(s, S) = 0, \tag{4.6}$$

where $\Theta^{(q)}$ is as defined in (2.9) with its antiderivative $\overline{\Theta}^{(q)}$ given by

$$\overline{\Theta}^{(q)}(x) := W^{(q)}(x) - \Phi(q)\overline{W}^{(q)}(x) > 0,$$

and

$$\Lambda(s, x) := \Phi(q)\Psi(s; \tilde{f})\overline{W}^{(q)}(x - s) + K - \varphi_s(x; \tilde{f}), \quad x, s \in \mathbb{R}, \tag{4.7}$$

$$\lambda(s, x) := \frac{\partial}{\partial x}\Lambda(s, x), \quad x > s. \tag{4.8}$$

Here, we shall confirm briefly how this is so. Note that when $\mathfrak{C}_s$ is satisfied, then $\mathfrak{C}_S$ is equivalent to the condition:

$$\mathfrak{C}'_S : \lambda(s, S) = 0. \tag{4.9}$$

Remark 4.2 We note the similarity between $\mathfrak{C}_s$ and $\mathfrak{C}_S$ (or $\mathfrak{C}'_S$) with the conditions $\mathfrak{C}_a$ and $\mathfrak{C}_b$ (or $\mathfrak{C}'_b$) as in (3.7), (3.10) (or (3.11)) in the two-sided singular control case.

First, by using the technique (using Eq. (4.3)) discussed above, we can compute (4.4): for all $s < S$,

$$\tilde{v}_{s,S}(S) = \frac{\Phi(q)}{q\overline{\Theta}^{(q)}(S - s)}\left[\overline{\Theta}^{(q)}(S - s)\left[\Psi(s; \tilde{f}) - \frac{q}{\Phi(q)}\left(K + \frac{C_U\psi'(0+)}{q}\right)\right] + \Lambda(s, S)\right],$$

$$\tilde{v}_{s,S}(x) = \begin{cases} -\frac{\overline{\Theta}^{(q)}(x-s)}{\overline{\Theta}^{(q)}(S-s)}\Lambda(s, S) + \Lambda(s, x) + \tilde{v}_{s,S}(S), & x \geq s, \\ K + \tilde{v}_{s,S}(S), & x < s. \end{cases} \tag{4.10}$$

Differentiating (4.10),

$$\tilde{v}'_{s,S}(x) = -\frac{\Theta^{(q)}(x - s)}{\overline{\Theta}^{(q)}(S - s)}\Lambda(s, S) + \lambda(s, x), \quad s < x < S. \tag{4.11}$$

From these expressions, we shall see that the conditions $\mathfrak{C}_s$ and $\mathfrak{C}_S$ as in (4.5) and (4.6) guarantee the desired smoothness/slope conditions described above: namely,

(1) $\tilde{v}_{s^*,S^*}(\cdot)$ is continuous (resp. differentiable) at s^* when X is of bounded (resp. unbounded) variation,
(2) $\tilde{v}'_{s^*,S^*}(S^*) = 0$.

(1) Regarding the continuity at s, by (4.10),

$$\tilde{v}_{s,S}(s+) = -\frac{\overline{\Theta}^{(q)}(0)}{\overline{\Theta}^{(q)}(S-s)}\Lambda(s,S) + K + \tilde{v}_{s,S}(S)$$
$$= -\frac{\overline{\Theta}^{(q)}(0)}{\overline{\Theta}^{(q)}(S-s)}\Lambda(s,S) + \tilde{v}_{s,S}(s-),$$

where $\overline{\Theta}^{(q)}(0) = 0$ if and only if X is of unbounded variation in view of (2.4). Hence, the continuity at $x = s$ holds if and only if $\mathfrak{C}_s$ holds for the case of bounded variation. On the other hand, it holds automatically for the unbounded variation case.

For the case of unbounded variation, we further pursue the differentiability at $x = s$. Equation (4.11) gives $\tilde{v}'_{s,S}(s+) = -\frac{\Theta^{(q)}(0)}{\overline{\Theta}^{(q)}(S-s)}\Lambda(s,S)$, and hence $\mathfrak{C}_s$ leads to the differentiability at s.

(2) Regarding the slope condition at S, we have $\tilde{v}'_{s,S}(S) = -\frac{\Theta^{(q)}(S-s)}{\overline{\Theta}^{(q)}(S-s)}\Lambda(s,S) + \lambda(s,S)$. Hence, given $\mathfrak{C}_s$, the condition $\mathfrak{C}_S$ guarantees $\tilde{v}'_{s,S}(S) = 0$ as desired.

Existence of (s^*, S^*) We now illustrate how the existence of (s^*, S^*) guaranteeing $\mathfrak{C}_s$ and $\mathfrak{C}_S$ can be shown. Here, as in Example 3.3, we shall assume Assumption 3.2: then,

$$\underline{a} \equiv \underline{a}(\tilde{f}') \quad \text{and} \quad \overline{a} \equiv \overline{a}(\tilde{f}')$$

are well-defined and finite as in the discussion given in Sect. 3.3.3.

We shall see that the desired s^* lies on the left of $\underline{a}$ while S^* lies on its right. As K decreases, the distance between s^* and S^* is expected to shrink and converge to $\underline{a}$, which is the optimal barrier in Example 3.3 for the case $b^* = \infty$.

To show the existence of (s^*, S^*), we shall first write

$$\Lambda(s,S) = \int_s^S \Psi(y;\tilde{f}')\overline{\Theta}^{(q)}(S-y)dy + K, \quad s, S \in \mathbb{R},$$
$$\lambda(s,S) = \Psi(S;\tilde{f}')W^{(q)}(0) + \int_s^S \Psi(y;\tilde{f}')\Theta^{(q)}(S-y)dy, \quad S > s. \tag{4.12}$$

In Fig. 6, we show sample plots of the functions $S \mapsto \Lambda(s,S)$ and $S \mapsto \lambda(s,S)$ for several values of starting points s, including $\underline{a}$ and a^*.

As can be confirmed in the figure and also clear from (4.12), by how $\underline{a}$ is chosen, we have the following properties:

(1) When $s > \underline{a}$, $\lambda(s,S) > 0$ for $S > s$ and hence $S \mapsto \Lambda(s,S)$ is monotonically increasing on $[s,\infty)$.

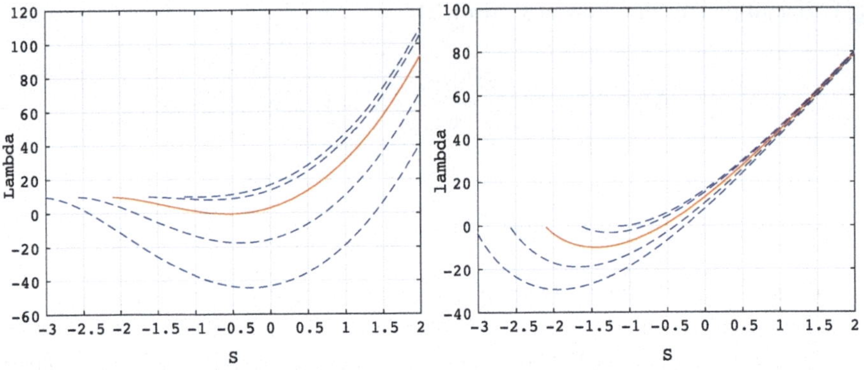

$S \mapsto \Lambda(s, S)$ $S \mapsto \lambda(s, S)$

Fig. 6 Existence of (s^*, S^*) for Example 4.3. Plots of $S \mapsto \Lambda(s, S)$ and $S \mapsto \lambda(s, S)$ on $[s, \infty)$ for five values of s are shown. The line in red corresponds to the one for $s = s^*$; the point at which $\Lambda(s^*, \cdot)$ is tangent to the x-axis becomes S^*. The rightmost curve corresponds to the one with $s = \underline{a}$; it is confirmed that $\Lambda(\underline{a}, \cdot)$ is monotonically increasing and $\lambda(\underline{a}, \cdot)$ is uniformly positive

(2) When $s < \underline{a}$, $\partial \Lambda(s, S)/\partial s = -\Psi(s; \tilde{f}')\overline{\Theta}^{(q)}(S - s) \geq 0$ by how $\underline{a}$ is chosen.
(3) For every fixed $s \in \mathbb{R}$, $\lim_{S \uparrow \infty} \Lambda(s, S) = \infty$.
(4) For every fixed $S \in \mathbb{R}$, $\lim_{s \downarrow -\infty} \Lambda(s, S) = -\infty$.
(5) For any $s \in \mathbb{R}$, $\Lambda(s, s) = K > 0$.

It is now clear how to obtain the desired (s^*, S^*). Similarly to Example 3.3, starting at $s = \underline{a}$, we decrease the value of s until we arrive at s^* such that $\inf_{S > s^*} \Lambda(s^*, S) = 0$. This exists because the function $s \mapsto \inf_{S > s} \Lambda(s, S)$, $s < \underline{a}$, is increasing by the property (2) above and goes to $-\infty$ as $s \downarrow -\infty$ by the property (4). Note that, because (4.12) implies $\lambda(s^*, S) < 0$ for $S \in (s^*, \underline{a})$, we must have $S^* > \underline{a}$. Because $\inf_{S > s^*} \Lambda(s^*, S) = 0$ attains a local minimum at $S = S^*$, we must have $\lambda(s^*, S^*) = \Lambda(s^*, S^*) = 0$, as desired.

4.2.2 Brief Remarks on the Cases of Examples 4.1 and 4.2

In [35] and [9], they use the first-order conditions to obtain (s^*, S^*) in Examples 4.1 and 4.2, respectively. To this end, they used the argument that the surface $(s, S) \mapsto v_{s,S}(x)$ has a global minimum (if formulated as a minimization problem).

The difficulty in their case is that because $\mathcal{I}$ has a boundary 0, it can happen that S^* (or both s^* and S^*) is zero. This means that the (s^*, S^*)-strategy, once activated, moves the controlled process to the default boundary. In Example 4.2 where 0 is regular for $\mathcal{I}^c = (0, \infty)$, ruin then occurs immediately. On the other hand, in Example 4.1, it is regular for $\mathcal{I}^c = (-\infty, 0)$ if and only if X is of unbounded variation. Hence, while ruin occurs immediately for the unbounded variation case, it stays above 0 for a positive amount of time a.s. This suggests one difficulty in solving the spectrally negative Lévy case.

If $S^* \neq 0$, the slope condition $v'_{s^*,S^*}(S^*) = -C_U = 1$ (resp. $v'_{s^*,S^*}(S^*) = C_U = -1$) is satisfied for Example 4.2 (resp. Example 4.1). Similarly, if $s^* \neq 0$, then the smoothness condition $v'_{s^*,S^*}(s^*) = -C_U = 1$ (resp. $v'_{s^*,S^*}(s^*) = C_U = -1$) is satisfied for Example 4.2 (resp. Example 4.1).

4.3 Quasi-Variational Inequalities and Verification

The verification of optimality asks that the candidate value function v_{s^*,S^*} satisfies the QVI (quasi-variational inequalities):

$$\begin{aligned}
&(\mathcal{L} - q)v_{s^*,S^*}(x) + f(x) \geq 0, \quad x \in \mathcal{I}^o \backslash \{s^*\}, \\
&v_{s^*,S^*}(x) \leq K + \inf_{u \in \mathcal{A}, x+u \in \mathcal{I}} \left[C_U|u| + v_{s^*,S^*}(x+u)\right], \quad x \in (-\infty, \overline{\mathcal{I}}], \\
&[(\mathcal{L} - q)v_{s^*,S^*}(x) + f(x)]\left[v_{s^*,S^*}(x) - K - \inf_{u \in \mathcal{A}, x+u \in \mathcal{I}} \left[C_U|u| + v_{s^*,S^*}(x+u)\right]\right] = 0, \\
&\quad x \in \mathcal{I}^o \backslash \{s^*\}.
\end{aligned} \tag{4.13}$$

Here, in the middle equality, if it is assumed for the case $\{u : u \in \mathcal{A}, x + u \in \mathcal{I}\}$ is empty, the right hand side is ∞.

For its proof, see [11, 12]. Similarly to the singular control case, in general we need additional assumptions on the tail growth of f and the Lévy measure. In particular, in [12, 49], it is assumed that the growth of f in the tail is at most polynomial.

4.3.1 The Case of Example 4.3

With (s^*, S^*) that satisfy $\mathfrak{C}_s$, the function (4.10) simplifies to, for $x \in \mathbb{R}$,

$$\tilde{v}_{s^*,S^*}(S^*) = \frac{\Phi(q)}{q}\Psi(s^*; \tilde{f}) - K - \frac{C_U\psi'(0+)}{q}, \tag{4.14}$$

$$\tilde{v}_{s^*,S^*}(x) = \Lambda(s^*, x) + \tilde{v}_{s^*,S^*}(S^*), \tag{4.15}$$

or equivalently

$$v_{s^*,S^*}(x) = \left(\frac{\Phi(q)}{q}\Psi(s^*; f) + \frac{C_U}{\Phi(q)}\right) Z^{(q)}(x - s^*) - C_U R^{(q)}(x - s^*) - \varphi_{s^*}(x; f). \tag{4.16}$$

See Fig. 7 for a sample plot of v_{s^*,S^*}.

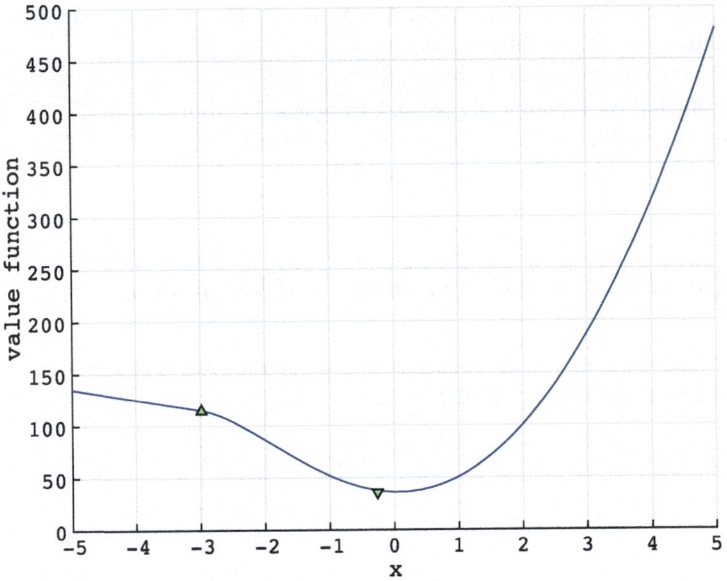

Fig. 7 A sample plot of the value function v_{s^*,S^*} for Example 4.3 when X is of unbounded variation. The up-pointing and down-pointing triangles show the points at s^* and S^*, respectively

Similarly to the singular control case (see Lemma 3.1), some inequalities of (4.13) are easily shown with minor assumptions on the function f.

Lemma 4.1 *Suppose $\mathfrak{C}_s$ holds.*

(1) *We have $(\mathcal{L} - q)v_{s^*,S^*}(x) + f(x) = 0$ for $x > s^*$.*
(2) *If Assumption 3.2 holds and $\underline{a}$ is well-defined and finite with $s^* \leq \underline{a} < \overline{a}$, then $(\mathcal{L} - q)v_{s^*,S^*}(x) + f(x) \geq 0$ on $(-\infty, s^*)$.*

Proof

(1) In view of (4.16), this is immediate by the results summarized in Sect. 2.7.3.
(2) Because $\tilde{v}_{s^*,S^*}(x) = K + \tilde{v}_{s^*,S^*}(S^*)$ for $x < s^*$ and by (4.14),

$$(\mathcal{L} - q)v_{s^*,S^*}(x) + f(x) = -q(K + \tilde{v}_{s^*,S^*}(S^*)) - C_U\psi'(0+) + C_U qx + f(x)$$
$$= \tilde{f}(x) - \tilde{f}(s^*) - \Psi(s^*; \tilde{f}').$$

This is positive by $x < s^* < \underline{a} \leq \overline{a}$ and how $\underline{a}$ and $\overline{a}$ are chosen. □

In view of Lemma 4.1, the remaining task is to show that

$$\begin{aligned} v_{s^*,S^*}(x) &= K + \inf_{u \geq 0}\left[C_U u + v_{s^*,S^*}(x+u)\right], \quad x \leq s^*, \\ v_{s^*,S^*}(x) &\leq K + \inf_{u \geq 0}\left[C_U u + v_{s^*,S^*}(x+u)\right], \quad x > s^*, \end{aligned} \tag{4.17}$$

or equivalently

$$\tilde{v}_{s^*,S^*}(x) = K + \inf_{u \geq 0} \tilde{v}_{s^*,S^*}(x+u), \quad x \leq s^*,$$

$$\tilde{v}_{s^*,S^*}(x) \leq K + \inf_{u \geq 0} \tilde{v}_{s^*,S^*}(x+u), \quad x > s^*.$$

These can be shown for $x \leq \underline{a}$ easily as follows. For $x \leq s^*$, in view of (4.15) and because S^* minimizes $\Lambda(s^*, x)$ over $x \in \mathbb{R}$, we must have

$$\tilde{v}_{s^*,S^*}(S^*) = \inf_{x \in \mathbb{R}} \tilde{v}_{s^*,S^*}(x). \tag{4.18}$$

Hence,

$$\tilde{v}_{s^*,S^*}(x) = \tilde{v}_{s^*,S^*}(s^*) = \tilde{v}_{s^*,S^*}(S^*) + K = K + \inf_{u \geq 0} \tilde{v}_{s^*,S^*}(x+u), \quad x \leq s^*. \tag{4.19}$$

The case $s^* \leq x \leq \underline{a}$ also holds by (4.18) and because $\tilde{v}'_{s^*,S^*}(x) = \lambda(s^*, x) < 0$ on $[s^*, \underline{a}]$ in view of how $\underline{a}$ is chosen and (4.12).

Unfortunately, the proof of (4.17) for $x > \underline{a}$ is difficult and, we need a nonstandard technique. As the fluctuation theory and scale function do not simplify the proof to our best knowledge, it is out of scope of this note. We refer the reader to the proof of Theorem 1(iii) of Benkherouf and Bensoussan [10].

Below, we summarize the functions and parameters that played important roles in characterizing the optimal solution in Examples 4.3.

4.3.2 Brief Remarks on the Cases of Examples 4.1 and 4.2

As in the singular control case, verification is in general harder for the spectrally negative case than for the spectrally positive case.

For Example 4.2, the variational inequalities (4.13) can be shown without much difficulty. Similarly to Example 4.3 above, the generator part of (4.13) holds trivially; this is due to the fact that in this case the controlling region is $(-\infty, s^*)$ and the waiting region is $(s^*, 0]$; the process does not jump from the former to the latter and hence the results similar to Lemma 4.1 hold. The other parts of (4.13) can be shown using the log-concavity of the scale function as in Sect. 2.7.2, which essentially shows that $-v'_{s^*,S^*}(x) < -C_U$ if and only if $x \in (s^*, S^*)$; see Lemma 5.3 of [9].

On the other hand, the verification for Example 4.1 can only be done for a subset of spectrally negative Lévy processes. This is again due to the fact, in this case, that the controlling region is (s^*, ∞) and the waiting region is $[0, s^*)$; the process can jump from the former to the latter, where the form of v_{s^*,S^*} changes.

5 Zero-Sum Games Between Two-Players

In this section, we consider optimal stopping games between two players: the *inf player* and the *sup player*, whose strategies are given by stopping times θ and τ, respectively. Here, a common expected payoff is minimized by the former and is maximized by the latter. The problem is terminated at the time either of the two players decides to stop or at the first exit time from some closed interval $\mathcal{I}$:

$$T_{\mathcal{I}^c} := \inf\{t > 0 : X_t \notin \mathcal{I}\}.$$

Without loss of generality, these can be assumed to satisfy

$$\theta, \tau \le T_{\mathcal{I}^c}, \quad a.s. \tag{5.1}$$

Let $q > 0$ be the discount factor and the terminal payoff be given by

(1) g_I: when the inf player stops first,
(2) g_S: when the sup player stops first,
(3) g: when both players stop simultaneously (including the case $\theta = \tau = T_{\mathcal{I}^c}$),

such that $g(x) = 0$ for $x \notin \mathcal{I}$. Then given any pair of strategies (θ, τ), the expected cost (resp. reward) for the inf (resp. sup) player is

$$v(x; \theta, \tau) := \mathbb{E}_x\Big[1_{\{\theta<\tau\}}e^{-q\theta}g_I(X_\theta) + 1_{\{\tau<\theta\}}e^{-q\tau}g_S(X_\tau) + 1_{\{\tau=\theta<\infty\}}e^{-q\tau}g(X_\tau)\Big]. \tag{5.2}$$

The objective is to determine, if it exists, a pair of stopping times $(\theta^*, \tau^*) \subset \mathcal{S}$, called the *saddle point*, that constitutes the *Nash equilibrium*:

$$v(x; \theta^*, \tau) \le v(x; \theta^*, \tau^*) \le v(x; \theta, \tau^*), \quad \forall \theta, \tau \in \mathcal{S}, \tag{5.3}$$

where $\mathcal{S}$ is the set of stopping times satisfying (5.1).

Example 5.1 Egami et al. [20] considered several games in the setting of a credit default swap (CDS) contract as extensions to the optimal stopping problem considered in Leung and Yamazaki [33].

As in a usual perpetual CDS contract, the sup player (protection buyer) pays premium continuously and whenever the default event $\{X < 0\}$ happens, the sup player receives from the inf player (seller) a fixed default payment 1, and the contract is terminated.

In their *cancellation game*, they added a feature that the sup player and inf player both have an option to cancel the contract before default for a fee, whoever cancels first. Specifically,

(1) the sup player begins by paying premium at rate p over time for a notional amount 1 to be paid at default;

(2) prior to default, the sup player and the inf player can select a time to cancel the contract;
(3) when the sup player cancels, he is incurred the fee γ_S to be paid to the inf player; when the inf player cancels, he is incurred γ_I to be paid to the sup player;
(4) if the sup player and the inf player exercise simultaneously, then both pay the fee upon exercise.

For the game to make sense, these parameters are assumed to satisfy

$$1 > \gamma_I \geq 0, \quad p > 0, \quad \gamma_S + \gamma_I > 0. \tag{5.4}$$

Namely, the inf player wants to minimize while the sup player wants to maximize the common expectation:

$$V(x; \theta, \tau) := \mathbb{E}_x \Bigg[- \int_0^{\tau \wedge \theta} e^{-qt} p \, \mathrm{d}t \\ + 1_{\{\tau \wedge \theta < \infty\}} \Big(e^{-qT_{(-\infty,0)}} 1_{\{\tau = \theta = T_{(-\infty,0)}\}} + 1_{\{\tau \wedge \theta < T_{(-\infty,0)}\}} e^{-q(\tau \wedge \theta)} \left(-\gamma_S 1_{\{\tau \leq \theta\}} + \gamma_I 1_{\{\tau \geq \theta\}} \right) \Big) \Bigg], \tag{5.5}$$

by choosing stopping times θ and τ, respectively.

Let

$$C(x; p) := \mathbb{E}_x \left[- \int_0^{T_{(-\infty,0)}} e^{-qt} p \, \mathrm{d}t + e^{-qT_{(-\infty,0)}} \right] = \left(\frac{p}{q} + 1 \right) \zeta(x) - \frac{p}{q}, \quad x > 0, \tag{5.6}$$

where, by (2.6),

$$\zeta(x) := \mathbb{E}_x \left[e^{-qT_{(-\infty,0)}} \right] = Z^{(q)}(x) - \frac{q}{\Phi(q)} W^{(q)}(x), \quad x \in \mathbb{R}.$$

Then, by the strong Markov property, (5.5) can be written

$$V(x; \theta, \tau) = C(x; p) + v(x; \theta, \tau), \quad x > 0,$$

where

$$v(x; \theta, \tau) := \mathbb{E}_x \left[e^{-q(\tau \wedge \theta)} \left(g_S(X_\tau) 1_{\{\tau < \theta\}} + g_I(X_\theta) 1_{\{\tau > \theta\}} + g(X_\tau) 1_{\{\tau = \theta\}} \right) 1_{\{\tau \wedge \theta < \infty\}} \right], \tag{5.7}$$

with, for $x \in \mathbb{R}$,

$$g_S(x) := 1_{\{x > 0\}} \Big[\Big(\frac{p}{q} - \gamma_S \Big) - \Big(\frac{p}{q} + 1 \Big) \zeta(x) \Big], \tag{5.8}$$

$$g_I(x) := 1_{\{x>0\}}\Big[\Big(\frac{p}{q} + \gamma_I\Big) - \Big(\frac{p}{q} + 1\Big)\zeta(x)\Big], \tag{5.9}$$

$$g(x) := 1_{\{x>0\}}\Big[\Big(\frac{p}{q} - \gamma_S + \gamma_I\Big) - \Big(\frac{p}{q} + 1\Big)\zeta(x)\Big]. \tag{5.10}$$

In other words, the problem is to identify the pair of strategies (θ^*, τ^*) such that (5.3) holds.

5.1 Threshold Strategies

If the (common) payoff functions have some monotonicity with respect to the position of X as in the examples given in Sect. 1.1, it is expected that both implement threshold strategies where one of them stops when X is sufficiently high while the other stops when it is sufficiently low. Hence, it is a reasonable conjecture that the equilibrium is characterized by two boundaries: $\alpha < \beta$ or $\beta < \alpha$.

We shall now consider a pair of strategies $(\theta_\alpha, \tau_\beta)$ such that

(1) if $\alpha < \beta$, then $\theta_\alpha := \inf\{t > 0 : X_t < \alpha\}$ and $\tau_\beta := \inf\{t > 0 : X_t > \beta\}$,
(2) if $\beta < \alpha$, then $\theta_\alpha := \inf\{t > 0 : X_t > \alpha\}$ and $\tau_\beta := \inf\{t > 0 : X_t < \beta\}$.

In order to satisfy the condition (5.1), we must have $\underline{I} \le \alpha < \beta \le \overline{I}$ and $\underline{I} \le \beta < \alpha \le \overline{I}$ for (1) and (2), respectively.

In this case, the players' expected NPVs of reward/cost (5.2) becomes

$$v_{\alpha,\beta}(x) := \mathbb{E}_x\Big[1_{\{\theta_\alpha < \tau_\beta\}} e^{-q\theta_\alpha} g_I(X_{\theta_\alpha}) + 1_{\{\tau_\beta < \theta_\alpha\}} e^{-q\tau_\beta} g_S(X_{\tau_\beta})\Big].$$

By the reviewed results in Sect. 2.4, this can be computed by the scale function and the Lévy measure.

Focusing on the strategy pairs given by $(\theta_\alpha, \tau_\beta)$, the first step again is to choose a candidate barrier pair (α^*, β^*) using two equations. The expected degree of smoothness is the same as the impulse control case (see Sect. 4.2) and is one less than the singular control case (see Sect. 3.2). More precisely, we have the following for the case $\alpha^* < \beta^*$ (the case $\beta^* < \alpha^*$ holds in the same way by swapping the roles of α^* and β^*):

(1) Regarding the smoothness of the value function at the lower barrier α^*,
   (a) if α^* is regular for $(-\infty, \alpha^*)$ (or equivalently X is of unbounded variation), then the continuous differentiability at α^* is expected;
   (b) if α^* is irregular for $(-\infty, \alpha^*)$ (or equivalently X is of bounded variation), then the continuity at α^* is expected.
(2) Regarding the smoothness at the upper barrier β^*, because it is always regular for (β^*, ∞), continuous differentiability is expected at β^* regardless of the path variation.

5.1.1 The Case of Example 5.1

In the cancellation game, the sup player has an incentive to cancel the contract when default is less likely, or equivalently when X is sufficiently high. On the other hand, the inf player tends to cancel it when default is likely to occur, or equivalently when X is sufficiently small. Because $\mathcal{I} = [0, \infty)$, we can conjecture that the sup player and the inf player choose the strategies τ_{β^*} and θ_{α^*} for some values $0 \leq \alpha^* < \beta^* \leq \infty$. Regarding the cases $\alpha^* = 0$ and $\beta^* = \infty$, see the interpretations given in Remark 5.2.

For $0 < \alpha < x < \beta < \infty$, it is straightforward to write

$$\begin{aligned} v_{\alpha,\beta}(x) - g_S(x) &= \Upsilon(x; \alpha, \beta) - \frac{p}{q} + \gamma_S, \\ v_{\alpha,\beta}(x) - g_I(x) &= \Upsilon(x; \alpha, \beta) - \frac{p}{q} - \gamma_I, \end{aligned} \tag{5.11}$$

where

$$\begin{aligned} \Upsilon(x; \alpha, \beta) := &-\gamma_S \mathbb{E}_x \left[e^{-q(\theta_\alpha \wedge \tau_\beta)} 1_{\{\tau_\beta < \theta_\alpha\}} \right] + \gamma_I \mathbb{E}_x \left[e^{-q(\theta_\alpha \wedge \tau_\beta)} 1_{\{\tau_\beta > \theta_\alpha \text{ or } \theta_\alpha = \tau_\beta = T_{(-\infty,0)}\}} \right] \\ &- \gamma_I \mathbb{E}_x \left[e^{-q(\theta_\alpha \wedge \tau_\beta)} 1_{\{\theta_\alpha = \tau_\beta = T_{(-\infty,0)}\}} \right]. \end{aligned} \tag{5.12}$$

By the results in Sect. 2.4 together with the compensation formula (see Theorem 4.4 of [30]), we can write

$$\Upsilon(x; \alpha, \beta) = W^{(q)}(x - \alpha) \frac{\Lambda(\alpha, \beta)}{W^{(q)}(\beta - \alpha)} - \Lambda(\alpha, x) + \frac{p}{q} - \gamma_S, \quad \beta > x > \alpha > 0, \tag{5.13}$$

where, for $0 < \alpha < \beta < \infty$,

$$\begin{aligned} \Lambda(\alpha, \beta) := &\frac{p}{q} - \gamma_S - \Big(\frac{p}{q} + \gamma_I\Big) Z^{(q)}(\beta - \alpha) \\ &+ \frac{1 - \gamma_I}{q} \int_{(-\infty,-\alpha)} \Big(Z^{(q)}(\beta - \alpha) - Z^{(q)}(\beta + u) \Big) \nu(\mathrm{d}u). \end{aligned} \tag{5.14}$$

We also define the derivative of (5.14) as, for $0 < \alpha < \beta < \infty$,

$$\begin{aligned} \lambda(\alpha, \beta) := &\frac{\partial}{\partial \beta} \Lambda(\alpha, \beta) = -\left(p + \gamma_I q\right) W^{(q)}(\beta - \alpha) \\ &+ (1 - \gamma_I) \int_{(-\infty,-\alpha)} \Big(W^{(q)}(\beta - \alpha) - W^{(q)}(\beta + u) \Big) \nu(\mathrm{d}u). \end{aligned}$$

We begin with establishing the continuous fit condition. First, by taking limits in (5.11), we have, for $0 < \alpha < \beta < \infty$

$$v_{\alpha,\beta}(\beta-) - g_S(\beta) = \Upsilon(\beta-;\alpha,\beta) + \gamma_S = 0, \tag{5.15}$$

$$v_{\alpha,\beta}(\alpha+) - g_I(\alpha) = W^{(q)}(0)\frac{\Lambda(\alpha,\beta)}{W^{(q)}(\beta-\alpha)}. \tag{5.16}$$

This means that continuous fit holds automatically at β. On the other hand, at α, while continuous fit holds automatically for the case of unbounded variation, it holds if and only if

$$\mathfrak{C}_\alpha : \frac{\Lambda(\alpha,\beta)}{W^{(q)}(\beta-\alpha)} = 0 \tag{5.17}$$

for the bounded variation case.

Now, by taking the derivative of (5.13), we obtain, for $\alpha < x < \beta$,

$$\begin{aligned} v'_{\alpha,\beta}(x+) - g'_S(x) &= v'_{\alpha,\beta}(x+) - g'_I(x) \\ &= \Upsilon'(x+;\alpha,\beta) = W^{(q)\prime}((x-\alpha)+)\frac{\Lambda(\alpha,\beta)}{W^{(q)}(\beta-\alpha)} - \lambda(\alpha,x). \end{aligned}$$

Hence, the smooth fit at β holds if and only if

$$\mathfrak{C}_\beta : W^{(q)\prime}((\beta-\alpha)-)\frac{\Lambda(\alpha,\beta)}{W^{(q)}(\beta-\alpha)} - \lambda(\alpha,\beta) = 0.$$

Assuming that it has paths of unbounded variation ($W^{(q)}(0) = 0$), then we obtain

$$v'_{\alpha,\beta}(\alpha+) - g'(\alpha) = W^{(q)\prime}(0+)\frac{\Lambda(\alpha,\beta)}{W^{(q)}(\beta-\alpha)}, \quad 0 < \alpha < \beta.$$

Therefore, $\mathfrak{C}_\alpha$ is also a sufficient condition for smooth fit at α for the unbounded variation case. In addition, if $\mathfrak{C}_\alpha$ holds, then $\mathfrak{C}_\beta$ simplifies to

$$\mathfrak{C}'_\beta : \lambda(\alpha,\beta) = 0.$$

We conclude that

(1) if (α^*, β^*) satisfy $\mathfrak{C}_\alpha$, then continuous fit at α^* holds for the bounded variation case and both continuous and smooth fit at α^* holds for the unbounded variation case;
(2) if (α^*, β^*) satisfy $\mathfrak{C}_\beta$, then both continuous and smooth fit conditions at β^* hold for all cases.

Remark 5.1 Note that, except that the form of Λ is different, the conditions $\mathfrak{C}_\alpha$ and $\mathfrak{C}_\beta$ (or $\mathfrak{C}'_\beta$) are the same as $\mathfrak{C}_a$ and $\mathfrak{C}_b$ (or $\mathfrak{C}'_b$) as in (3.7) and (3.10) (or (3.11)) in the two-sided singular control case and are similar to $\mathfrak{C}_s$ and $\mathfrak{C}_S$ (or $\mathfrak{C}'_S$) as in (4.5) and (4.6) (or (4.9)) in the impulse control case.

In order to show the existence of a pair that satisfy $\mathfrak{C}_\alpha$ and $\mathfrak{C}_\beta$, consider the function, for $0 < \alpha < \beta$,

$$\widehat{\lambda}(\alpha,\beta) := \frac{\lambda(\alpha,\beta)}{W^{(q)}(\beta-\alpha)} = -(p+q\gamma_I) + (1-\gamma_I)\int_{(-\infty,-\alpha)}\left(1-\frac{W^{(q)}(\beta+u)}{W^{(q)}(\beta-\alpha)}\right)\nu(du).$$

By using the log-concavity of the scale function as in Sect. 2.7.2, the following can be easily derived.

Lemma 5.1

(1) *For fixed* $0 < \beta < \infty$, $\alpha \mapsto \widehat{\lambda}(\alpha,\beta)$ *is decreasing on* $(0,\beta)$.
(2) *For fixed* $\alpha > 0$, $\beta \mapsto \widehat{\lambda}(\alpha,\beta)$ *is decreasing on* (α,∞).

Using Lemma 5.1(2) and (2.13), for $\alpha > 0$, we can extend $\widehat{\lambda}(\alpha,\beta)$ to the cases $\beta = \alpha$ and $\beta = \infty$ with

$$\widehat{\lambda}(\alpha) \equiv \widehat{\lambda}(\alpha,\alpha+) := \lim_{\beta\downarrow\alpha}\widehat{\lambda}(\alpha,\beta) = -(p+q\gamma_I) + (1-\gamma_I)\bar{\nu}(\alpha),$$

$$\widehat{\lambda}(\alpha,\infty) := \lim_{\beta\to\infty}\widehat{\lambda}(\alpha,\beta) = -(p+q\gamma_I) + (1-\gamma_I)\Phi(q)\Psi(\alpha;\bar{\nu}) = \Phi(q)\Psi(\alpha;\widehat{\lambda}),$$

where

$$\bar{\nu}(x) := \nu(-\infty,-x), \quad x > 0.$$

We shall see that the function $\widehat{\lambda}(\cdot)$ plays the same role as $\tilde{f}'(\cdot)$ in Examples 3.3 and 4.3. Because $\widehat{\lambda}(\cdot)$ and $\Psi(\cdot;\widehat{\lambda})$ are monotonically decreasing, we can define $\overline{\alpha} := \overline{a}(-\widehat{\lambda})$ and $\underline{\alpha} := \underline{a}(-\widehat{\lambda})$ as in Definitions 2.1 and 2.2, respectively. These will serve as bounds on α^* and we will have $\underline{\alpha} \le \alpha^* < \overline{\alpha}$.

Egami et al. [20] show that there always exists a pair (α^*,β^*) belonging to one of the following four cases:

case 1: $0 < \alpha^* < \beta^* < \infty$;
case 2: $0 < \alpha^* < \beta^* = \infty$;
case 3: $0 = \alpha^* < \beta^* < \infty$;
case 4: $0 = \alpha^* < \beta^* = \infty$;

which satisfy $\mathfrak{C}_\alpha$ when $\alpha^* > 0$ and $\mathfrak{C}_\beta$ when $\beta^* < \infty$.

Here, we only give a brief sketch of the proof that if

$$\underline{\alpha} > 0 \quad \text{and} \quad \sup_{\beta > \underline{\alpha}}\Lambda(\underline{\alpha},\beta) > 0, \tag{5.18}$$

then **case 1** holds. (If these are violated, $\alpha^* = 0$ and/or $\beta^* = \infty$; see Remark 5.2 below.) To this end, observe that

$$\frac{\partial}{\partial\alpha}\Lambda(\alpha,\beta) = -W^{(q)}(\beta-\alpha)\widehat{\lambda}(\alpha) \tag{5.19}$$

is negative for every $\alpha \in (0, \overline{\alpha})$ by how $\overline{\alpha}$ is chosen as in Definition 2.1. Hence, the function $\alpha \mapsto \sup_{\beta>\alpha}\Lambda(\alpha,\beta)$ is monotonically decreasing on $(0, \overline{\alpha})$. Thanks to the continuity of $\Lambda(\alpha,\beta)$ and (5.18), if we can show that $\sup_{\beta>\overline{\alpha}}\Lambda(\overline{\alpha},\beta) < 0$, then there must exist $\alpha^* \in (\underline{\alpha}, \overline{\alpha})$ such that $\sup_{\beta>\alpha^*}\Lambda(\alpha^*,\beta) = 0$ with its local maximum attained at β^*. Indeed, by Lemma 5.1(2) and how $\overline{\alpha}$ is chosen, $\widehat{\lambda}(\overline{\alpha},\beta) \le 0$ or equivalently $\lambda(\overline{\alpha},\beta) \le 0$ for $\beta \in (\overline{\alpha}, \infty)$ and hence $\sup_{\beta>\overline{\alpha}}\Lambda(\overline{\alpha},\beta) = \Lambda(\overline{\alpha},\overline{\alpha}+) = -(\gamma_I+\gamma_S) < 0$.

These properties of the shapes of λ and Λ can be confirmed by the numerical plots given in Fig. 8.

Remark 5.2 While the details are omitted in this note, when (5.18) does not hold, necessarily $\alpha^* = 0$ and/or $\beta^* = \infty$. In the latter case, it can be shown that the sup player never stops in the equilibrium.

In the case $\alpha^* = 0$, it may not yield the Nash equilibrium for the unbounded variation case. To see this, we notice that a default happens as soon as X goes below zero. Therefore, in the event that X continuously passes (creeps) through zero, the inf player would optimally seek to exercise at a level as close to zero as possible. Nevertheless, this timing strategy is not admissible, though it can be approximated arbitrarily closely by admissible stopping times. It can be shown that $\alpha^* = 0$ is possible only if the jump part X^d of X is of bounded variation.

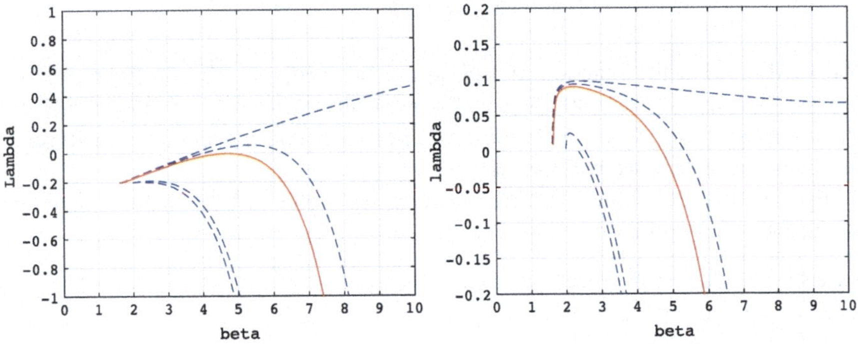

$\beta \mapsto \Lambda(\alpha,\beta)$ $\beta \mapsto \lambda(\alpha,\beta)$

Fig. 8 Existence of (α^*, β^*) for Example 5.1. Plots of $\beta \mapsto \Lambda(\alpha,\beta)$ on $[\alpha,\infty)$ for the starting values $\alpha = \underline{\alpha}, (\underline{\alpha}+\alpha^*)/2, \alpha^*, (\alpha^*+\overline{\alpha})/2, \overline{\alpha}$. The solid curve in red corresponds to the one for $\alpha = \alpha^*$; the point at which $\Lambda(\alpha^*,\cdot)$ is tangent to the x-axis (or $\lambda(\alpha^*,\cdot)$ vanishes) becomes β^*. The function $\Lambda(\underline{\alpha},\cdot)$ is monotonically increasing while $\Lambda(\overline{\alpha},\cdot)$ is monotonically decreasing. Equivalently, $\lambda(\underline{\alpha},\cdot)$ is uniformly positive while $\lambda(\overline{\alpha},\cdot)$ is uniformly negative

5.2 Variational Inequalities and Verification

The verification of optimality (for both players) require that, when $\alpha^* < \beta^*$,

$$\begin{aligned} g_S(x) \le v_{\alpha^*,\beta^*}(x) \le g_I(x), &\quad x \in \mathcal{I}, \\ (\mathcal{L} - q)v_{\alpha^*,\beta^*}(x) \ge 0, &\quad x \in (-\infty, \alpha^*) \cap \mathcal{I}^o, \\ (\mathcal{L} - q)v_{\alpha^*,\beta^*}(x) = 0, &\quad x \in (\alpha^*, \beta^*) \cap \mathcal{I}^o, \\ (\mathcal{L} - q)v_{\alpha^*,\beta^*}(x) \le 0, &\quad x \in (\beta^*, \infty) \cap \mathcal{I}^o. \end{aligned} \tag{5.20}$$

On the other hand, when $\alpha^* > \beta^*$, it requires that

$$\begin{aligned} g_S(x) \le v_{\alpha^*,\beta^*}(x) \le g_I(x), &\quad x \in \mathcal{I}, \\ (\mathcal{L} - q)v_{\alpha^*,\beta^*}(x) \le 0, &\quad x \in (-\infty, \beta^*) \cap \mathcal{I}^o, \\ (\mathcal{L} - q)v_{\alpha^*,\beta^*}(x) = 0, &\quad x \in (\beta^*, \alpha^*) \cap \mathcal{I}^o, \\ (\mathcal{L} - q)v_{\alpha^*,\beta^*}(x) \ge 0, &\quad x \in (\alpha^*, \infty) \cap \mathcal{I}^o. \end{aligned}$$

Suppose $\alpha^* < \beta^*$. From the inf player's perspective, assuming that the sup player's strategy is given by τ_{β^*} (so that the state space for the inf player is $\mathcal{I}_{\beta^*} := (-\infty, \beta^*) \cap \mathcal{I}$), the above variational inequalities satisfy those for the minimization problem for the inf player that

$$\begin{aligned} v_{\alpha^*,\beta^*}(x) \le g_I(x), &\quad x \in \mathcal{I}_{\beta^*}, \\ (\mathcal{L} - q)v_{\alpha^*,\beta^*}(x) \ge 0, &\quad x \in (-\infty, \alpha^*) \cap \mathcal{I}^o_{\beta^*}, \\ (\mathcal{L} - q)v_{\alpha^*,\beta^*}(x) = 0, &\quad x \in (\alpha^*, \beta^*). \end{aligned}$$

Similarly, from the sup player's perspective, assuming that the inf player's strategy is given by θ_{α^*} (so that the state space of the sup player is $\mathcal{I}_{\alpha^*} := (\alpha^*, \infty) \cap \mathcal{I}$), the above variational inequalities satisfy those for the maximization problem for the sup player that

$$\begin{aligned} v_{\alpha^*,\beta^*}(x) \ge g_S(x), &\quad x \in \mathcal{I}_{\alpha^*}, \\ (\mathcal{L} - q)v_{\alpha^*,\beta^*}(x) \le 0, &\quad x \in (\beta^*, \infty) \cap \mathcal{I}^o_{\alpha^*}, \\ (\mathcal{L} - q)v_{\alpha^*,\beta^*}(x) = 0, &\quad x \in (\alpha^*, \beta^*). \end{aligned}$$

The case $\alpha^* > \beta^*$ is similar, and hence we omit the details.

This is a rough illustration on why these conditions are imposed for verification. We refer the reader to [20] and also [21, 40] for more rigorous arguments. In general, if v_{α^*,β^*} is unbounded or $\mathcal{I}$ has a finite boundary at which v_{α^*,β^*} fails to be smooth/continuous, some localizing arguments are necessary.

5.2.1 Verification for Example 5.1

Here we shall illustrate a proof technique on how the candidate value function v_{α^*,β^*} solves the variational inequalities, focusing on Example 5.1 in the case $0 < \alpha^* < \beta^* < \infty$.

By (5.11), we can write

$$v_{\alpha^*,\beta^*}(x) = \left\{ \begin{array}{ll} g_S(x), & x \geq \beta^* \\ g_S(x) + (v_{\alpha^*,\beta^*}(x) - g_S(x)), & \alpha^* < x < \beta^* \\ g_I(x), & x \leq \alpha^* \end{array} \right\} = -\Big(\frac{p}{q} + 1\Big)\zeta(x) + J(x) \tag{5.21}$$

where

$$J(x) := \begin{cases} \frac{p}{q} - \gamma_S, & x \geq \beta^*, \\ \Upsilon(x; \alpha^*, \beta^*), & \alpha^* \leq x < \beta^*, \\ \frac{p}{q} + \gamma_I, & 0 \leq x < \alpha^*, \\ \frac{p}{q} + 1 & x < 0. \end{cases} \tag{5.22}$$

Here, by (5.17),

$$\Upsilon(x; \alpha^*, \beta^*) = \Big(\frac{p}{q} + \gamma_I\Big) Z^{(q)}(x - \alpha^*) \\ - \frac{1 - \gamma_I}{q} \int_{(-\infty, -\alpha^*)} \Big(Z^{(q)}(x - \alpha^*) - Z^{(q)}(x + u)\Big)\, \nu(\mathrm{d}u). \tag{5.23}$$

See Fig. 9 for a sample plot of the value function along with the stopping values.

Below, we show briefly that v_{α^*,β^*} solves (5.20) when $0 < \alpha^* < \beta^* < \infty$.

Lemma 5.2 *Suppose $W^{(q)}$ is sufficiently smooth on $(0, \infty)$ (i.e. C^1 when X is of bounded variation and C^2 when it is of unbounded variation). Then we have the following:*

(1) $g_S(x) \leq v_{\alpha^*,\beta^*}(x) \leq g_I(x), \quad x \in [0, \infty)$,
(2) $(\mathcal{L} - q)v_{\alpha^*,\beta^*}(x) \geq 0, \quad x \in (0, \alpha^*)$,
(3) $(\mathcal{L} - q)v_{\alpha^*,\beta^*}(x) = 0, \quad x \in (\alpha^*, \beta^*)$,
(4) $(\mathcal{L} - q)v_{\alpha^*,\beta^*}(x) \leq 0, \quad x \in (\beta^*, \infty)$.

Brief sketch of proof

(1) We show for $x \in (\alpha^*, \beta^*)$; the other cases are immediate.

The proof is relatively straightforward by the log-concavity of the scale function as in Sect. 2.7.2 and the shapes of Λ and λ given by

$$\Lambda(\alpha^*, \beta) \leq 0 \quad \text{and} \quad \lambda(\alpha^*, \beta) \geq 0, \quad \alpha^* < \beta < \beta^*. \tag{5.24}$$

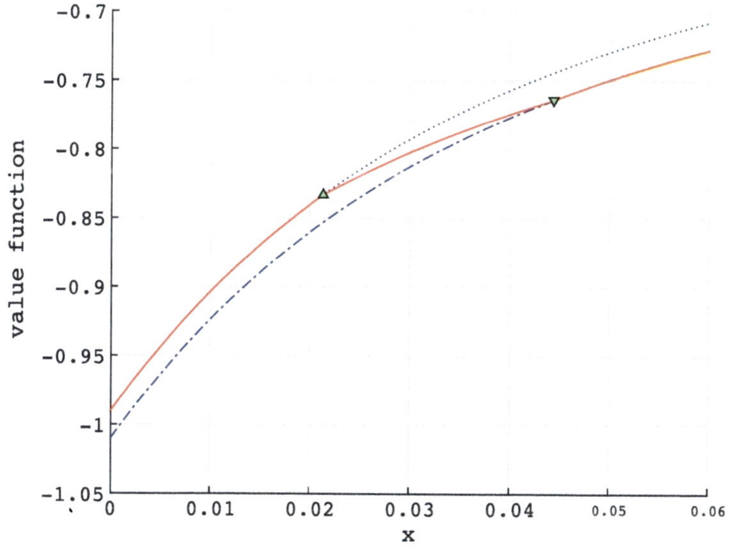

Fig. 9 A sample plot of the value function v_{α^*,β^*} (solid red line) for Example 5.1 when X is of unbounded variation. The up-pointing and down-pointing triangles show the points at α^* and β^*, respectively. The two dotted lines show the stopping values g_S and g_I

Here (5.24) holds because, by Lemma 5.1, $\beta \mapsto \Lambda(\alpha^*, \beta)$ increases on (α^*, β^*) and decreases on (β^*, ∞) with its peak given at $\Lambda(\alpha^*, \beta^*) = 0$ (see Fig. 8).

Now, with the help of (5.19) and the log-concavity,

$$\frac{\partial_+}{\partial_+ \alpha}(v_{\alpha,\beta^*}(x) - g_I(x)) = \Big[\frac{\partial_+}{\partial_+ \alpha}\frac{W^{(q)}(x-\alpha)}{W^{(q)}(\beta^*-\alpha)}\Big]\Lambda(\alpha, \beta^*) > 0, \quad \alpha^* < \alpha < x < \beta^*.$$

Hence, by this, (5.16) and (5.24), $0 \geq W^{(q)}(0)\Lambda(x, \beta^*)/W^{(q)}(\beta^* - x) = v_{x,\beta^*}(x+) - g_I(x) \geq v_{\alpha^*,\beta^*}(x) - g_I(x)$ for $\alpha^* < x < \beta^*$.

On the other hand, by (5.24),

$$\frac{\partial_+}{\partial_+ \beta}(v_{\alpha^*,\beta}(x) - g_S(x)) = \frac{W^{(q)}(x-\alpha^*)}{(W^{(q)}(\beta-\alpha^*))^2}\Big[\lambda(\alpha^*, \beta)W^{(q)}(\beta-\alpha^*) - \Lambda(\alpha^*, \beta)W^{(q)\prime}((\beta-\alpha^*)+)\Big] > 0, \quad \alpha^* < x < \beta < \beta^*.$$

Therefore, by this and (5.15), $0 = v_{\alpha^*,x}(x-) - g_S(x) \leq v_{\alpha^*,\beta^*}(x) - g_S(x)$ for $\alpha^* < x < \beta^*$.

(2) By the assumption that $W^{(q)}$ is sufficiently smooth, the identity (2.17) holds, and therefore

$$(\mathcal{L} - q)\zeta(x) = 0, \quad x > 0. \tag{5.25}$$

Hence,

$$(\mathcal{L} - q)v_{\alpha^*,\beta^*}(x) = (1 - \gamma_I)\bar{v}(x) - (q\gamma_I + p) = \widehat{\lambda}(x). \tag{5.26}$$

Because $x < \alpha^* < \overline{\alpha}$, this must be positive by how $\overline{\alpha}$ is chosen.

(3) In view of (5.21), (5.22), and (5.23), it is immediate by (2.16) together with (5.25).

(4) This is as usual the hardest part because the process can jump from the stopping region of the sup player (β^*, ∞) to the other two regions $(-\infty, \alpha^*)$ and (α^*, β^*), where the form of v_{α^*,β^*} changes. However, it is more straightforward than the two-sided singular control case that we studied in Sect. 3.

In Egami et al. [20], they first show that $(\mathcal{L} - q)v_{\alpha^*,\beta^*}(\beta^*+) \leq (\mathcal{L} - q)v_{\alpha^*,\beta^*}(\beta^*-) = 0$ using how α^* and β^* are chosen so that v_{α^*,β^*} gets smooth/continuous at β^*. It then remains to show that $x \mapsto (\mathcal{L} - q)v_{\alpha^*,\beta^*}(x)$ is decreasing on (β^*, ∞). In view of the decomposition (5.21) and also (5.25), it is equivalent to showing that $(\mathcal{L} - q)J(x)$ is decreasing on (β^*, ∞). Indeed, because $J' = J'' = 0$ on $x > \beta^*$,

$$(\mathcal{L} - q)J(x) = \int_{(-\infty,\beta^*-x)} \left[J(x+u) - \left(\frac{p}{q} - \gamma_S\right) \right] \nu(du) - (p - q\gamma_S), \quad x > \beta^*,$$

where the integrand is nonnegative and monotonically decreasing in x and the set $(-\infty, \beta^* - x)$ is decreasing in x as well. □

In Table 3, we summarize the functions and parameters that played major roles in the above analysis for Examples 5.1.

Table 3 Summary of the key functions and parameters in Example 5.1

$\Lambda(\alpha,\beta)$	$:= \frac{p}{q} - \gamma_S - (\frac{p}{q} + \gamma_I) Z^{(q)}(\beta-\alpha) + \frac{1-\gamma_I}{q} \int_{(-\infty,-\alpha)} \left[Z^{(q)}(\beta-\alpha) - Z^{(q)}(\beta+u) \right] \nu(du)$
$\widehat{\lambda}(\alpha)$	$:= -(p + q\gamma_I) + (1-\gamma_I)\bar{v}(\alpha)$
$\underline{\alpha}$	$:= \underline{a}(-\widehat{\lambda})$
$\leq \alpha^*$	$:= \alpha$ of (α, β) such that $\mathfrak{C}_\alpha$ and $\mathfrak{C}_\beta$ hold simultaneously
$< \overline{\alpha}$	$:= \overline{a}(-\widehat{\lambda})$
$< \beta^*$	$:= \beta$ of (α, β) such that $\mathfrak{C}_\alpha$ and $\mathfrak{C}_\beta$ hold simultaneously

It can be shown that $\alpha^* = \underline{\alpha}$ when $\beta^* = \infty$

5.3 Other Optimal Stopping Games

There are many other existing games studied for a spectrally one-sided Lévy process. The following problems can be formulated as (5.3). However, there are clear differences with the problem considered above.

Example 5.2 The McKean optimal stopping game corresponds to the case $\mathcal{I} = \mathbb{R}$ with $g_S(x) = g(x) = (K - e^x) \vee 0$ and $g_I = (K - e^x) \vee 0 + \delta$ for some $K, \delta > 0$. In other words, this is an extension of the American put option where the seller (inf player) can also exercise with an additional fee δ. This problem was solved by Baurdoux and Kyprianou [5] for a spectrally negative Lévy process. It is required that $0 \leq \psi(1) \leq q$ for the solution to be nontrivial.

Example 5.3 As a way to model a version of the convertible bond, Gapeev and Kühn [23] and Baurdoux et al. [7] considered the problem where the cost (resp. reward) for the inf (resp. sup) player is given by

$$V(x; \theta, \tau) := \mathbb{E}_x\Big[\int_0^{\tau\wedge\theta} e^{-qt}(C_1 + C_2 e^{X_t})\mathrm{d}t + 1_{\{\theta\leq\tau\}}e^{-q\theta}(e^{X_\theta} \vee K) + 1_{\{\tau<\theta\}}e^{-q\tau+X_\tau}\Big],$$

for $C_1 \geq 0$ and $C_2, K > 0$. This can be easily transformed to the formulation given in the beginning of this section. Indeed, by the strong Markov property, we can write $V(x; \theta, \tau) = v(x; \theta, \tau) + F(x)$ where

$$F(x) := \mathbb{E}_x\Big[\int_0^{\infty} e^{-rt}(C_1 + C_2 e^{X_t})\mathrm{d}t\Big],$$

$$v(x; \theta, \tau) := \mathbb{E}_x\Big[1_{\{\theta\leq\tau\}}e^{-q\theta}\big(e^{X_\theta} \vee K - F(X_\theta)\big) + 1_{\{\tau<\theta\}}e^{-q\tau}(e^{X_\tau} - F(X_\tau))\Big].$$

Hence, solving this is equivalent to solving (5.2) with $g_I(x) = g(x) = e^x \vee K - F(x)$, $g_S(x) = e^x - F(x)$, and $\mathcal{I} = \mathbb{R}$.

Gapeev and Kühn [23] considered the case of a Brownian motion plus i.i.d. exponential jumps. Baurdoux et al. [7] studied for a spectrally positive Lévy process.

In these examples, while the fluctuation theory and scale function can be used as main tools, the above techniques described in this section may not be directly used.

In Example 5.2, Baurdoux and Kyprianou [5] showed that the equilibrium is given by either $\tau^* := \inf\{t > 0 : X_t < k^*\}$ and $\sigma^* = \infty$, or $\tau^* := \inf\{t > 0 : X_t < x^*\}$ and $\sigma^* := \inf\{t > 0 : X_t \in [\log K, y^*]\}$ for some thresholds k^*, x^* and y^*. While continuous/smooth fit can be used to identify these values, due to the critical barrier $\log K$, one does not observe the dependency between the two parameters that we have seen in this section.

In Example 5.3, as shown in [23] and [7], the equilibrium is given by two up-crossing times where at least one of them is the first time X goes above the critical barrier $\log K$. Therefore, again one does not observe the dependency between the two parameters.

5.4 *When a Stopper Is Replaced with a Controller*

One can naturally consider the case where the stopper(s) are replaced with singular controller(s).

The game between a controller and a stopper has been studied by Hernández-Hernández et al. [26] for the case driven by a diffusion process, where they obtained general results on the verification lemma and gave some explicitly solvable examples.

The case driven by a spectrally one-sided Lévy process is studied by Hernández-Hernández and Yamazaki [25], where they considered the problem where a stopper maximizes and a controller minimizes the expected value of some monotone payoff. They considered both the spectrally negative and positive cases. Not surprisingly, the solution procedures are similar to the ones illustrated in this note: the candidate barriers (a^*, b^*), which separate the state space into the stopping, waiting, and controlling regions, are chosen by continuous/smooth fit so that

(1) the value function at the boundary for the controller is continuously differentiable (resp. twice continuously differentiable) if it is irregular (resp. regular) for the controlling region;
(2) the value function at the boundary for the stopper is continuous (resp. continuously differentiable) if it is irregular (resp. regular) for the stopping region.

The verification of optimality can be carried out by showing the verification lemma as in the one given in Sect. 5.2. As we have seen, many parts of the verification can be carried out without much effort. However, the difficulty is again to show the sub/super harmonicity at the region where the process can jump instantaneously to the other regions. To deal with this, Hernández-Hernández and Yamazaki [25] applied similar techniques as the ones discussed in Sects. 3.4 and 5.2.1.

The game between two singular controllers is also of great interest. Under a certain monotonicity assumption on the payoff function, it is expected that the optimally controlled process becomes the doubly reflected Lévy process similarly to the two-sided singular control case we studied in Sect. 3. Hence, the candidate value function can be computed again using the scale function and is expected to preserve the same smoothness as those observed in Sect. 3. Consequently, the two boundaries can be chosen in essentially the same way. The verification lemma can be easily obtained by modifying (3.19). It is expected that many of the techniques used in Sect. 3 can be recycled.

Acknowledgements The author thanks the anonymous referee for constructive comments and suggestions. K. Yamazaki is supported by MEXT KAKENHI Grant Number 26800092 and 17K05377.

References

1. D. Applebaum, *Lévy Processes and Stochastic Calculus* (Cambridge University Press, Cambridge, 2009)
2. S. Asmussen, F. Avram, M.R. Pistorius, Russian and American put options under exponential phase-type Lévy models. Stoch. Process. Appl. **109**(1), 79–111 (2004)
3. F. Avram, A.E. Kyprianou, M.R. Pistorius, Exit problems for spectrally negative Lévy processes and applications to (Canadized) Russian options. Ann. Appl. Probab. **14**(1), 215–238 (2004)
4. F. Avram, Z. Palmowski, M.R. Pistorius, On the optimal dividend problem for a spectrally negative Lévy process. Ann. Appl. Probab. **17**(1), 156–180 (2007)
5. E. Baurdoux, A.E. Kyprianou, The McKean stochastic game driven by a spectrally negative Lévy process. Electron. J. Probab. **13**(8), 173–197 (2008)
6. E.J. Baurdoux, K. Yamazaki, Optimality of doubly reflected Lévy processes in singular control. Stoch. Process. Appl. **125**(7), 2727–2751 (2015)
7. E.J. Baurdoux, A.E. Kyprianou, J.C. Pardo, The Gapeev-Kühn stochastic game driven by a spectrally positive Lévy process. Stoch. Process. Appl. **121**(6), 1266–1289 (2008)
8. E. Bayraktar, A.E. Kyprianou, K. Yamazaki, On optimal dividends in the dual model. Astin Bull. **43**(3), 359–372 (2013)
9. E. Bayraktar, A.E. Kyprianou, K. Yamazaki, Optimal dividends in the dual model under transaction costs. Insur. Math. Econom. **54**, 133–143 (2014)
10. L. Benkherouf, A. Bensoussan, Optimality of an (s, S) policy with compound Poisson and diffusion demands: a quasi-variational inequalities approach. SIAM J. Control Optim. **48**(2), 756–762 (2009)
11. A. Bensoussan, J.-L. Lions, *Impulse Control and Quasi-Variational Inequalities* (Wiley, London, 1984)
12. A. Bensoussan, R.H. Liu, S.P. Sethi, Optimality of an (s, S) policy with compound Poisson and diffusion demands: a quasi-variational inequalities approach. SIAM J. Control Optim. **44**(5), 1650–1676 (2005)
13. J. Bertoin, *Lévy Processes*. Cambridge Tracts in Mathematics, vol. 121 (Cambridge University Press, Cambridge, 1996)
14. J. Bertoin, Exponential decay and ergodicity of completely asymmetric Lévy processes in a finite interval. Ann. Appl. Probab. **7**(1), 156–169 (1997)
15. T. Chan, A.E. Kyprianou, M. Savov, Smoothness of scale functions for spectrally negative Lévy processes. Probab. Theory Relat. Fields **150**, 691–708 (2011)
16. R.A. Doney, *Fluctuation Theory for Lévy Processes*. Lecture Notes in Mathematics, vol. 1897 (Springer, Berlin, 2007)
17. M. Egami, K. Yamazaki, Precautional measures for credit risk management in jump models. Stochastics **85**(1), 111–143 (2013)
18. M. Egami, K. Yamazaki, On the continuous and smooth fit principle for optimal stopping problems in spectrally negative Lévy models. Adv. Appl. Probab. **46**(1), 139–167 (2014)
19. M. Egami, K. Yamazaki, Phase-type fitting of scale functions for spectrally negative Lévy processes. J. Comput. Appl. Math. **264**, 1–22 (2014)
20. M. Egami, T. Leung, K. Yamazaki, Default swap games driven by spectrally negative Lévy processes. Stoch. Process. Appl. **123**(2), 347–384 (2013)
21. E. Ekström, G. Peskir, Optimal stopping games for Markov processes. SIAM J. Control Optim. **47**(2), 684–702 (2008)
22. D.J. Emery, Exit problem for a spectrally positive process. Adv. Appl. Probab. **5**, 498–520 (1973)
23. P.V. Gapeev, C.Kühn, Perpetual convertible bonds in jump-diffusion models. Stat. Decis. **23**(1), 15–31 (2005)
24. J.M. Harrison, M.I. Taksar, Instantaneous control of Brownian motion. Math. Oper. Res. **8**(3), 439–453 (1983)

25. D. Hernández-Hernández, K. Yamazaki, Games of singular control and stopping driven by spectrally one-sided Lévy processes. Stoch. Process. Appl. **125**(1), 1–38 (2015)
26. D. Hernández-Hernández, R.S. Simon, M. Zervos, A zero-sum game between a singular stochastic controller and a discretionary stopper. Ann. Appl. Probab. **25**(1), 46–80 (2015)
27. D. Hernández-Hernández, J.-L. Pérez, K. Yamazaki, Optimality of refraction strategies for spectrally negative Lévy processes. SIAM J. Control Optim. **54**(3), 1126–1156 (2016)
28. M. Jeanblanc-Picqué, Impulse control method and exchange rate. Math. Financ. **3**(2), 161–177 (1993)
29. A. Kuznetsov, A. Kyprianou, V. Rivero, *The Theory of Scale Functions for Spectrally Negative Lévy Processes*. Springer Lecture Notes in Mathematics, vol. 2061. (Springer, Berlin, 2013), pp. 97–186
30. A.E. Kyprianou, *Introductory Lectures on Fluctuations of Lévy Processes with Applications* (Springer, Berlin, 2006)
31. A.E. Kyprianou, B.A. Surya, Principles of smooth and continuous fit in the determination of endogenous bankruptcy levels. Financ. Stoch. **11**(1), 131–152 (2007)
32. A.E. Kyprianou, R. Loeffen, J.-L. Pérez, et al., Optimal control with absolutely continuous strategies for spectrally negative Lévy processes. J. Appl. Probab. **49**(1), 150–166 (2012)
33. T. Leung, K. Yamazaki, American step-up and step-down credit default swaps under levy models. Quant. Financ. **13**(1), 137–157 (2013)
34. R.L. Loeffen, On optimality of the barrier strategy in de Finetti's dividend problem for spectrally negative Lévy processes. Ann. Appl. Probab. **18**(5), 1669–1680 (2008)
35. R.L. Loeffen, An optimal dividends problem with transaction costs for spectrally negative Lévy processes. Insur. Math. Econom. **45**(1), 41–48 (2009)
36. P. Matomäki, On solvability of a two-sided singular control problem. Math. Method Oper. Res. **76**(3), 239–271 (2012)
37. E. Mordecki, Optimal stopping and perpetual options for Lévy processes. Financ. Stoch. **6**(4), 473–493 (2002)
38. G. Mundaca, B. Øksendal, Optimal stochastic intervention control with application to the exchange rate. J. Math. Econ. **29**(2), 225–243 (1998)
39. B. Øksendal, A. Sulem, *Applied Stochastic Control of Jump Diffusions*, 2nd edn. (Springer, Berlin, 2007)
40. G. Peskir, Optimal stopping games and Nash equilibrium. Theory Probab. Appl. **53**(3), 558–571 (2009)
41. G. Peskir, A. Shiryaev, *Optimal Stopping and Free-Boundary Problems*. Lectures in Mathematics ETH Zürich (Birkhäuser, Basel, 2006)
42. M.R. Pistorius, On doubly reflected completely asymmetric Lévy processes. Stoch. Process. Appl. **107**(1), 131–143 (2003)
43. M.R. Pistorius, On exit and ergodicity of the spectrally one-sided Lévy process reflected at its infimum. J. Theor. Probab. **17**(1), 183–220 (2004)
44. P.E. Protter, *Stochastic Integration and Differential Equations*. Stochastic Modelling and Applied Probability, vol. 21, 2nd edn. (Springer, Berlin, 2005). Version 2.1
45. A. Shiryaev, The problem of the most rapid detection of a disturbance in a stationary process. Sov. Math. Dokl. **2**, 795–799 (1961)
46. V. Suprun, Problem of destruction and resolvent of terminating process with independent increments. Ukr. Math. J. **28**, 39–45 (1976)
47. B.A. Surya, K. Yamazaki, Optimal capital structure with scale effects under spectrally negative Lévy models. Int. J. Theor. Appl. Financ. **17**(2), 1450013 (2014)
48. K. Yamazaki, Cash management and control band policies for spectrally one-sided levy processes, in *Recent Advances in Financial Engineering 2014*, pp. 199–215 (2016)
49. K. Yamazaki, Inventory control for spectrally positive Lévy demand processes. Math. Oper. Res. **42**(1), 212–237 (2017).
50. C. Yin, Y. Wen, Y. Zhao, On the optimal dividend problem for a spectrally positive Lévy process. Astin Bull. **44**(3), 635–651 (2014)

Part II
Research Articles

Asymptotic Results for the Severity and Surplus Before Ruin for a Class of Lévy Insurance Processes

Ekaterina T. Kolkovska and Ehyter M. Martín-González

Abstract We investigate a classical two-sided jumps risk process perturbed by a spectrally negative α-stable process, in which the gain size distribution has a rational Laplace transform. We consider three classes of light- and heavy-tailed claim size distributions. We obtain the asymptotic behaviors of the ruin probability and of the joint tail of the surplus prior to ruin and the severity of ruin, for large values of the initial capital. We also show that our asymptotic results are sharp. This extends our previous work (Kolkovska and Martín-González, Gerber-Shiu functionals for classical risk processes perturbed by an α-stable motion. Insur Math Econ 66:22–28, 2016).

Keywords Two-sided risk process · Stable process · Ruin probability · Severity of ruin · Surplus before ruin · Asymptotic ruin probability

Mathematics Subject Classification 60G51

1 Introduction

For a given risk process $X = \{X(t),\ t \geq 0\}$, the expected discounted penalty function, named also the Gerber-Shiu functional, is defined by

$$\phi(u) = \mathbb{E}\left[e^{-\delta\tau_0}\omega\left(|X(\tau_0)|, X(\tau_0-)\right) 1_{\{\tau_0<\infty\}} \,\middle|\, X(0) = u\right],$$

E. T. Kolkovska (✉)
Área de Probabilidad y Estadística, Centro de Investigación en Matemáticas, Guanajuato, Mexico
e-mail: todorova@cimat.mx

E. M. Martín-González
División de Ciencias Naturales y Exactas, Departamento de Matemáticas, Universidad de Guanajuato, Guanajuato, Mexico

© Springer International Publishing AG, part of Springer Nature 2018
D. Hernández-Hernández et al. (eds.), *XII Symposium of Probability and Stochastic Processes*, Progress in Probability 73,
https://doi.org/10.1007/978-3-319-77643-9_3

where $\tau_0 = \inf\{t \geq 0 : X(t) < 0\}$ is the ruin time, $\delta \geq 0$ is a constant representing a discounting factor, and $\omega : \mathbb{R}_+ \times \mathbb{R}_+ \to \mathbb{R}_+$ is a nonnegative penalty function. The random variables $|X(\tau_0)|$ and $X(\tau_0-)$ are known respectively as the severity of ruin and the surplus immediately before ruin. The functional ϕ was introduced in [12] as a generalization of the concept of ruin probability, which is obtained as a particular case when $\delta = 0$ and $\omega \equiv 1$, and has been investigated intensively since then. Many other important risk measures arise as particular cases of the Gerber-Shiu functional, including the distribution of the claim that causes the ruin given that $\tau_0 < \infty$, the joint Laplace transform of the severity of ruin and the surplus prior to ruin, the Laplace transform of the time to ruin and the joint tail distribution of the severity of ruin and the surplus prior to ruin.

The classical two-sided jumps risk process is given by

$$X(t) = u + ct + \sum_{j=1}^{N_1(t)} Y_{j1} - \sum_{j=1}^{N_2(t)} Y_{j2} := u + ct + Z_1(t) - Z_2(t), \tag{1.1}$$

where $u \geq 0$ and $c > 0$ are constants representing, respectively, the initial capital of the insurance company and the prime per unit time that the company receives, and $Z_1 = \{Z_1(t), t \geq 0\}$, $Z_2 = \{Z_2(t), t \geq 0\}$ are two independent compound Poisson processes with respective intensities and jump distributions, λ_i and F_i, $i = 1, 2$, where $\lambda_i \geq 0$ for $i = 1, 2$. Here $Z_1(t)$ and $Z_2(t)$ model respectively the accumulated random gains and random claims at time t. In the case when $\lambda_1 = 0$ the resulting process is called the classical risk process.

In a previous paper [15] we investigated a perturbed two-sided jumps classical risk process $V_\alpha = \{V_\alpha(t), t \geq 0\}$, given by

$$V_\alpha(t) = X(t) - \eta W_\alpha(t), \; \eta > 0, t \geq 0, \tag{1.2}$$

where X is the risk process defined in (1.1) and $\{W_\alpha(t), t \geq 0\}$ is an independent standard α-stable process with index of stability $1 < \alpha < 2$ and skewness parameter $\beta = 1$. Moreover, F_1 possesses a density f_1 whose Laplace transform $\widehat{f_1}$ is a rational function of the form

$$\widehat{f_1}(r) = \frac{Q(r)}{\prod_{i=1}^{N}(q_i + r)^{m_i}}, \qquad r \geq 0, \tag{1.3}$$

where $N, m_i \in \mathbb{N}$ with $m_1 + m_2 + \cdots + m_N = m$, $0 < q_1 < q_2 < \cdots < q_m$ and Q is a polynomial function of degree at most $m - 1$. The family of distributions satisfying (1.3) is widely used in probability applications. This is a wide class of light-tailed distributions which includes Coxian distributions, combinations of exponential distribution, phase-type distributions, combinations of Erlang distributions and many others. It is dense in the class of general nonnegative distributions (see e.g. [7] and [16, Theorem 8.2.8].) and this property allows for numerical approximations for ϕ in the case of general gain distributions. Under some additional assumptions on the claim size distribution function F_2 and the penalty function ω, in [15] we obtained a formula for the Laplace transform of ϕ and an expression for ϕ as an infinite series of convolutions of given functions. However, such infinite sums of convolutions are hard to work with in practice, and therefore, it is of interest to study the asymptotic behavior of such expressions.

In this paper we investigate the same model as in [15], to which we refer the reader for motivation and explanations about the meaning of the model parameters. Based upon the results obtained in [15], here we obtain an asymptotic formula for the ruin probability $\psi(u) := \mathbb{P}[\tau_0 < \infty | V_\alpha(0) = u]$ as $u \to \infty$, see Theorem 1 below. In Theorem 2 we obtain an asymptotic formula, as $u \to \infty$, of the joint tail distribution

$$\Upsilon_{a,b}(u) := \mathbb{P}[|V_\alpha(\tau_0)| > a, V_\alpha(\tau_0-) > b, \tau_0 < \infty | V(0) = u], \quad a > 0, \ b > 0. \tag{1.4}$$

In Theorem 3 we show that such asymptotic formula holds uniformly in the parameters a and b. These results extend our previous work [14], where we investigated similar behaviors for the classical risk process perturbed by W_α. Other asymptotic results for the ruin probability and the asymptotic distribution of the overshoot of the process about high levels are obtained by Klüppelberg et al. [13] in the case when the Lévy risk process is spectrally positive or spectrally negative. In Doney et al. [6] asymptotic results for the time of ruin, the surplus before the time of ruin and the overshoot at ruin time are obtained for Lévy risk processes under the assumptions that the positive part of the Lévy measure of the process is heavy tailed, and the renewal measure of the descending ladder process is of regular variation. In the case we study here the risk process V_α has two-sided jumps distribution, such that the upward-jump distribution is light-tailed. Therefore, our results complement the investigation in [13] and [6].

We remark that expressions for Gerber-Shiu functionals of a more general class of Lévy risk processes than the one we treat here are given in Biffis and Morales [2] in terms of infinite series of convolutions of integral functions. However, the integrals involved in such convolution formula are not easy to calculate in general, since they are integrals with respect to pure jumps measures and require Laplace transform inversion techniques. In [1] the authors give an expression for a generalized version of the Gerber-Shiu functional for spectrally negative Lévy risk processes in terms of integrals of the associated scale functions of the processes. However, in most cases the scale functions are difficult to obtain explicitly.

The paper is organized as follows: in Sect. 2 we give additional assumptions on the process V_α that we need, as well as several definitions and preliminary results that we use in the sequel. In Sect. 3 we obtain asymptotics for the ruin probability of the process V_α, using Karamata's theorem combined with certain results from [8]. The final Sect. 4 contains our main results, Theorems 2 and 3, and their proofs.

2 Definitions and Preliminary Results

In what follows we consider the process V_α and denote by ψ the corresponding ruin probability $\psi(u) = \mathbb{P}[\tau_0 < \infty | V_\alpha(0) = u]$ starting with an initial capital $u \geq 0$. As above, we write $\Upsilon_{a,b}(u) = \mathbb{P}[|V_\alpha(\tau_0)| > a,\, V_\alpha(\tau_0-) > b,\, \tau_0 < \infty | V(0) = u]$, $u \geq 0$, for the joint tail of the severity of ruin and surplus prior to ruin, where a and b are fixed positive numbers. These two functions ψ and $\Upsilon_{a,b}$ are particular cases of ϕ respectively, when $\omega(x, y) = 1$ and when $\omega(x, y) = 1_{\{x>a,y>b\}}$. We recall that the survival probability $\Phi(u) = 1 - \psi(u)$, $u \in \mathbb{R}$, is a distribution function.

We consider the Generalized Lundberg equation

$$L(r) := cr + \eta^\alpha r^\alpha + \lambda_1 \widehat{F}_1(-r) + \lambda_2 \widehat{F}_2(r) - (\lambda_1 + \lambda_2) = 0.$$

In [15, Proposition 3.6] it is proved that L has exactly $m + 1$ roots in the right-half complex plane $\mathbb{C}_+ = \{z \in \mathbb{C} : Re(z) \geq 0\}$, and when $\delta = 0$, 0 is a root of the above equation with multiplicity 1. We denote the roots of L by $\rho_1, \ldots, \rho_{m+1}$, with $\rho_1 = 0$ when $\delta = 0$. We assume that the following conditions hold.

(a) The upward distribution F_1 has a density f_1, whose Laplace transform has the form (1.3).
(b) The Net Profit Condition $\mathbb{E}[V_\alpha(1) - u] = c + \lambda_1\mu_1 - \lambda_2\mu_2 > 0$ holds, where $\mu_j = \mathbb{E}[X_{1j}] < \infty$, $j = 1, 2$.
(c) The roots $\rho_1, \ldots, \rho_{m+1}$, are all different.

Notice that assumption (b) implies that $\lim_{t\to\infty} V_\alpha(t) = +\infty$ with probability 1. For $a > 0$ we denote by $z_{\alpha,a}$ the density of the extremal stable distribution $\zeta_{\alpha,a}$; see e.g. [15, page 376] for the definition of $\zeta_{\alpha,a}$. It is known [11, Lemma 1] that the Laplace transform of $z_{\alpha,a}$ exists for all $r \geq 0$ and is given by $\widehat{z}_{\alpha,a}(r) = \frac{a}{a+r^{\alpha-1}}$. We set $E(\rho_j) = \frac{\prod_{l=1}^{N}(q_l-\rho_j)^{m_l}}{\prod_{l\neq j}(\rho_l-\rho_j)}$ and denote by T_r the Dickson-Hipp operator introduced in [5], which is defined by $T_r f(x) = \int_x^\infty e^{-r(y-x)} f(y)\, dy$ for any $x \geq 0$, all complex number $r = r_1 + ir_2$ with $r_1 \geq 0$, and all integrable nonnegative functions f. We define the function

$$g_0(x) = \lambda_2 \sum_{j=1}^{m+1} E(\rho_j) T_{\rho_j} f_2(x), \quad x > 0,$$

and for $\alpha < 2$ and $u > 0$ we denote $l_\alpha(u) = \frac{(\alpha-1)u^{-\alpha}}{\Gamma(2-\alpha)}$ and $f_\alpha(u) = \sum_{j=2}^{m+1} E(\rho_j)\rho_j T_{\rho_j} l_\alpha(u)$. It is easily shown that $\widehat{f_\alpha}(r) = \sum_{j=2}^{m+1} E(\rho_j)\rho_j \frac{\rho_j^{\alpha-1} - r^{\alpha-1}}{\rho_j - r}$. From [15, Lemma 5.3] it follows that f_α and g_0 are real valued functions. In the sequel we will assume that these two functions are nonnegative. This assumption holds at least in the case when F_2 is a convex sum of exponential distribution functions with positive coefficients, since in this case it follows similarly as in [4] that the roots ρ_j, $j = 1, \ldots, m+1$, of the Lundberg equation $L(r) = 0$ are nonnegative real numbers. This implies, due to the definition of $E(\rho_j)$, that also $E(\rho_j)$ are nonnegative numbers.

Now we define the distribution functions

$$F_\alpha(x) = \frac{1}{\mathcal{C}_F} \int_{0+}^{x} f_\alpha(y)dy, \; G_0(x) = \frac{1}{\mathcal{C}_G} \int_{0+}^{x} g_0(y)dy,$$

$$U_\alpha(x) = \frac{1}{\mathcal{C}_U} \int_{0+}^{x} \nu_\alpha(y)dy, \quad x > 0, \qquad (2.1)$$

and $F_{2,I}(x) = \frac{1}{\mu_2}\int_0^x \overline{F}_2(y)\,dy$, $x \geq 0$. Here $\mathcal{C}_F = \int_{0+}^{\infty} f_\alpha(x)\,dx$, $\mathcal{C}_G = \int_{0+}^{\infty} g_0(x)\,dx$ and $\mathcal{C}_U = \int_{0+}^{\infty} \nu_\alpha(x)\,dx$. The functions $\nu : \mathbb{R}_+ \to \mathbb{R}_+$ and $W_\alpha : \mathbb{R}_+ \to \mathbb{R}_+$ are defined by their Laplace transforms

$$\widehat{\nu}_\alpha(r)\left(1 + \frac{1}{\theta}\widehat{f}_\alpha(r)\widehat{z}_{\alpha,\theta}(r)\right) = \widehat{z}_{\alpha,\theta}(r), \qquad (2.2)$$

where $\theta = c/\eta^\alpha + \kappa$ and $\kappa = \frac{1}{\eta^\alpha}\widehat{g}_0(0) + \widehat{f}_\alpha(0)$, and

$$\widehat{W}_\alpha(r) = \frac{\frac{1}{\eta^\alpha \theta}\widehat{\nu}_\alpha(r)}{1 - \frac{1}{\theta}\left[\kappa\widehat{\nu}_\alpha(r) + \frac{1}{\eta^\alpha}\widehat{g}_0(r)\widehat{\nu}_\alpha(r)\right]}. \qquad (2.3)$$

In [15, Proposition 5.6] we give representations of ν and W_α as series of convolutions of given functions.

We recall [15, Proposition 5.4 b)] that the Laplace transform of the ruin probability ψ satisfies the equality

$$\widehat{\psi}(r) = \frac{1}{r} - \frac{(c + \lambda_1\mu_1 - \lambda_2\mu_2)}{r}\frac{\prod_{j=1}^{N} q_j^{m_j}}{\prod_{j=2}^{m+1}\rho_j}\widehat{W}_\alpha(r), \quad r > 0. \qquad (2.4)$$

Notice that the roots of Lundberg's equation appear in conjugate pairs because the equation coefficients are real, hence $\frac{\prod_{j=1}^{N} q_j^{m_j}}{\prod_{j=2}^{m+1}\rho_j} > 0$.

We also recall the following definitions: Let F be a distribution function such that $F(0) = 0$ with tail $\overline{F} = 1 - F$. If there exist numbers $c_1, c_2 > 0$ such that $\overline{F}(x) \leq c_1 e^{-c_2 x}$ for all $x > 0$, then F is called light-tailed distribution function. Otherwise

F is a heavy-tailed distribution function and in such a case we write $F \in \mathcal{H}$. In case that $\lim_{x\to\infty} \overline{F^{*2}}(x)/\overline{F}(x) = 2$ we say that F belongs to the class of subexponential distributions and write $F \in \mathcal{S}$. The distribution function F belongs to the class $\mathcal{L}$ if for any $y \geq 0$ there holds $\lim_{x\to\infty} \overline{F}(x-y)/\overline{F}(x) = 1$. Finally, F belongs to the class $\mathcal{R}_c$ for $c \geq 0$ if F has a density f such that $\lim_{x\to\infty} f(x)/\overline{F}(x) = c$. We say that $f : \mathbb{R}_+ \to \mathbb{R}_+$ is a regularly varying function of x at ∞, with order $a \in \mathbb{R}$, if $\lim_{x\to\infty} f(xt)/f(x) = t^a$ for $t > 0$, and write $f \in RV_a$. In the particular case when $a = 0$, we say that f is a slowly varying function of x at ∞. If f is regularly varying of order a, then it can be written as $f(x) = x^a L(x)$, where L is a slowly varying function. We define $f \sim g$ if $\lim_{x\to\infty} f(x)/g(x) = 1$. We write $F \in \overline{RV}_a$ if F is such that $\overline{F}(x) \sim x^a L(x)$. The following inclusions hold (see [9]):

$$\overline{RV}_a \subset \mathcal{S} \subset \mathcal{L} \subset \mathcal{H} \text{ and } \mathcal{R}_0 \subset \mathcal{L}. \tag{2.5}$$

Lemma 1 *Let F_1, F_2 be two distribution functions such that $F_i(0) = 0$, $i = 1, 2$, and let $H = F_1 * F_2$ be their convolution.*

a) If $F_2 \in \mathcal{S}$ and $\overline{F}_1(x) = o(\overline{F}_2(x))$ as $x \to \infty$, then $H \in \mathcal{S}$. Moreover, $\overline{H}(x) \sim \overline{F}_2(x)$.
b) If $\overline{F}_i(x) \sim x^{-\delta} L_i(x)$ for $i = 1, 2$, where L_1 and L_2 are slowly varying functions, then $\overline{H}(x) \sim x^{-\delta}(L_1(x) + L_2(x))$ as $x \to \infty$.
c) If $\overline{F}_2(x) \sim c\overline{F}_1(x)$ for some $c \in (0, \infty)$, then $F_1 \in \mathcal{S}$ if and only if $F_2 \in \mathcal{S}$ and $\overline{H} \sim (1 + c)\overline{F}_2(x)$.
d) If $\beta \in (0, 1)$ and $K(x) = (1 - \beta) \sum_{n=0}^{\infty} \beta^n F_1^{*n}(x)$ then the following three conditions are equivalent:

$$K \in \mathcal{S}, \quad F_1 \in \mathcal{S}, \quad \overline{K}(x) \sim \tfrac{\beta}{1-\beta} \overline{F}_1(x).$$

Proof For a) and d) see, respectively, [8, Proposition 1a) and Theorem 3]. For b) see [10, page 278]. The proof of c) is given in [16, lemmas 2.5.2 and 2.5.4]. ■

3 Asymptotic Behavior of the Ruin Probability

In what follows we will use the elementary identities

$$\widehat{F}(r) = \frac{\widehat{f}(r)}{r} \text{ and } \widehat{\overline{F}}(r) = \frac{1 - \widehat{f}(r)}{r}, \quad r > 0, \tag{3.1}$$

valid for any distribution function F with $F(0) = 0$ and having density f. First we state the following auxiliary result.

Proposition 1 *The following asymptotics hold:*

a) $\lim_{x\to\infty} \frac{\overline{F}_\alpha(x)}{\frac{x^{1-\alpha}}{\Gamma(2-\alpha)}} = \frac{1}{\mathcal{C}_F}\left(1 - \frac{\prod_{i=1}^N q_i^{m_i}}{\prod_{j=2}^{m+1}\rho_j}\right)$*, hence* $F_\alpha \in \mathcal{S}$.

b) If $F_2 \in \mathcal{R}_0$*, then* $\lim_{x\to\infty} \frac{\overline{G}_0(x)}{\overline{F}_{2,I}(x)} = \frac{\lambda_2\mu_2}{\mathcal{C}_G}\frac{\prod_{i=1}^N q_i^{m_i}}{\prod_{j=2}^{m+1}\rho_j}$*. If in addition* $F_{2,I} \in \mathcal{S}$*, then* $G_0 \in \mathcal{S}$.

c) If $\overline{F}_2(x) = o(x^{-\alpha})$*, then* $\overline{G}_0(x) = o(x^{1-\alpha})$.

d) $\lim_{x\to\infty} \frac{\overline{U}_\alpha(x)}{\overline{\zeta}_{\alpha,\theta}(x)} = \mathcal{C}_U\frac{\prod_{i=1}^N q_i^{m_i}}{\prod_{j=2}^{m+1}\rho_j}$*, hence* $U_\alpha \in \mathcal{S}$.

Proof

a) Let us define $F^*(u) = \int_0^u \overline{F}_\alpha(x)dx$. From (3.1) we obtain $\widehat{F}^*(r) = (1 - \frac{1}{\mathcal{C}_F}\widehat{f}_\alpha(r))/r^2$, hence $\lim_{r\downarrow 0}\frac{r\widehat{F}^*(r)}{r^{\alpha-2}} = \lim_{r\downarrow 0}\frac{1-\frac{1}{\mathcal{C}_F}\widehat{f}_\alpha(r)}{r^{\alpha-1}}$. From the definition of $\mathcal{C}_F$ it follows that $1 - \frac{1}{\mathcal{C}_F}\widehat{f}_\alpha(0) = 0$. Using L'Hospital's rule gives

$$\lim_{r\downarrow 0}\frac{1-\frac{1}{\mathcal{C}_F}\widehat{f}_\alpha(r)}{r^{\alpha-1}} = \lim_{r\downarrow 0}\frac{1+\frac{1}{\mathcal{C}_F}\sum_{j=2}^{m+1}E(\rho_j)\rho_j\frac{\rho_j^{\alpha-1}-r^{\alpha-1}}{\rho_j - r}}{r^{\alpha-1}}$$

$$= \lim_{r\downarrow 0}\frac{\frac{1}{\mathcal{C}_F}\sum_{j=2}^{m+1}E(\rho_j)\rho_j\left(\frac{\rho_j^{\alpha-1}-r^{\alpha-1}}{(\rho_j-r)^2} - \frac{(\alpha-1)r^{\alpha-2}}{\rho_j - r}\right)}{(\alpha-1)r^{\alpha-2}}$$

$$= -\frac{1}{\mathcal{C}_F}\sum_{j=2}^{m+1}E(\rho_j) = \frac{1}{\mathcal{C}_F}\left(1 - \frac{\prod_{i=1}^N q_i^{m_i}}{\prod_{j=2}^{m+1}\rho_j}\right), \tag{3.2}$$

where the last equality follows by [15, Lemma 5.3]. From [10, Theorem 1, page 443] we obtain the limit in part a), which implies that $\overline{F}_\alpha$ is regularly varying. Using (2.5) we also obtain that $F_\alpha \in \mathcal{S}$.

b) Notice that

$$\lim_{x\to\infty}\left|\frac{\int_x^\infty T_{\rho_j}f_2(y)dy}{\overline{F}_{2,I}(x)}\right|$$

$$\le \lim_{x\to\infty}\frac{\int_x^\infty\int_y^\infty e^{-Re(\rho_j)(z-y)}f_2(z)dzdy}{\overline{F}_{2,I}(x)}, \quad j = 2, 3, \ldots, m+1. \tag{3.3}$$

Taking limits when $x \to \infty$ in the right-hand side of (3.3) yields

$$\lim_{x\to\infty} \frac{\int_x^\infty e^{Re(\rho_j)y} \int_y^\infty e^{-Re(\rho_j)z} f_2(z)dzdy}{\overline{F}_{2,I}(x)} = \lim_{x\to\infty} \frac{\int_x^\infty e^{-Re(\rho_j)z} f_2(z)dzdy}{e^{-Re(\rho_j)x}\overline{F}_2(x)}$$

$$= \lim_{x\to\infty} \frac{e^{-Re(\rho_j)x} f_2(x)}{Re(\rho_j)e^{-Re(\rho_j)x}\overline{F}_2(x) + e^{-Re(\rho_j)x} f_2(x)} = \lim_{x\to\infty} \frac{\frac{f_2(x)}{\overline{F}_2(x)}}{Re(\rho_j) + \frac{f_2(x)}{\overline{F}_2(x)}}, \tag{3.4}$$

where the first and second equalities follow by L'Hospital's rule. Using the assumption that $F_2 \in \mathcal{R}_0$, we obtain from (3.4) and (3.3) that

$$\lim_{x\to\infty} \left| \frac{\int_x^\infty T_{\rho_j} f_2(y)\, dy}{\overline{F}_{2,I}(x)} \right| = 0. \tag{3.5}$$

Since $\int_x^\infty g_0(y)dy = \lambda_2 \frac{\prod_{i=1}^N q_i^{m_i}}{\prod_{j=2}^{m+1} \rho_j} \mu_2 \overline{F}_{2,I}(x) - \lambda_2 \sum_{j=2}^{m+1} E(\rho_j) \int_x^\infty T_{\rho_j} f_2(y)$, the triangle inequality yields

$$\frac{\lambda_2\mu_2 \prod_{i=1}^N q_i^{m_i}}{C_G \prod_{j=2}^{m+1} \rho_j} - \left| \frac{-\frac{\lambda_2}{C_G} \sum_{j=2}^{m+1} E(\rho_j) \int_x^\infty T_{\rho_j} f_2(y)dy}{\overline{F}_{2,I}(x)} \right| \leq \left| \frac{\overline{G}_0(x)}{\overline{F}_{2,I}(x)} \right|, \tag{3.6}$$

and

$$\left| \frac{\overline{G}_0(x)}{\overline{F}_{2,I}(x)} \right| \leq \frac{\lambda_2\mu_2 \prod_{i=1}^N q_i^{m_i}}{C_G \prod_{j=2}^{m+1} \rho_j} + \left| \frac{-\frac{\lambda_2}{C_G} \sum_{j=2}^{m+1} E(\rho_j) \int_x^\infty T_{\rho_j} f_2(y)dy}{\overline{F}_{2,I}(x)} \right|. \tag{3.7}$$

The limit in part b) follows from letting $x \to \infty$ in (3.6) and (3.7) and using (3.5). Assuming that $F_{2,I} \in \mathcal{S}$, the relation $G_0 \in \mathcal{S}$ follows from part c) of Lemma 1.

c) Let us assume that $\overline{F}_2(x) = o(x^{-\alpha})$, hence L'Hospital's rule implies that

$$\lim_{x\to\infty} \frac{\overline{F}_{2,I}(x)}{x^{1-\alpha}} = 0. \tag{3.8}$$

This yields

$$\lim_{x\to\infty} \frac{\int_x^\infty \int_y^\infty e^{-Re(\rho_j)(z-y)} f_2(z)dzdy}{x^{1-\alpha}} \leq \lim_{x\to\infty} \frac{\int_x^\infty \int_y^\infty f_2(z)dzdy}{x^{1-\alpha}}$$

$$= \mu_2 \lim_{x\to\infty} \frac{\overline{F}_{2,I}(x)}{x^{1-\alpha}} = 0,$$

and from (3.8) we obtain

$$\lim_{x\to\infty}\left|\frac{\int_x^{\infty} T_{\rho_j} f_2(y)dy}{x^{1-\alpha}}\right| = 0. \tag{3.9}$$

Using (3.6) and (3.7) we obtain the inequalities

$$\frac{\mu_2 \prod_{i=1}^{N} q_i^{m_i}}{\mathcal{C}_G \prod_{j=2}^{m+1} \rho_j}\frac{\overline{F}_{2,I}(x)}{x^{1-\alpha}} - \left|\frac{-\frac{1}{\mathcal{C}_G}\sum_{j=2}^{m+1} E(\rho_j)\int_x^{\infty} T_{\rho_j} f_2(y)dy}{x^{1-\alpha}}\right| \leq \left|\frac{\overline{G}_0(x)}{x^{1-\alpha}}\right| \tag{3.10}$$

and

$$\left|\frac{\overline{G}_0(x)}{x^{1-\alpha}}\right| \leq \frac{\mu_2 \prod_{i=1}^{N} q_i^{m_i}}{\mathcal{C}_G \prod_{j=2}^{m+1} \rho_j}\frac{\overline{F}_{2,I}(x)}{x^{1-\alpha}} + \left|\frac{-\frac{1}{\mathcal{C}_G}\sum_{j=2}^{m+1} E(\rho_j)\int_x^{\infty} T_{\rho_j} f_2(y)dy}{x^{1-\alpha}}\right|. \tag{3.11}$$

The result now follows by letting $x \to \infty$ in (3.10) and (3.11), and using (3.8) and (3.9).

d) Putting $r = 0$ in (2.2) gives $\mathcal{C}_U = (1 + \frac{\mathcal{C}_F}{\theta})^{-1}$. Dividing both sides of (2.2) by $\mathcal{C}_U$ yields $\dfrac{\widehat{\nu}_\alpha(r)}{\mathcal{C}_U}\left(1 + \frac{1}{\theta}\widehat{f}_\alpha(r)\widehat{z}_{\alpha,\theta}(r)\right) = \dfrac{\widehat{z}_{\alpha,\theta}(r)}{\mathcal{C}_U}$, hence:

$$\begin{aligned}\left(1-\frac{\widehat{\nu}_\alpha(r)}{\mathcal{C}_U}\right)\left(1+\frac{1}{\theta}\widehat{f}_\alpha(r)\widehat{z}_{\alpha,\theta}(r)\right) &= 1+\frac{1}{\theta}\widehat{f}_\alpha(r)\widehat{z}_{\alpha,\theta}(r) - \frac{\widehat{z}_{\alpha,\theta}(r)}{\mathcal{C}_U}\\ &= 1+\frac{1}{\theta}\widehat{f}_\alpha(r)\widehat{z}_{\alpha,\theta}(r) - \left(1+\frac{\mathcal{C}_F}{\theta}\right)\widehat{z}_{\alpha,\theta}(r)\\ &= 1-\widehat{z}_{\alpha,\theta}(r) - \frac{\mathcal{C}_F}{\theta}\widehat{z}_{\alpha,\theta}(r)\left(1-\frac{1}{\mathcal{C}_F}\widehat{f}_\alpha(r)\right).\end{aligned} \tag{3.12}$$

We define the function $U_\alpha^*(x) = \int_0^x \overline{U}_\alpha(y)dy$, $x > 0$. From (3.1) we get

$$\widehat{U}_\alpha^*(r) = \frac{\widehat{\overline{U}}_\alpha(r)}{r} = \frac{1-\frac{\widehat{\nu}_\alpha(r)}{\mathcal{C}_U}}{r^2}. \tag{3.13}$$

It follows from (3.12) that

$$\frac{r\widehat{U}_\alpha^*(r)}{r^{\alpha-2}} = \frac{\frac{1-\widehat{z}_{\alpha,\theta}(r)}{r^{\alpha-1}} - \frac{\mathcal{C}_F}{\theta}\frac{\widehat{z}_{\alpha,\theta}(r)\left(1-\frac{1}{\mathcal{C}_F}\widehat{f}_\alpha(r)\right)}{r^{\alpha-1}}}{1+\frac{1}{\theta}\widehat{f}_\alpha(r)\widehat{z}_{\alpha,\theta}(r)}. \tag{3.14}$$

Since $\widehat{z}_{\alpha,\theta}(r) = \frac{\theta}{\theta+r^{\alpha-1}}$ we obtain $\lim_{r\downarrow 0} \frac{1-\widehat{z}_{\alpha,\theta}(r)}{r^{\alpha-1}} = \frac{1}{\theta}$. Using this equality together with (3.2) and letting $r \downarrow 0$ in (3.14), we obtain

$$\lim_{r\downarrow 0} \frac{r\widehat{U}^*_\alpha(r)}{r^{\alpha-2}} = \frac{\frac{1}{\theta} - \frac{1}{\theta} + \frac{1}{\theta}\frac{\prod_{i=1}^N q_i^{m_i}}{\prod_{j=2}^{m+1}\rho_j}}{1+\frac{\mathcal{C}_F}{\theta}} = \frac{\mathcal{C}_U}{\theta}\frac{\prod_{i=1}^N q_i^{m_i}}{\prod_{j=2}^{m+1}\rho_j}, \tag{3.15}$$

where in the last equality we used that $\mathcal{C}_U = (1+\frac{\mathcal{C}_F}{\theta})^{-1}$. Since U^*_α has the monotone density $\overline{U}_\alpha$, Theorem 1 (page 443) in [10] gives the limit in part d). This implies that the tail of U_α is asymptotically regularly varying with index $1-\alpha$, hence from (2.5) we conclude that $U_\alpha \in \mathcal{S}$.

■

Now we are ready to obtain the main result in this section.

Theorem 1 *Consider the following three cases for the claim size distribution F_2. As $x \to \infty$,*

$$\begin{aligned} &\textit{Case 1}: \overline{F}_2(x) = o(x^{-\alpha}),\\ &\textit{Case 2}: \overline{F}_2(x) \sim \kappa x^{-\alpha} \textit{ for some } \kappa > 0,\\ &\textit{Case 3}: F_{2,I} \in \mathcal{S},\ F_2 \in \mathcal{R}_0 \textit{ and } x^{-\alpha} = o\left(\overline{F}_2(x)\right). \end{aligned} \tag{3.16}$$

Then, as $u \to \infty$, we have:

a) In case 1:

$$\psi(u) \sim \frac{\eta^\alpha}{(c+\lambda_1\mu_1 - \lambda_2\mu_2)\,\Gamma(2-\alpha)} u^{1-\alpha}, \tag{3.17}$$

b) In case 2:

$$\psi(u) \sim \frac{1}{c+\lambda_1\mu_1-\lambda_2\mu_2}\left[\frac{\eta^\alpha}{\Gamma(2-\alpha)} + \frac{\lambda_2\kappa}{\alpha-1}\right] u^{1-\alpha}, \tag{3.18}$$

c) In case 3:

$$\psi(u) \sim \frac{\lambda_2\mu_2}{c+\lambda_1\mu_1-\lambda_2\mu_2}\overline{F}_{2,I}(u), \tag{3.19}$$

and in all cases $\Phi \in \mathcal{S}$.

Proof

Case 1. We define the function $G_0^*(x) = \int_0^x \overline{G}_0(y)dy, x > 0$. Due to (3.1) we have $\widehat{G}_0^*(r) = \frac{1-\widehat{g}_0(r)}{r^2}$. From Proposition 1 c) and the assumption that $\overline{F}_2(x) = o(x^{-\alpha})$ we obtain $\overline{G}_0(x) = o\left(x^{1-\alpha}\right)$, hence Theorem 1 (page 443) in [10] and the equality $\widehat{G}_0^*(r) = \frac{1-\widehat{g}_0(r)}{r^2}$ imply

$$0 = \lim_{r\downarrow 0} \frac{r\widehat{G}_\alpha^*(r)}{r^{\alpha-2}} = \lim_{r\downarrow 0} \frac{1-\widehat{g}_0(r)\mathcal{C}_G^{-1}}{r^{\alpha-1}}. \tag{3.20}$$

Using that $0 = \psi(\infty) = \lim_{u\to\infty}\psi(u)$, the final value theorem for Laplace transforms $\psi(\infty) = \lim_{r\downarrow 0} r\widehat{\psi}(r)$ and (2.4) we obtain $\widehat{W}_\alpha(0) = \left((c+\lambda_1\mu_1-\lambda_2\mu_2)\frac{\prod_{j=2}^{m+1}\rho_j}{\prod_{i=1}^N q_i^{m_i}}\right)^{-1}$. Setting $r = 0$ in (2.3) yields

$$\frac{1}{(c+\lambda_1\mu_1-\lambda_2\mu_2)\frac{\prod_{j=2}^{m+1}\rho_j}{\prod_{i=1}^N q_i^{m_i}}} = \frac{\frac{1}{\eta^\alpha\theta}\widehat{\nu}_\alpha(0)}{1-\frac{1}{\theta}\left[\kappa\widehat{\nu}_\alpha(0)+\frac{1}{\eta^\alpha}\widehat{g}_0(0)\widehat{\nu}_\alpha(0)\right]}$$

$$= \frac{\frac{1}{\eta^\alpha\theta}\mathcal{C}_U}{1-\frac{1}{\theta}\left[\kappa\mathcal{C}_U+\frac{1}{\eta^\alpha}\mathcal{C}_G\mathcal{C}_U\right]},$$

or equivalently

$$(c+\lambda_1\mu_1-\lambda_2\mu_2)\frac{\prod_{j=2}^{m+1}\rho_j}{\prod_{i=1}^N q_i^{m_i}} = \frac{1-\frac{1}{\theta}\left[\kappa\mathcal{C}_U+\frac{1}{\eta^\alpha}\mathcal{C}_G\mathcal{C}_U\right]}{\frac{1}{\eta^\alpha\theta}\mathcal{C}_U}. \tag{3.21}$$

Now we set $\psi^*(u) = \int_0^u \psi(y)dy$. Due to (3.1), (2.4), (2.3) and (3.21) we have

$$\widehat{\psi}^*(r) = \frac{1-\left[\frac{1-\frac{1}{\theta}\left[\kappa\mathcal{C}_U+\frac{1}{\eta^\alpha}\mathcal{C}_G\mathcal{C}_U\right]}{\frac{1}{\eta^\alpha\theta}\mathcal{C}_U}\right]\frac{\frac{1}{\eta^\alpha\theta}\widehat{\nu}_\alpha(r)}{1-\frac{1}{\theta}\left[\kappa\widehat{\nu}_\alpha(r)+\frac{1}{\eta^\alpha}\widehat{g}_0(r)\widehat{\nu}_\alpha(r)\right]}}{r^2}$$

$$= \frac{1-\frac{1}{\theta}\left[\kappa\widehat{\nu}_\alpha(r)+\frac{1}{\eta^\alpha}\widehat{g}_0(r)\widehat{\nu}_\alpha(r)\right]-\left[\frac{1-\frac{1}{\theta}\left[\kappa\mathcal{C}_U+\frac{1}{\eta^\alpha}\mathcal{C}_G\mathcal{C}_U\right]}{\mathcal{C}_U}\right]\widehat{\nu}_\alpha(r)}{r^2\left(1-\frac{1}{\theta}\left[\kappa\widehat{\nu}_\alpha(r)+\frac{1}{\eta^\alpha}\widehat{g}_0(r)\widehat{\nu}_\alpha(r)\right]\right)}.$$

It follows that

$$\lim_{r\downarrow 0}\frac{r\widehat{\psi}^*(r)}{r^{\alpha-2}}=\lim_{r\downarrow 0}\frac{1}{r^{\alpha-1}}\left(\frac{1-\dfrac{\widehat{\nu}_\alpha(r)}{\mathcal{C}_U}+\frac{c_G}{\eta^\alpha\theta}\left[1-\dfrac{\widehat{g}_0(r)}{\mathcal{C}_G}\right]\widehat{\nu}_\alpha(r)}{1-\frac{1}{\theta}\left[\kappa\widehat{\nu}_\alpha(r)+\eta^{-\alpha}\widehat{g}_0(r)\widehat{\nu}_\alpha(r)\right]}\right)$$

$$=\frac{\dfrac{1}{\theta}\dfrac{\prod_{i=1}^N q_i^{m_i}}{\prod_{j=2}^{m+1}\rho_j}\mathcal{C}_U}{1-\frac{1}{\theta}\left[\kappa\mathcal{C}_U+\eta^{-\alpha}\mathcal{C}_G\mathcal{C}_U\right]},\tag{3.22}$$

where the last equality follows from (3.13), (3.15) and (3.20). From (3.21) we obtain

$$\frac{\dfrac{1}{\theta}\dfrac{\prod_{i=1}^N q_i^{m_i}}{\prod_{j=2}^{m+1}\rho_j}\mathcal{C}_U}{\dfrac{c+\lambda_1\mu_1-\lambda_2\mu_2}{\eta^\alpha\theta}\dfrac{\prod_{i=1}^N q_i^{m_i}}{\prod_{j=2}^{m+1}\rho_j}\mathcal{C}_U}=\frac{\eta^\alpha}{c+\lambda_1\mu_1-\lambda_2},$$

hence from (3.22), $\lim_{r\downarrow 0}\dfrac{r\widehat{\psi}^*(r)}{r^{\alpha-2}}=\dfrac{\eta^\alpha}{c+\lambda_1\mu_1-\lambda_2}$. The asymptotic formula (3.17) now follows from [10, Theorem 1, page 443]. Since (3.17) implies that Φ has a regularly varying tail, from (2.5) we conclude $\Phi\in\mathcal{S}$.

Case 2. We work again with the functions ψ^* and G_0^* defined before. Due to $F_2\in\overline{RV}_{-\alpha}$ and $F_2\in\mathcal{R}_0$, from part b) of Proposition 1 we obtain $\overline{G}_0(x)\sim\dfrac{\lambda_2\mu_2}{\mathcal{C}_G}\dfrac{\prod_{i=1}^N q_i^{m_i}}{\prod_{j=2}^{m+1}\rho_j}\overline{F}_{2,I}(x)$. Since $\overline{F}_2(x)\sim\kappa x^{-\alpha}$, an application of L' Hospital's rule to $\lim_{x\to\infty}\frac{\overline{F}_{2,I}(x)}{x^{1-\alpha}}$ yields $\overline{F}_{2,I}(x)\sim\kappa\frac{x^{1-\alpha}}{(\alpha-1)\mu_2}$. Hence $\overline{G}_0(x)\sim\dfrac{\lambda_2\kappa}{\mathcal{C}_G(\alpha-1)}\dfrac{\prod_{i=1}^N q_i^{m_i}}{\prod_{j=2}^{m+1}\rho_j}x^{1-\alpha}$. Applying [10, Theorem 1, page 443] to $G_0^*(x)$ gives

$$\frac{\lambda_2\mu_2\kappa}{\mathcal{C}_G(\alpha-1)}\frac{\prod_{i=1}^N q_i^{m_i}}{\prod_{j=2}^{m+1}\rho_j}=\lim_{r\downarrow 0}\frac{r\widehat{G}_\alpha^*(r)}{r^{\alpha-2}}=\lim_{r\downarrow 0}\frac{1-\widehat{g}_0(r)\mathcal{C}_G^{-1}}{r^{\alpha-1}}.\tag{3.23}$$

From the last equality we see, as in case 1, that the limit (3.22) remains valid also in this case. Therefore

$$\lim_{r\downarrow 0}\frac{r\widehat{\psi}^*(r)}{r^{\alpha-2}}=\lim_{r\downarrow 0}\frac{1}{r^{\alpha-1}}\left(\frac{1-\dfrac{\widehat{\nu}_\alpha(r)}{\mathcal{C}_U}+\frac{c_G}{\eta^\alpha\theta}\left[1-\dfrac{\widehat{g}_0(r)}{\mathcal{C}_G}\right]\widehat{\nu}_\alpha(r)}{1-\frac{1}{\theta}\left[\kappa\widehat{\nu}_\alpha(r)+\eta^{-\alpha}\widehat{g}_0(r)\widehat{\nu}_\alpha(r)\right]}\right),$$

hence, substituting (3.13), (3.15) and (3.23) in the above equality gives

$$\lim_{r\downarrow 0}\frac{r\widehat{\psi}^*(r)}{r^{\alpha-2}} = \frac{\frac{1}{\theta}\frac{\prod_{i=1}^N q_i^{m_i}}{\prod_{j=2}^{m+1}\rho_j}C_U}{1-\frac{1}{\theta}\left[\kappa C_U+\eta^{-\alpha}C_G C_U\right]}\left[1+\frac{\lambda_2\kappa\Gamma(2-\alpha)}{\eta^\alpha(\alpha-1)}\right]$$
$$= \frac{\eta^\alpha(\alpha-1)+\lambda_2\kappa\Gamma(2-\alpha)}{(c+\lambda_1\mu_1-\lambda_2\mu_2)(\alpha-1)}.$$

The asymptotic formula (3.18) follows from [10, Theorem 1, page 443]. Since the right-hand side of (3.18) is a regularly varying function, it follows that Φ has a regularly varying tail. This finishes the proof of case 2.

Case 3. The equality $W_\alpha(x) = \frac{1}{\eta^\alpha\theta}\nu_\alpha * \sum_{n=0}^\infty \frac{1}{\theta^n}\left[\kappa\nu_\alpha+\frac{1}{\eta^\alpha}g_0*\nu_\alpha\right]^{*n}(x)$ is proved in [15, Proposition 5.6]. From (2.4) we note that $(c+\lambda_1\mu_1-\lambda_2\mu_2)\frac{\prod_{j=2}^{m+1}\rho_j}{\prod_{i=1}^N q_i^{m_i}}W_\alpha$ is the density function of the probability of survival Φ, hence using the above equality and the definitions of U_α and G_0 in (2.1), it follows that

$$\Phi(x) = \frac{1}{\eta^\alpha\theta}(c+\lambda_1\mu_1-\lambda_2\mu_2)\frac{\prod_{j=2}^{m+1}\rho_j}{\prod_{i=1}^N q_i^{m_i}}C_U U_\alpha$$
$$*\sum_{n=0}^\infty\frac{1}{\theta^n}\left(\kappa C_U U_\alpha+\frac{1}{\eta^\alpha}C_G C_U G_0 * U_\alpha\right)^{*n}(x).$$

Now we define $\beta = \frac{1}{\theta}\left[\kappa C_U+\eta^{-\alpha}C_G C_U\right]$. Using (3.21), we obtain from the last equality that

$$\Phi(x) = (1-\beta)U_\alpha * \sum_{n=0}^\infty\frac{\beta^n}{\theta^n}\left[\frac{1}{\beta}\left(\kappa C_U U_\alpha+\frac{1}{\eta^\alpha}C_G C_U G_0 * U_\alpha\right)\right]^{*n}(x)$$
$$= U_\alpha * K(x), \tag{3.24}$$

where $K(x) = (1-\beta)\sum_{n=0}^\infty\beta^n K_0^{*n}(x)$ with $K_0(x) = \left(\frac{1}{\theta\beta}\left[\kappa C_U U_\alpha + \eta^{-\alpha}C_G C_U G_0 * U_\alpha\right]\right)(x)$, $x>0$. Since U_α and G_0 are distribution functions, using the definition of β we see that K_0 and K are distribution functions as well, and Φ is the convolution of the distribution functions U_α and K. In view of this, we need to study the asymptotic behaviour of $\overline{K}$.

The assumption that $x^{-\alpha} = o(\overline{F}_2(x))$, together with an application of L'Hospital's rule, imply that $x^{1-\alpha} = o\left(\overline{F}_{2,I}(x)\right)$. Since by assumption $F_2\in\mathcal{R}_0$, part b) of Proposition 1 yields $\overline{G}_0(x)\sim\frac{\lambda_2\mu_2}{C_G}\frac{\prod_{i=1}^N q_i^{m_i}}{\prod_{j=2}^{m+1}\rho_j}\overline{F}_{2,I}(x)$, hence $x^{1-\alpha} = o\left(\overline{G}_0(x)\right)$, and due to part d) of Proposition 1 we get

$\overline{U}_\alpha(x) = o\big(\overline{G}_0(x)\big)$. It follows from the definition of K_0 and Lemma 1 a) that $1 - K_0(x) \sim \frac{\eta^{-\alpha}C_G C_U}{\theta\beta}\overline{G}_0(x)$. Since by assumption $F_{2,I} \in \mathcal{S}$, from part b) of Proposition 1 we obtain $G_0 \in \mathcal{S}$. It follows from Lemma 1 d) that

$$\overline{K}(x) \sim \frac{\beta}{1-\beta}\frac{\eta^{-\alpha}C_G C_U}{\theta\beta}\overline{G}_0(x)$$

$$\sim \frac{\lambda_2\mu_2}{\dfrac{c+\lambda_1\mu_1-\lambda_2\mu_2\frac{\prod_{i=1}^N q_i^{m_i}}{\prod_{j=2}^{m+1}\rho_j}C_U}{\eta^\alpha\theta}}\frac{C_U}{\eta^\alpha\theta}\frac{\prod_{i=1}^N q_i^{m_i}}{\prod_{j=2}^{m+1}\rho_j}\overline{F}_{2,I}(x),$$

which reduces to $\overline{K}(x) \sim \frac{\lambda_2\mu_2}{c+\lambda_1\mu_1-\lambda_2\mu_2}\overline{F}_{2,I}(x)$ after simplifying the coefficient in the right-hand side of the asymptotic expression above. From here we obtain (3.19) using (3.24), Lemma 1 a) and the relation $\overline{U}_\alpha(x) = o(\overline{F}_{2,I}(x))$ as $x \to \infty$. Hence $\Phi \in \mathcal{S}$.

■

Corollary 1 *For the three cases in (3.16) the ruin probability $\psi(u)$ admits the asymptotic expression*

$$\psi(u) \sim \frac{\eta^\alpha}{(c+\lambda_1\mu_1-\lambda_2\mu_2)\,\Gamma(2-\alpha)}u^{1-\alpha} + \frac{\lambda_2\mu_2}{c+\lambda_1\mu_1-\lambda_2\mu_2}\overline{F}_{2,I}(u) \text{ as } u \to \infty. \tag{3.25}$$

In particular, if $\overline{F}_2(u) \sim L_1(u)u^{-\alpha}$ for some slowly varying function L_1 and $\overline{F}_2$ belongs to any of the cases in (3.16), then

$$\psi(u) \sim \frac{1}{c+\lambda_1\mu_1-\lambda_2\mu_2}\left[\frac{\eta^\alpha}{\Gamma(2-\alpha)} + \frac{\lambda_2}{\alpha-1}L_1(u)\right]u^{1-\alpha}. \tag{3.26}$$

Proof The estimate (3.25) follows directly from Theorem 1. To obtain (3.26) we consider the three cases in (3.16).

Case 1. We have $\lim_{u\to\infty}\frac{L_1(u)u^{-\alpha}}{u^{-\alpha}} = \lim_{u\to\infty}\frac{\overline{F}_2(u)}{u^{-\alpha}}\frac{L_1(u)u^{-\alpha}}{\overline{F}_2(u)} = 0$. Hence

$$\lim_{u\to\infty}\frac{\psi(u)}{\frac{1}{c+\lambda_1\mu_1-\lambda_2\mu_2}\left[\frac{\eta^\alpha}{\Gamma(2-\alpha)} + \frac{\lambda_2\mu_2}{\alpha-1}L_1(u)\right]u^{1-\alpha}}$$

$$= \lim_{u\to\infty}\frac{\frac{\psi(u)}{u^{1-\alpha}}}{\frac{1}{c+\lambda_1\mu_1-\lambda_2\mu_2}\left[\frac{\eta^\alpha}{\Gamma(2-\alpha)} + \frac{\lambda_2\mu_2}{\alpha-1}\frac{L_1(u)u^{1-\alpha}}{u^{1-\alpha}}\right]} = 1,$$

where we used (3.17) to obtain the last equality.

Case 2. We set $C = \frac{1}{c+\lambda_1\mu_1-\lambda_2\mu_2}\left[\frac{\eta^\alpha}{\Gamma(2-\alpha)} + \frac{\lambda_2\kappa}{\alpha-1}\right]$.
Using the equality $\lim_{u\to\infty} L_1(u) = \kappa$ and (3.18) we obtain that

$$\lim_{u\to\infty} \frac{\psi(u)}{\frac{1}{c+\lambda_1\mu_1-\lambda_2\mu_2}\left[\frac{\eta^\alpha}{\Gamma(2-\alpha)} + \frac{\lambda_2}{\alpha-1}L_1(u)\right]u^{1-\alpha}}$$

$$= \lim_{u\to\infty} \frac{\frac{\psi(u)}{Cu^{1-\alpha}}}{\frac{1}{C}\frac{1}{c+\lambda_1\mu_1-\lambda_2\mu_2}\left[\frac{\eta^\alpha}{\Gamma(2-\alpha)} + \frac{\lambda_2}{\alpha-1}\frac{L_1(u)u^{1-\alpha}}{u^{1-\alpha}}\right]} = 1.$$

Case 3. Notice that $u^{-\alpha} = o\left(\overline{F}_2(u)\right)$ implies $u^{1-\alpha} = o\left(\overline{F}_{2,I}(u)\right)$. Using now Karamata's theorem (see e.g. [3, Proposition 1.5.10]) we obtain that $\lim_{u\to\infty} \frac{\overline{F}_{2,I}(u)}{L_1(u)u^{1-\alpha}} = \frac{\alpha-1}{\mu_2}$. Hence

$$\lim_{u\to\infty} \frac{\psi(u)}{\frac{1}{c+\lambda_1\mu_1-\lambda_2\mu_2}\left[\frac{\eta^\alpha}{\Gamma(2-\alpha)} + \frac{\lambda_2\mu_2}{\alpha-1}L_1(u)\right]u^{1-\alpha}}$$

$$= \lim_{u\to\infty} \frac{\frac{\psi(u)}{\frac{\lambda_2\mu_2}{c+\lambda_1\mu_1-\lambda_2\mu_2}\overline{F}_{2,I}(u)}}{\frac{\eta^\alpha}{\Gamma(2-\alpha)}\frac{u^{1-\alpha}}{\lambda_2\mu_2\overline{F}_{2,I}(u)} + \frac{1}{\alpha-1}\frac{L_1(u)u^{1-\alpha}}{\overline{F}_{2,I}(u)}} = 1.$$

■

4 Asymptotic Behavior of the Joint Tail of the Severity of Ruin and the Surplus Prior to Ruin

For fixed $\beta > 0$ and $a \geq 0$, we define the function

$$B(x;\beta,a) := \int_x^\infty e^{-\beta(y-x)}\left(\lambda_2\overline{F}_2(y+a) + \frac{\eta^\alpha(\alpha-1)}{\Gamma(2-\alpha)}(y+a)^{-\alpha}\right)dy, \quad x \geq 0. \tag{4.1}$$

In order to obtain asymptotic expressions for $\Upsilon_{a,b}(u)$ as $u \to \infty$ in such a way that $\Delta := \max\{u, b\} \to \infty$, we establish some preliminary lemmas.

Lemma 2

a) For all $x \geq 0$ and $\beta > 0$, $B(x;\beta,a) \leq \lambda_2\mu_2 + \frac{\eta^\alpha}{\Gamma(2-\alpha)}a^{1-\alpha}$ for any $a \geq 0$.
b) The asymptotic relation $B(x;\beta,a) = o(\psi(x+a))$ as $x \to \infty$, holds in any of the cases in (3.16).

Proof

a) Since $e^{-\beta(y-x)} \leq 1$ when $y \geq x$, and $\overline{F}_2(y+a) \leq \overline{F}_2(y)$, we have

$$
\begin{aligned}
&B(x;\beta,a)\\
&\leq \int_x^\infty \left(\lambda_2 \overline{F}_2(y+a) + \frac{\eta^\alpha(\alpha-1)}{\Gamma(2-\alpha)}(y+a)^{-\alpha}\right) dy\\
&\leq \int_0^\infty \left(\lambda_2 \overline{F}_2(y) + \frac{\eta^\alpha(\alpha-1)}{\Gamma(2-\alpha)}(y+a)^{-\alpha}\right) dy,
\end{aligned}
$$

which implies a).

b) Using that $\overline{F}_2(y+a) \leq \overline{F}_2(x+a)$ and $(y+a)^{-\alpha} \leq (x+a)^{-\alpha}$ for all $y \geq x$, we see that

$$
\begin{aligned}
B(x;\beta,a) &\leq \int_x^\infty e^{-\beta(y-x)}\left(\lambda_2 \overline{F}_2(x+a) + \frac{\eta^\alpha(\alpha-1)}{\Gamma(2-\alpha)}(x+a)^{-\alpha}\right) dy\\
&= \frac{1}{\beta}\left(\lambda_2 \overline{F}_2(x+a) + \frac{\eta^\alpha(\alpha-1)}{\Gamma(2-\alpha)}(x+a)^{-\alpha}\right). \qquad (4.2)
\end{aligned}
$$

For the first two cases in (3.16), the limit $\lim\limits_{x\to\infty} \frac{\overline{F}_2(x)}{x^{-\alpha}}$ exists and is finite, hence in any of these two cases we obtain

$$
\begin{aligned}
&\lim_{x\to\infty} \frac{\frac{1}{\beta}\left(\lambda_2 \overline{F}_2(x+a) + \frac{\eta^\alpha(\alpha-1)}{\Gamma(2-\alpha)}(x+a)^{-\alpha}\right)}{(x+a)^{1-\alpha}}\\
&= \lim_{x\to\infty} \frac{\frac{1}{\beta}\left(\lambda_2 \frac{\overline{F}_2(x+a)}{(x+a)^{-\alpha}} + \frac{\eta^\alpha(\alpha-1)}{\Gamma(2-\alpha)}\right)}{x+a} = 0. \qquad (4.3)
\end{aligned}
$$

Due to (3.17) and (3.18) we obtain, again in cases 1 and 2 of (3.16), that $\psi(u) \sim Au^{1-\alpha}$ for some constant $A > 0$. This and (4.3) imply

$$
\begin{aligned}
&\lim_{x\to\infty} \frac{\frac{1}{\beta}\left(\lambda_2 \overline{F}_2(x+a) + \frac{\eta^\alpha(\alpha-1)}{\Gamma(2-\alpha)}(x+a)^{-\alpha}\right)}{\psi(x+a)}\\
&= \lim_{x\to\infty} \frac{\dfrac{\frac{1}{\beta}\left(\lambda_2 \overline{F}_2(x+a) + \frac{\eta^\alpha(\alpha-1)}{\Gamma(2-\alpha)}(x+a)^{-\alpha}\right)}{(x+a)^{1-\alpha}}}{\dfrac{\psi(x+a)}{(x+a)^{1-\alpha}}} = 0.
\end{aligned}
$$

Hence we obtain the result in these two cases by dividing by $\psi(x+a)$ both sides of (4.2) and making $x \to \infty$ afterwards.

In the remaining case 3, the assumption that $F_2 \in \mathcal{R}_0$ and L'Hospital's rule imply that $\overline{F}_{2,I} \in \mathcal{R}_0$. From (3.19) we obtain that $\psi(u) \sim A_2\overline{F}_{2,I}(u)$ for some constant $A_2 > 0$. Moreover, from the proof of Theorem 1 c) we see that $x^{1-\alpha} = o(\overline{F}_{2,I}(x))$. Using these two results together with $\overline{F}_{2,I} \in \mathcal{R}_0$, it follows that

$$\lim_{x\to\infty} \frac{\frac{1}{\beta}\left(\lambda_2\overline{F}_2(x+a) + \frac{\eta^\alpha(\alpha-1)}{\Gamma(2-\alpha)}(x+a)^{-\alpha}\right)}{\psi(x+a)}$$

$$= \lim_{x\to\infty} \frac{\dfrac{\frac{1}{\beta}\left(\lambda_2\overline{F}_2(x+a) + \frac{\eta^\alpha(\alpha-1)}{\Gamma(2-\alpha)}(x+a)^{-\alpha}\right)}{\overline{F}_{2,I}(x+a)}}{\dfrac{\psi(x+a)}{\overline{F}_{2,I}(x+a)}} = 0.$$

Again, the result follows dividing both sides of (4.2) by $\psi(x+a)$ and making $x \to \infty$.

■

Recall the definition of the joint tail distribution $\Upsilon_{a,b}$ given in (1.4).

Lemma 3 *The joint tail distribution admits the representation*

$$\Upsilon_{a,b}(u) = h_\alpha * W_\alpha(u), \quad u > 0, \tag{4.4}$$

where

$$h_\alpha(u) = \frac{\prod_{i=1}^{N} q_i^{m_i}}{\prod_{j=2}^{m+1}\rho_j}\int_u^\infty \left[\lambda_2\overline{F}_2(a+z) + \frac{\eta^\alpha(\alpha-1)}{\Gamma(2-\alpha)}(a+z)^{-\alpha}\right] 1_{\{z>b\}}dz + I_{a,b}(u),$$

and $I_{a,b}(x) = \sum_{j=2}^{m+1} E(\rho_j)\int_x^\infty e^{-\rho_j(y-x)}\left(\lambda_2\overline{F}_2(y+a) + \frac{\eta^\alpha(\alpha-1)}{\Gamma(2-\alpha)}(y+a)^{-\alpha}\right)$ $1_{\{y>b\}}dy$. *Moreover, if F_2 belongs to any of the cases in (3.16), then for fixed $a, b > 0$,*

$$\int_0^u I_{a,b}(u-y)\Phi(dy) = o(\psi(u)) \text{ as } u \to \infty, \tag{4.5}$$

and the following limit holds:

$$\lim_{u\to\infty}\int_0^u I_{a,b}(u-y)\Phi(dy) = 0, \tag{4.6}$$

uniformly on the sets $\{a \geq \xi, b \geq \eta\}$ for all fixed $\xi, \eta > 0$.

Proof Formula (4.4) follows directly from [15, Corollary 5.1]. To prove (4.5) we first note that

$$\begin{aligned}&\left|I_{a,b}(x)\right|\\&\quad\leq\sum_{j=2}^{m+1}|E(\rho_j)|\int_x^\infty e^{-Re(\rho_j)(y-x)}\left(\lambda_2\overline{F}_2(y+a)+\frac{\eta^\alpha(\alpha-1)}{\Gamma(2-\alpha)}(y+a)^{-\alpha}\right)dy,\end{aligned}$$

which due to (4.1) is equivalent to

$$\left|I_{a,b}(x)\right|\leq\sum_{j=2}^{m+1}|E(\rho_j)|B\left[x;\,Re(\rho_j),a\right]. \tag{4.7}$$

Let $\varepsilon > 0$ be given. From Lemma 2 b) there exists $u_0 > 0$ such that $\sum_{j=2}^{m+1}|E(\rho_j)|B\left[u;\,Re(\rho_j),a\right] < \varepsilon\psi(u)$ for all $u > u_0$. It follows from (4.7) that

$$\begin{aligned}\left|\frac{\int_0^u I_{a,b}(u-y)\Phi(dy)}{\psi(u)}\right| &< \frac{\varepsilon\int_0^{u-u_0}(1-\Phi(u-y))\Phi(dy)}{\psi(u)}+\frac{\int_{u-u_0}^u|I_{a,b}(u-y)|\Phi(dy)}{\psi(u)}\\&\leq\frac{\varepsilon\int_0^u(1-\Phi(u-y))\Phi(dy)}{\psi(u)}+\frac{\int_{u-u_0}^u|I_{a,b}(u-y)|\Phi(dy)}{\psi(u)}\\&\leq\sum_{j=2}^{m+1}|E(\rho_j)|\left(\lambda_2\mu_2+\frac{\eta^\alpha}{\Gamma(2-\alpha)}a^{1-\alpha}\right):=c_0,\end{aligned}$$

where in the last equality we used Lemma 2 a). Hence

$$\begin{aligned}\left|\frac{\int_0^u I_{a,b}(u-y)\Phi(dy)}{\psi(u)}\right| &< \frac{\varepsilon\left(\Phi(u)-\Phi*\Phi(u)\right)}{\psi(u)}+c_0\frac{\Phi(u)-\Phi(u-u_0)}{\psi(u)}\\&=\frac{\varepsilon\left(1-\Phi*\Phi(u))-\psi(u)\right)}{\psi(u)}+c_0\frac{\psi(u-u_0)-\psi(u)}{\psi(u)}.\end{aligned}$$

The estimate in (4.5) follows from the last inequality and the fact that $\Phi\in\mathcal{S}$. Since $I_{a,b}$ is, by its definition, nonincreasing in a and b, it follows that

$$\int_0^u I_{a,b}(u-y)\Phi(dy)\leq\int_0^u I_{\xi,\eta}(u-y)\Phi(dy), \tag{4.8}$$

for all $a\geq\xi$ and $b\geq\eta$.

Since $\lim_{u\to\infty} \psi(u) = 0$, using (4.5) we obtain that $\lim_{u\to\infty} \int_0^u I_{a,b}(u-y)\Phi(dy) = 0$. Hence the result follows from (4.8) by making $u \to \infty$. ■

We now obtain the main results of this section.

Theorem 2 *Let F_2 belong to any of the three cases given in (3.16). Then, for fixed $a > 0$, the joint tail of the severity of ruin and the surplus prior to ruin, $\Upsilon_{a,b}$, admits the following asymptotic expressions as $u \to \infty$ in such a way that $\Delta = \max\{u, b\} \to \infty$:*

a) in case 1, $\Upsilon_{a,b}(u) \sim \dfrac{\eta^\alpha}{(c+\lambda_1\mu_1-\lambda_2\mu_2)\,\Gamma(2-\alpha)}(a+\Delta)^{1-\alpha}$,

b) in case 2, $\Upsilon_{a,b}(u) \sim \dfrac{1}{c+\lambda_1\mu_1-\lambda_2\mu_2}\left[\dfrac{\eta^\alpha}{\Gamma(2-\alpha)} + \dfrac{\lambda_2\kappa}{\alpha-1}\right](a+\Delta)^{1-\alpha}$,

c) in case 3, $\Upsilon_{a,b}(u) \sim \dfrac{\lambda_2\mu_2}{c+\lambda_1\mu_1-\lambda_2\mu_2}\overline{F}_{2,I}(a+\Delta)$.

Proof From (4.4) and [15, Corollary 5.5] it follows that

$$
\begin{aligned}
\Upsilon_{a,b}(u) = {} & \frac{1}{c+\lambda_1\mu_1-\lambda_2\mu_2} \\
& \times \int_0^u \int_{u-y}^\infty \left[\lambda_2\overline{F}_2(a+z) + \frac{\eta^\alpha(\alpha-1)}{\Gamma(2-\alpha)}(a+z)^{-\alpha}\right] 1_{\{z>b\}}\, dz\, \Phi(dy) \\
& + \frac{\frac{\prod_{i=1}^N q_i^{m_i}}{\prod_{j=2}^{m+1} \rho_j}}{c+\lambda_1\mu_1-\lambda_2\mu_2} \int_0^u I_{a,b}(u-y)\, \Phi(dy).
\end{aligned}
\tag{4.9}
$$

In view of (4.5) we need only to study the asymptotic behavior of

$$
\begin{aligned}
\Upsilon^*(u, a, b) := {} & \frac{1}{c+\lambda_1\mu_1-\lambda_2\mu_2} \\
& \times \int_0^u \int_{u-y}^\infty \left[\lambda_2\overline{F}_2(a+z) + \frac{\eta^\alpha(\alpha-1)}{\Gamma(2-\alpha)}(a+z)^{-\alpha}\right] 1_{\{z>b\}}\, dz\, \Phi(dy)
\end{aligned}
$$

as $u \to \infty$ in such a way that $\Delta = \max\{u, b\} \to \infty$. First we suppose that $\Delta = u$ and define

$$
\begin{aligned}
& \Upsilon_0(u, a) \\
& := \frac{1}{c+\lambda_1\mu_1-\lambda_2\mu_2} \int_0^u \left[\lambda_2\mu_2\overline{F}_{2,I}(a+u-y) + \frac{\eta^\alpha}{\Gamma(2-\alpha)}(a+u-y)^{1-\alpha}\right] \Phi(dy).
\end{aligned}
\tag{4.10}
$$

Therefore

$$\begin{aligned}\Upsilon^*(u,a,b) & \\ &\leq \frac{1}{c+\lambda_1\mu_1-\lambda_2\mu_2}\int_0^u\int_{u-y}^\infty\left[\lambda_2\overline{F}_2(a+z)+\frac{\eta^\alpha(\alpha-1)}{\Gamma(2-\alpha)}(a+z)^{-\alpha}\right]dz\,\Phi(dy)\\ &=\Upsilon_0(u,a)\end{aligned} \tag{4.11}$$

and

$$\begin{aligned}\Upsilon^*(u,a,b) & \\ &\geq \frac{1}{c+\lambda_1\mu_1-\lambda_2\mu_2}\int_0^u\int_{u}^\infty\left[\lambda_2\overline{F}_2(a+z)+\frac{\eta^\alpha(\alpha-1)}{\Gamma(2-\alpha)}(a+z)^{-\alpha}\right]dz\,\Phi(dy)\\ &=\frac{\lambda_2\mu_2\overline{F}_{2,I}(a+u)+\frac{\eta^\alpha}{\Gamma(2-\alpha)}(a+u)^{1-\alpha}}{c+\lambda_1\mu_1-\lambda_2\mu_2}\Phi(u).\end{aligned} \tag{4.12}$$

The above inequality and Corollary 1 imply that

$$\liminf_{u\to\infty}\frac{\Upsilon^*(u,a,b)}{\psi(u+a)}\geq 1 \tag{4.13}$$

because $\lim_{u\to\infty}\Phi(u)=1$. To finish the proof it suffices to show that

$$\lim_{u\to\infty}\frac{\Upsilon_0(u,a)}{\psi(u+a)}=1 \tag{4.14}$$

for any of the claim size distributions in (3.16). Indeed, the asymptotics in the three cases follow from (4.14) together with (4.11), (4.13), (4.9) and (4.5).

We note that

$$\begin{aligned}\Upsilon_0(u,a) &= \frac{1}{c+\lambda_1\mu_1-\lambda_2\mu_2}\\ &\times\left[\lambda_2\mu_2\overline{F}_{2,I}(a)\int_0^u\left(1-F_{a,I}(u-y)\right)\Phi(dy)+\frac{a^{1-\alpha}\eta^\alpha}{\Gamma(2-\alpha)}\int_0^u\left(1-P_{a,\alpha}(u-y)\right)\Phi(dy)\right],\end{aligned}$$

where we define for $a > 0$ the functions $F_{a,I}(u) = 1 - \dfrac{\overline{F}_{2,I}(u+a)}{\overline{F}_{2,I}(a)}$ and $P_{a,\alpha}(u) = 1 - \left(\dfrac{au}{a+u}\right)^{\alpha-1} u^{1-\alpha},\ u \geq 0$. Hence,

$$\begin{aligned}
\Upsilon_0(u,a) &= \frac{1}{c+\lambda_1\mu_1-\lambda_2\mu_2} \\
&\quad \times \left[\lambda_2\mu_2\overline{F}_{2,I}(a)\left(\Phi(u) - F_{a,I} * \Phi(u)\right) + \frac{a^{1-\alpha}\eta^\alpha}{\Gamma(2-\alpha)}\left(\Phi(u) - P_{a,\alpha} * \Phi(u)\right)\right] \\
&= \frac{1}{c+\lambda_1\mu_1-\lambda_2\mu_2}\Big(\lambda_2\mu_2\overline{F}_{2,I}(a)\left[\Phi(u) - 1 + 1 - F_{a,I} * \Phi(u)\right] \\
&\quad + \frac{a^{1-\alpha}\eta^\alpha}{\Gamma(2-\alpha)}\left[\Phi(u) - 1 + 1 - P_{a,\alpha} * \Phi(u)\right]\Big) \\
&= \frac{1}{c+\lambda_1\mu_1-\lambda_2\mu_2} \\
&\quad \times \Big(\lambda_2\mu_2\overline{F}_{2,I}(a)\left[1 - F_{a,I} * \Phi(u) - \Psi(u)\right] \\
&\quad + \frac{a^{1-\alpha}\eta^\alpha}{\Gamma(2-\alpha)}\left[1 - P_{a,\alpha} * \Phi(u) - \Psi(u)\right]\Big).
\end{aligned} \tag{4.15}$$

Case 1. Due to Theorem 1 a) we have $\Phi \in \mathcal{S}$ and $\psi(u) \sim \dfrac{\eta^\alpha}{(c+\lambda_1\mu_1-\lambda_2\mu_2)\Gamma(2-\alpha)} u^{1-\alpha}$, hence Lemma 1 c) and the assumption $\overline{F}_{2,I}(u) = o(u^{1-\alpha})$ as $u \to \infty$, imply

$$1 - F_{a,I} * \Phi(u) \sim \frac{\eta^\alpha}{(c+\lambda_1\mu_1-\lambda_2\mu_2)\Gamma(2-\alpha)} u^{1-\alpha}.$$

This shows that

$$\frac{\lambda_2\mu_2\overline{F}_{2,I}(a)}{c+\lambda_1\mu_1-\lambda_2\mu_2}\left[1 - F_{a,I} * \Phi(u) - \psi(u)\right] = o(u^{1-\alpha}) \text{ as } u \to \infty. \tag{4.16}$$

From Lemma 1 b), as $u \to \infty$,

$$1 - P_{a,\alpha} * \Phi(u) \sim \left[\left(\frac{au}{a+u}\right)^{\alpha-1} + \frac{\eta^\alpha}{(c+\lambda_1\mu_1-\lambda_2\mu_2)\Gamma(2-\alpha)}\right] u^{1-\alpha},$$

which due to (3.17) implies

$$1 - P_{a,\alpha} * \Phi(u) - \psi(u) \sim \left(\frac{a}{a+u}\right)^{\alpha-1}. \tag{4.17}$$

Using the expression for $\Upsilon_0(u, a)$ given in (4.15), together with (4.16) and (4.17) we obtain $\Upsilon_0(u, a) \sim \frac{1}{c + \lambda_1\mu_1 - \lambda_2\mu_2}\Big[\lambda_2\mu_2\overline{F}_{2,I}(a + \Delta) + \frac{\eta^\alpha}{\Gamma(2-\alpha)}(a + \Delta)^{1-\alpha}\Big]$, and (4.14) follows.

Case 2. Since by assumption $\overline{F}_2(u) \sim \kappa u^{1-\alpha}$, L'Hospital's rule gives $\overline{F}_{2,I}(u) \sim \frac{\kappa}{\mu_2(\alpha-1)}u^{1-\alpha}$. Hence $\overline{F}_{a,I}(u) \sim \frac{\kappa}{\mu_2(\alpha-1)\overline{F}_{2,I}(a)}u^{1-\alpha}$. From (3.18) we have $\Psi(u) \sim Cu^{1-\alpha}$, where the constant C is given by $C = \frac{1}{c+\lambda_1\mu_1-\lambda_2\mu_2}\Big[\frac{\eta^\alpha}{\Gamma(2-\alpha)} + \frac{\lambda_2\kappa}{\alpha-1}\Big]$. Using this and Lemma 1 b) gives $1 - F_{a,I} * \Phi(u) \sim \Big[C + \frac{\kappa}{\mu_2(\alpha-1)}\Big]u^{1-\alpha}$. It follows that

$$1 - F_{a,I} * \Phi(u) - \psi(u) \sim \frac{\kappa}{\mu_2(\alpha-1)}u^{1-\alpha} \sim \frac{\kappa}{\mu_2(\alpha-1)}(a+u)^{1-\alpha}. \tag{4.18}$$

From Lemma 1 b) and (3.18),

$$1 - P_{a,\alpha} * \Phi(u) \sim \left[\left(\frac{au}{a+u}\right)^{\alpha-1} + \frac{\eta^\alpha}{(c+\lambda_1\mu_1-\lambda_2\mu_2)\Gamma(2-\alpha)} + \frac{\lambda_2\kappa}{c+\lambda_1\mu_1-\lambda_2\mu_2}\right]u^{1-\alpha}.$$

This together with (3.18) yields

$$1 - P_{a,\alpha} * \Phi(u) - \psi(u) \sim \left(\frac{a}{a+u}\right)^{\alpha-1}. \tag{4.19}$$

Now using (4.18) and (4.19), we obtain

$$\Upsilon_0(u, a) \sim \frac{1}{c+\lambda_1\mu_1-\lambda_2\mu_2}\left[\frac{\eta^\alpha}{\Gamma(2-\alpha)} + \frac{\lambda_2\kappa}{\alpha-1}\right](a+u)^{1-\alpha}.$$

Case 3. Using the assumption $u^{-\alpha} = o\big(\overline{F}_2(u)\big)$ and L'Hospital's rule we get $u^{1-\alpha} = o\big(\overline{F}_{2,I}(u)\big)$. Since $\overline{P}_{a,\alpha}(u) = \left(\frac{au}{a+u}\right)^{\alpha-1}u^{1-\alpha}$ and $\lim_{u\to\infty}\frac{\left(\frac{auy}{a+uy}\right)^{\alpha-1}}{\left(\frac{au}{a+u}\right)^{\alpha-1}} = 1$ for all $y > 0$, we have $\overline{P}_{a,\alpha}(u) \sim u^{1-\alpha}$. Hence $\overline{P}_{a,\alpha}(u) = o\big(\overline{F}_{2,I}(u)\big)$, and from Corollary 1 and (3.19) we obtain $1 - P_{a,\alpha} * \Phi(u) \sim \frac{\lambda_2\mu_2}{c+\lambda_1\mu_1-\lambda_2\mu_2}\overline{F}_{2,I}(u)$. Using (3.19) again we conclude that $1 - P_{a,\alpha} * \Phi(u) - \psi(u) = o(\overline{F}_{2,I}(u))$. Due to Lemma 1 b),

$$1 - F_{a,I} * \Phi(u) \sim \left(\frac{1}{\overline{F}_{2,I}(a)} + \frac{\lambda_2\mu_2}{c + \lambda_1\mu_1 - \lambda_2\mu_2}\right) \overline{F}_{2,I}(a+u),$$

which implies $1 - F_{a,I} * \Phi(u) - \psi(u) \sim \frac{\overline{F}_{2,I}(u+a)}{\overline{F}_{2,I}(a)}$. In this way we obtain (4.14).

In the case of $\Delta = b$ we have

$$\begin{aligned}
&\Upsilon^*(u,a,b)\\
&= \frac{1}{c + \lambda_1\mu_1 - \lambda_2\mu_2} \int_0^u \int_{u-y}^{\infty} \left[\lambda_2 \overline{F}_2(a+z) + \frac{\eta^\alpha(\alpha-1)}{\Gamma(2-\alpha)}(a+z)^{-\alpha}\right] 1_{\{z>b\}}\, dz\, \Phi(dy)\\
&= \frac{1}{c + \lambda_1\mu_1 - \lambda_2\mu_2} \int_0^u \left[\lambda_2\mu_2 \overline{F}_{2,I}(a+b) + \frac{\eta^\alpha}{\Gamma(2-\alpha)}(a+b)^{1-\alpha}\right] \Phi(dy)\\
&= \frac{\lambda_2\mu_2\overline{F}_{2,I}(a+b) + \frac{\eta^\alpha}{\Gamma(2-\alpha)}(a+b)^{1-\alpha}}{c + \lambda_1\mu_1 - \lambda_2\mu_2} \Phi(u).
\end{aligned}$$

The asymptotics for $\Upsilon_{a,b}$ follow by dividing $\Upsilon_{a,b}(u)$ by $\dfrac{\lambda_2\mu_2\overline{F}_{2,I}(a+b) + \frac{\eta^\alpha}{\Gamma(2-\alpha)}(a+b)^{1-\alpha}}{c + \lambda_1\mu_1 - \lambda_2\mu_2}$, letting afterward $u \to \infty$, and proceeding as in the cases 1, 2 and 3 above with u replaced by b. ■

Corollary 2 *For any of the cases in (3.16), the joint tail $\Upsilon_{a,b}$ has the asymptotic expression when $u \to \infty$ and $\Delta = \max\{u, b\} \to \infty$:*

$$\Upsilon_{a,b}(u) \sim \frac{\eta^\alpha}{(c + \lambda_1\mu_1 - \lambda_2\mu_2)\,\Gamma(2-\alpha)}(a+\Delta)^{1-\alpha} + \frac{\lambda_2\mu_2}{c + \lambda_1\mu_1 - \lambda_2\mu_2}\overline{F}_{2,I}(a+\Delta).$$

In particular, if $\overline{F}_2(u) \sim L_1(u)u^{-\alpha}$ for some slowly varying function L_1, and $\overline{F}_2$ satisfies any of the cases in (3.16), it follows

$$\Upsilon_{a,b}(u) \sim \frac{1}{c + \lambda_1\mu_1 - \lambda_2\mu_2}\left[\frac{\eta^\alpha}{\Gamma(2-\alpha)} + \frac{\lambda_2}{\alpha-1}L_1(a+\Delta)\right](a+\Delta)^{1-\alpha}.$$

We have the following sharper result, which shows that the asymptotics of $\Upsilon_{a,b}$ given in Theorem 2, hold uniformly on the parameters a and b.

Theorem 3 *Let F_2 belong to any of the three cases given in (3.16). The following limits hold, when $u \to \infty$, uniformly on the sets $A_{\xi,\eta} = \{a \geq \xi, b \geq \eta\}$, for fixed $\xi, \eta > 0$.*

1. *In case 1:*

$$\lim_{u\to\infty}\left|\Upsilon_{a,b}(u)-\frac{\eta^\alpha}{(c+\lambda_1\mu_1-\lambda_2\mu_2)\,\Gamma(2-\alpha)}(a+u)^{1-\alpha}\right|=0.$$

2. *In case 2:*

$$\lim_{u\to\infty}\left|\Upsilon_{a,b}(u)-\frac{1}{c+\lambda_1\mu_1-\lambda_2\mu_2}\left[\frac{\eta^\alpha}{\Gamma(2-\alpha)}+\frac{\lambda_2\kappa}{\alpha-1}\right](a+u)^{1-\alpha}\right|=0.$$

3. *In case 3:*

$$\lim_{u\to\infty}\left|\Upsilon_{a,b}(u)-\frac{\lambda_2\mu_2}{c+\lambda_1\mu_1-\lambda_2\mu_2}\overline{F}_{2,I}(a+u)\right|=0.$$

Proof By (4.9) and (4.6), we only need to study the uniform convergence of $\Upsilon^*(u,a,b) = \frac{1}{c+\lambda_1\mu_1-\lambda_2\mu_2}\int_0^u\int_{u-y}^\infty\left[\lambda_2\overline{F}_2(a+z)+\frac{\eta^\alpha(\alpha-1)}{\Gamma(2-\alpha)}(a+z)^{-\alpha}\right]$ $1_{\{z>b\}}dz\Phi(dy)$ on the sets $A_{\xi,\eta}$.

Using (4.12) we obtain

$$\begin{aligned}\Upsilon^*(u,a,b)-\frac{\eta^\alpha(a+u)^{1-\alpha}}{(c+\lambda_1\mu_1-\lambda_2\mu_2)\Gamma(2-\alpha)}&\geq\frac{\eta^\alpha(a+u)^{1-\alpha}(\Phi(u)-1)}{(c+\lambda_1\mu_1-\lambda_2\mu_2)\Gamma(2-\alpha)}\\&\quad+\frac{\lambda_2\mu_2\overline{F}_{2,I}(a+u)\Phi(u)}{c+\lambda_1\mu_1-\lambda_2\mu_2}\\&\geq\frac{\eta^\alpha(a+u)^{1-\alpha}(\Phi(u)-1)}{(c+\lambda_1\mu_1-\lambda_2\mu_2)\Gamma(2-\alpha)}.\end{aligned}$$

Since $(a+u)^{1-\alpha}(\Phi(u)-1)$ is nonincreasing as a function of a, we have

$$(a+u)^{1-\alpha}(\Phi(u)-1)\leq(\xi+u)^{1-\alpha}(\Phi(u)-1),$$

and since $\lim_{u\to\infty}(\xi+u)^{1-\alpha}(\Phi(u)-1)=0$, the convergence

$$\lim_{u\to\infty}(a+u)^{1-\alpha}(\Phi(u)-1)=0 \tag{4.20}$$

is uniform on $\{a\geq\xi\}$. Hence for all $\varepsilon>0$ and $\xi>0$ there exists $A>0$ such that for $u\geq A$ we have

$$\Upsilon^*(u,a,b)-\frac{\eta^\alpha(a+u)^{1-\alpha}}{(c+\lambda_1\mu_1-\lambda_2\mu_2)\Gamma(2-\alpha)}\geq-\varepsilon. \tag{4.21}$$

We will use that, by the definition (4.10) and the equivalent formula (4.15),

$$\overline{F}_{2,I}(a)\left(1 - F_{a,I} * \Phi(u) - \psi(u)\right) = \int_0^u \overline{F}_{2,I}(a+u-y)\Phi(dy)$$

and

$$a^{1-\alpha}\left(1 - P_{a,\alpha} * \Phi(u) - \psi(u)\right) = \int_0^u (a+u-y)^{1-\alpha}\Phi(dy),$$

hence

$$\begin{aligned} &\overline{F}_{2,I}(a)\left(1 - F_{a,I} * \Phi(u) - \psi(u)\right) \text{ and} \\ &a^{1-\alpha}\left(1 - P_{a,\alpha} * \Phi(u) - \psi(u)\right) \text{ are nonincreasing in } a. \end{aligned} \tag{4.22}$$

Case 1: We have, by (4.12):

$$\Upsilon^*(u,a,b) - \frac{\eta^\alpha (a+u)^{1-\alpha}}{(c+\lambda_1\mu_1 - \lambda_2\mu_2)\Gamma(2-\alpha)} \le \Upsilon_0(a,b) - \frac{\eta^\alpha (a+u)^{1-\alpha}}{(c+\lambda_1\mu_1 - \lambda_2\mu_2)\Gamma(2-\alpha)}.$$

Hence, from the definition of $\Upsilon_0(a,b)$ in (4.10) and the equality (4.15) we obtain:

$$\begin{aligned} &\Upsilon^*(u,a,b) - \frac{\eta^\alpha (a+u)^{1-\alpha}}{(c+\lambda_1\mu_1 - \lambda_2\mu_2)\Gamma(2-\alpha)} \\ &\le \frac{1}{c+\lambda_1\mu_1-\lambda_2\mu_2}\Bigg[\lambda_2\mu_2\overline{F}_{2,I}(a)\left(1 - F_{a,I} * \Phi(u) - \psi(u)\right) \\ &\quad + \frac{\eta^\alpha a^{1-\alpha}}{\Gamma(2-\alpha)}\left(1 - P_{a,\alpha} * \Phi(u) - \psi(u) - \left(\frac{a+u}{a}\right)^{1-\alpha}\right)\Bigg]. \end{aligned} \tag{4.23}$$

We know from (4.22) that $1 - P_{a,\alpha} * \Phi(u) - \psi(u)$ nonincreasing in a. Since $\left(\frac{a+u}{a}\right)^{1-\alpha} = \left(\frac{1}{1+u/a}\right)^{\alpha-1}$ and $1+u/a$ is decreasing in a, it follows that $-\left(\frac{a+u}{a}\right)^{1-\alpha}$ is decreasing in a. Hence $1 - P_{a,\alpha} * \Phi(u) - \psi(u) - \left(\frac{a+u}{a}\right)^{1-\alpha}$ is decreasing in a.

From this and (4.17) we obtain, similarly as in (4.20), that

$$\lim_{u\to\infty} \frac{a^{1-\alpha}\eta^\alpha}{\Gamma(2-\alpha)}\left|1 - P_{a,\alpha} * \Phi(u) - \psi(u) - \left(\frac{a+u}{a}\right)^{1-\alpha}\right| = 0, \tag{4.24}$$

uniformly on $\{a \ge \xi\}$.

For the remaining term in (4.23), from (4.16) and (4.22) we obtain using the same argument

$$\lim_{u\to\infty} \lambda_2\mu_2\overline{F}_{2,I}(a)\left|1 - F_{a,I} * \Phi(u) - \psi(u)\right| = 0, \tag{4.25}$$

uniformly on $\{a \geq \xi\}$. Due to (4.23), (4.24) and (4.25) it follows that, for all $\varepsilon > 0$, $\xi > 0$ and $\eta > 0$ there exists $A > 0$ such that for all $u > A$ and $a > \xi, b > \eta$ we have

$$\Upsilon^*(u, a, b) - \frac{\eta^\alpha (a+u)^{1-\alpha}}{(c + \lambda_1\mu_1 - \lambda_2\mu_2)\Gamma(2-\alpha)} \leq \varepsilon. \tag{4.26}$$

Hence, the result follows from (4.21) and (4.26).

Case 2: Similarly as in the previous case, we obtain from (4.22) and (4.19)

$$\lim_{u\to\infty} \left|\frac{a^{1-\alpha}\eta^\alpha}{\Gamma(2-\alpha)}\left(1 - P_{a,\alpha} * \Phi(u) - \psi(u) - \left(\frac{a+u}{a}\right)^{1-\alpha}\right)\right| = 0, \tag{4.27}$$

uniformly in $\{a \geq \xi\}$. It also follows from (4.18) that

$$\lim_{u\to\infty} \left|\overline{F}_2(a)\left(1 - F_{a,I} * \Phi(u) - \psi(u)\right)\right| = 0, \tag{4.28}$$

uniformly in $\{a \geq \xi\}$. Hence, for all $\varepsilon, \xi, \eta > 0$ there exists an $A > 0$ such that for all $u > A$ and $a > \xi, b > \eta$ we have

$$\Upsilon^*(u, a, b) - \frac{\eta^\alpha (a+u)^{1-\alpha}}{(c + \lambda_1\mu_1 - \lambda_2\mu_2)\Gamma(2-\alpha)} \leq \varepsilon. \tag{4.29}$$

The result follows now from (4.21) and (4.29).

Case 3: By (4.12) it holds

$$\begin{aligned}\Upsilon^*(u, a, b) &- \frac{\lambda_2\mu_2}{c + \lambda_1\mu_1 - \lambda_2\mu_2}\overline{F}_{2,I}(u+a)\\ &\geq \frac{\lambda_2\mu_2}{c + \lambda_1\mu_1 - \lambda_2\mu_2}\overline{F}_{2,I}(u+a)\left(\Phi(u) - 1\right)\\ &+ \frac{\eta^\alpha}{\Gamma(2-\alpha)}(a+u)^{1-\alpha}\Phi(u).\end{aligned}$$

As in the above cases we obtain that, for all ε, ξ, $\eta > 0$ there exists an $A > 0$ such that for all $u > A$ and $a > \xi$, $b > \eta$ we have

$$\Upsilon^*(u, a, b) - \frac{\lambda_2\mu_2}{c + \lambda_1\mu_1 - \lambda_2\mu_2}\overline{F}_{2,I}(u+a) \geq -\varepsilon. \tag{4.30}$$

On the other hand,

$$\begin{aligned}
&\Upsilon^*(u,a,b) - \frac{\lambda_2\mu_2}{c+\lambda_1\mu_1-\lambda_2\mu_2}\overline{F}_{2,I}(u+a) \le \Upsilon_0(u,a) - \frac{\lambda_2\mu_2}{c+\lambda_1\mu_1-\lambda_2\mu_2}\overline{F}_2(a+u)\\
&= \frac{1}{c+\lambda_1\mu_1-\lambda_2\mu_2}\Big\{\lambda_2\mu_2\left[\overline{F}_{2,I}(a)\left(1-F_{a,I}*\Phi(u)-\psi(u)\right)-\overline{F}_{2,I}(a+u)\right]\\
&+\frac{a^{1-\alpha}\eta^\alpha}{\Gamma(2-\alpha)}\left(1-P_{a,\alpha}*\Phi(u)-\psi(u)\right)\Big\}. \qquad (4.31)
\end{aligned}$$

Since in this case we have $\lim_{u\to\infty}(1-P_{a,\alpha}*\Phi(u)-\psi(u))=0$, using that $1-P_{a,\alpha}*\Phi(u)$ is decreasing in a, it follows that

$$\lim_{u\to\infty}\left|\frac{a^{1-\alpha}\eta^\alpha}{\Gamma(2-\alpha)}\left(1-P_{a,\alpha}*\Phi(u)-\psi(u)\right)\right| = 0, \qquad (4.32)$$

uniformly on $\{a \ge \xi\}$, for any $\xi > 0$. For the remaining term in (4.31), there holds

$$\begin{aligned}
&\overline{F}_{2,I}(a)\left(1-F_{a,I}*\Phi(u)-\psi(u)\right)\\
&= \overline{F}_{2,I}(a)\left[\Phi(u)-F_{a,I}*\Phi(u)\right] = \overline{F}_{2,I}(a)\int_0^u\left(1-F_{a,I}(u-y)\right)\Phi(dy)\\
&= \overline{F}_{2,I}(a)\int_0^u\overline{F}_{a,I}(u-y)\Phi(dy) = \overline{F}_{2,I}(a)\int_0^u\frac{\overline{F}_{2,I}(a+u-y)}{\overline{F}_{2,I}(a)}\Phi(dy)\\
&= (\overline{F}_{2,I}(a+\cdot)*\Phi)(u). \qquad (4.33)
\end{aligned}$$

Hence $\lim_{u\to\infty}\left(\overline{F}_{2,I}(a+\cdot)*\Phi\right)(u)=0$ uniformly on $\{a\ge\xi\}$ for $\xi>0$. Since

$$\begin{aligned}
&\overline{F}_{2,I}(a)\left|1-F_{a,I}*\Phi(u)-\psi(u)-\overline{F}_{2,I}(a+u)\right|\\
&\le\left[(\overline{F}_{2,I}(a+\cdot)*\Phi)(u)+\overline{F}_{2,I}(a)\overline{F}_{2,I}(a+u)\right],
\end{aligned}$$

and $\lim_{u\to\infty}\left[(\overline{F}_{2,I}(a+\cdot)*\Phi)(u)+\overline{F}_{2,I}(a)\overline{F}_{2,I}(a+u)\right]=0$ uniformly on $\{a\ge\xi\}$ for $\xi>0$, we obtain that

$$\lim_{u\to\infty}\overline{F}_{2,I}(a)\left|1-F_{a,I}*\Phi(u)-\psi(u)-\overline{F}_{2,I}(a+u)\right|=0, \qquad (4.34)$$

uniformly on $\{a\ge\xi\}$, $\xi>0$. Using (4.31), (4.32) and (4.34) we obtain that, for all $\varepsilon,\xi,\eta>0$ there exists $A>0$ such that for all $u>A$ and $a>\xi, b>\eta$, it follows

$$\Upsilon^*(u,a,b)-\frac{\lambda_2\mu_2}{c+\lambda_1\mu_1-\lambda_2\mu_2}\overline{F}_{2,I}(u+a)\le\varepsilon. \qquad (4.35)$$

The result follows now from (4.30) and (4.35). ■

Acknowledgements The authors express their gratitude to an anonymous referee whose careful revision and suggestions greatly improved the presentation of the paper. This work was partially supported by CONACyT (Mexico) Grant No. 257867.

References

1. E. Biffis, A. Kyprianou, A note on scale functions and the time value of ruin for Lévy insurance risk processes. Insur. Math. Econ. **46**(1), 85–91 (2010)
2. E. Biffis, M. Morales, On a generalization of the Gerber-Shiu function to path-dependent penalties. Insur. Math. Econ. **46**(1), 92–97 (2010)
3. N.H. Bingham, C.M. Goldie, J.L. Teugels, *Regular Variation* (Cambridge University Press, Cambridge, 1987)
4. N. Bowers, H. Gerber, J. Hickman, D. Jones, C. Nesbitt, *Actuarial Mathematics* (Society of Actuaries, Schaumburg, 1986)
5. D. Dickson, C. Hipp, On the time to ruin for Erlang(2) risk processes. 4th IME Conference (Barcelona, 2000). Insur. Math. Econ. **29**(3), 333–344 (2001)
6. R. Doney, C. Klüppelberg, R. Maller, Passage time and fluctuation calculations for subexponential Lévy processes. *Bernoulli* **22**(3), 1491–1519 (2016)
7. D. Dufresne, Fitting combinations of exponentials to probability distributions. Appl. Stoch. Models Bus. Ind. **23**(1), 23–48 (2007)
8. P. Embrechts, C. Goldie, N. Veraverbeke, Subexponentiality and infinite divisibility. Z. Wahrsch. Verw. Gebiete **49**(3), 335–347 (1979)
9. P. Embrechts, C. Klüppelberg, T. Mikosch, *Modelling Extremal Events for Insurance and Finance* (Springer, Berlin, 1997)
10. W. Feller, *An Introduction to Probability Theory and Its Applications*, vol. II (Wiley, New York, 1971)
11. H. Furrer, Risk processes perturbed by α-stable Lévy motion. Scand. Actuar. J. **1998**(1), 59–74 (1998)
12. H.U. Gerber, E.S.W. Shiu, On the time value of ruin. With discussion and a reply by the authors. North Am. Actuar. J. **2**(1), 48–78 (1998)
13. C. Klüppelberg, A. Kyprianou, M. Maller, Ruin probabilities and overshoots for general Lévy insurance risk processes. Ann. Appl. Probab. **14**(4), 1766–1801 (2004)
14. E.T. Kolkovska, E.M. Martín-González, Gerber-Shiu functionals for classical risk processes perturbed by an α-stable motion. Insur. Math. Econ. **66**, 22–28 (2016)
15. E.T. Kolkovska, E.M. Martín-González, Path functionals of a class of Lévy insurance risk processes. Commun. Stoch. Anal. **10**(3), 363–387 (2016)
16. T. Rolski, H. Schmidli, V. Schmidt, J. Teugels, *Stochastic Processes for Insurance and Finance* (Wiley, New York, 1999)

Characterization of the Minimal Penalty of a Convex Risk Measure with Applications to Robust Utility Maximization for Lévy Models

Daniel Hernández-Hernández and Leonel Pérez-Hernández

Abstract The minimality of the penalty function associated with a convex risk measure is analyzed in this paper. First, in a general static framework, we provide necessary and sufficient conditions for a penalty function defined in a convex and closed subset of the absolutely continuous measures with respect to some reference measure $\mathbb{P}$ to be minimal on this set. When the probability space supports a Lévy process, we establish results that guarantee the minimality property of a penalty function described in terms of the coefficients associated with the density processes. These results are applied in the solution of the robust utility maximization problem for a market model based on Lévy processes.

Keywords Convex risk measures · Fenchel-Legendre transformation · Minimal penalization · Lévy process · Robust utility maximization

Mathematics Subject Classification 91B30, 46E30

1 Introduction

The definition of coherent risk measure was introduced by Artzner et al. in their fundamental works [1, 2] for finite probability spaces, giving an axiomatic characterization that was extended later by Delbaen [3] to general probability spaces. In the papers mentioned above one of the fundamental axioms was the positive homogeneity, and in further works it was removed, defining the concept of

D. Hernández-Hernández
Centro de Investigaci6n en Matemáticas, Guanajuato, Mexico
e-mail: dher@cimat.mx

L. Pérez-Hernández (✉)
Departamento de Economía y Finanzas, Universidad de Guanajuato, Guanajuato, Mexico

© Springer International Publishing AG, part of Springer Nature 2018

D. Hernández-Hernández et al. (eds.), *XII Symposium of Probability and Stochastic Processes*, Progress in Probability 73,
https://doi.org/10.1007/978-3-319-77643-9_4

convex risk measure introduced by Föllmer and Schied [4, 5], Frittelli and Rosazza Gianin [7, 8] and Heath [10].

This is a rich area that has received a lot of attention and much work has been developed. There exists by now a well established theory in the static and dynamic cases, but there are still many questions unanswered in the static framework that need to be analyzed carefully. The one we focus on in this paper is the characterization of the penalty functions that are minimal for the corresponding static risk measure. Up to now, there are mainly two ways to deal with minimal penalty functions, namely the definition or the biduality relation. With the results presented in this paper we can start with a penalty function, which essentially discriminate models within a convex closed subset of absolutely continuous probability measures with respect to (w.r.t.) the market measure, and then guarantee that it corresponds to the minimal penalty of the corresponding convex risk measure on this subset. This property is, as we will see, closely related with the lower semicontinuity of the penalty function, and the complications to prove this property depend on the structure of the probability space.

We first provide a general framework, within a measurable space with a reference probability measure $\mathbb{P}$, and show necessary and sufficient conditions for a penalty function defined in a convex and closed subset of the absolutely continuous measures with respect to the reference measure to be minimal within this subset. The characterization of the form of the penalty functions that are minimal when the probability space supports a Lévy process is then studied. This requires to characterize the set of absolutely continuous measures for this space, and it is done using results that describe the density process for spaces which support semimartingales with the weak predictable representation property. Roughly speaking, using the weak representation property, every density process splits in two parts, one is related with the continuous local martingale part of the decomposition and the other with the corresponding discontinuous one. It is shown some kind of continuity property for the quadratic variation of a sequence of densities converging in L^1. From this characterization of the densities, a family of penalty functions is proposed, which turned out to be minimal for the risk measures generated by duality.

The previous results are applied to the solution of the robust utility maximization problem. The formulation of this problem, described formally in Sect. 6, is justified by the axiomatic system proposed by Maccheroni et al. [17], which led to utility functionals of the form

$$X \longrightarrow \inf_{\mathbb{Q}\in\mathcal{Q}'} \left\{\mathbb{E}_{\mathbb{Q}}\left[U\left(X\right)\right] + \vartheta\left(\mathbb{Q}\right)\right\}. \tag{1.1}$$

The elements of this display will be described in detail in the last section. For previous works on this direction we refer the interested reader to the works of Quenez [18], Schied [19] and Hernández-Hernández and Schied [11], and references therein.

The paper is organized as follows. Section 2 contains the description of the minimal penalty functions for a general probability space, providing necessary and sufficient conditions, the last one restricted to a subset of equivalent probability measures. Section 3 reports the structure of the densities for a probability space that supports a Lévy processes and the convergence properties needed to prove the lower semicontinuity of the set of penalty functions defined in Sect. 4. In this section we show that these penalty functions are minimal. The description of the market model is presented in Sect. 5, together with the characterization of the equivalent martingale measures and, finally, in the last section we solve the robust utility maximization problem using duality theory.

2 Minimal Penalty Function of Risk Measures Concentrated in $\mathcal{Q}_{\ll}(\mathbb{P})$

Given a penalty function ψ, it is possible to induce a convex risk measure ρ, which in turn has a representation by means of a minimal penalty function ψ_ρ^*. Starting with a penalty function ψ, we give in this section necessary and sufficient conditions in order to guarantee that it is the minimal penalty within the set of absolutely continuous probability measures. We begin recalling briefly some known results from the theory of static risk measures, and then a characterization for minimal penalties is presented.

2.1 *Preliminaries from Static Measures of Risk*

Let $X : \Omega \to \mathbb{R}$ be a mapping from a set Ω of possible market scenarios, representing the discounted net worth of the position. Uncertainty is represented by the measurable space $(\Omega, \mathcal{F})$, and we denote by $\mathcal{X}$ the linear space of bounded financial positions, including constant functions.

Definition 2.1

(i) The function $\rho : \mathcal{X} \to \mathbb{R}$, quantifying the risk of X, is a *monetary risk measure* if it satisfies the following properties:

$$\text{Monotonicity: If } X \leq Y \text{ then } \rho(X) \geq \rho(Y) \ \forall X, Y \in \mathcal{X}. \tag{2.1}$$

$$\text{Translation Invariance: } \rho(X + a) = \rho(X) - a \ \forall a \in \mathbb{R} \ \forall X \in \mathcal{X}. \tag{2.2}$$

(ii) When this function satisfies also the convexity property

$$\rho(\lambda X + (1-\lambda) Y) \leq \lambda \rho(X) + (1-\lambda) \rho(Y) \ \forall \lambda \in [0,1] \ \forall X, Y \in \mathcal{X}, \tag{2.3}$$

it is said that ρ is a convex risk measure.

(iii) The function ρ is called normalized if $\rho(0) = 0$, and sensitive, with respect to a measure $\mathbb{P}$, when for each $X \in L_{+}^{\infty}(\mathbb{P})$ with $\mathbb{P}[X > 0] > 0$ we have that $\rho(-X) > \rho(0)$.

We say that a set function $\mathbb{Q} : \mathcal{F} \to [0, 1]$ is a *probability content* if it is finitely additive and $\mathbb{Q}(\Omega) = 1$. The set of *probability contents* on this measurable space is denoted by $\mathcal{Q}_{cont}$. From the general theory of static convex risk measures [6], we know that any map $\psi : \mathcal{Q}_{cont} \to \mathbb{R} \cup \{+\infty\}$, with $\inf_{\mathbb{Q}\in\mathcal{Q}_{cont}} \psi(\mathbb{Q}) \in \mathbb{R}$, induces a static convex measure of risk as a mapping $\rho : \mathfrak{M}_b \to \mathbb{R}$ given by

$$\rho(X) := \sup_{\mathbb{Q}\in\mathcal{Q}_{cont}} \left\{\mathbb{E}_{\mathbb{Q}}[-X] - \psi(\mathbb{Q})\right\}. \tag{2.4}$$

Here $\mathfrak{M}$ denotes the class of measurable functions and $\mathfrak{M}_b$ the subclass of bounded measurable functions. The function ψ will be referred as a *penalty function*. Föllmer and Schied [5, Theorem 3.2] and Frittelli and Rosazza Gianin [7, Corollary 7] proved that any convex risk measure is essentially of this form.

More precisely, a convex risk measure ρ on the space $\mathfrak{M}_b(\Omega, \mathcal{F})$ has the representation

$$\rho(X) = \sup_{\mathbb{Q}\in\mathcal{Q}_{cont}} \left\{\mathbb{E}_{\mathbb{Q}}[-X] - \psi_\rho^*(\mathbb{Q})\right\}, \tag{2.5}$$

where

$$\psi_\rho^*(\mathbb{Q}) := \sup_{X\in\mathcal{A}_\rho} \mathbb{E}_{\mathbb{Q}}[-X], \tag{2.6}$$

and $\mathcal{A}_\rho := \{X \in \mathfrak{M}_b : \rho(X) \leq 0\}$ is the *acceptance set* of ρ.

Remark 2.1 The penalty ψ_ρ^* is called the *minimal penalty function* associated to ρ because, for any other penalty function ψ fulfilling (2.4), $\psi(\mathbb{Q}) \geq \psi_\rho^*(\mathbb{Q})$, for all $\mathbb{Q} \in \mathcal{Q}_{cont}$. Furthermore, for the minimal penalty function, the next biduality relation is satisfied

$$\psi_\rho^*(\mathbb{Q}) = \sup_{X\in\mathfrak{M}_b(\Omega,\mathcal{F})} \left\{\mathbb{E}_{\mathbb{Q}}[-X] - \rho(X)\right\}, \quad \forall\mathbb{Q} \in\mathcal{Q}_{cont}. \tag{2.7}$$

Let $\mathcal{Q}(\Omega, \mathcal{F})$ be the family of probability measures on the measurable space $(\Omega, \mathcal{F})$. Among the measures of risk, the class of them which representation in (2.5) is concentrated on the set of probability measures $\mathcal{Q} \subset \mathcal{Q}_{cont}$ are of special interest. Recall that a function $I : E \subset \mathbb{R}^\Omega \to \mathbb{R}$ is *sequentially continuous from below (above)* when $\{X_n\}_{n\in\mathbb{N}} \uparrow X \Rightarrow \lim_{n\to\infty} I(X_n) = I(X)$ (respectively $\{X_n\}_{n\in\mathbb{N}} \downarrow X \Rightarrow \lim_{n\to\infty} I(X_n) = I(X)$). Föllmer and Schied [6] proved that any sequentially continuous from below convex measure of risk is concentrated on the set $\mathcal{Q}$. Later, Krätschmer [15, Prop. 3 p. 601] established that the sequential continuity from below is not only a sufficient but also a necessary condition in order to have a representation, by means of the minimal penalty function in terms of probability measures.

We denote by $\mathcal{Q}_{\ll}(\mathbb{P})$ the subclass of absolutely continuous probability measure with respect to $\mathbb{P}$ and by $\mathcal{Q}_{\approx}(\mathbb{P})$ the subclass of equivalent probability measure. Of course, $\mathcal{Q}_{\approx}(\mathbb{P}) \subset \mathcal{Q}_{\ll}(\mathbb{P}) \subset \mathcal{Q}(\Omega, \mathcal{F})$.

Remark 2.2 When a convex risk measures in $\mathcal{X} := L^{\infty}(\mathbb{P})$ satisfies the property

$$\rho(X) = \rho(Y) \text{ if } X = Y \ \mathbb{P}\text{-a.s.} \tag{2.8}$$

and is represented by a penalty function ψ as in (2.4), we have that

$$\mathbb{Q} \in \mathcal{Q}_{cont} \setminus \mathcal{Q}_{cont}^{\ll} \Longrightarrow \psi(\mathbb{Q}) = +\infty, \tag{2.9}$$

where $\mathcal{Q}_{cont}^{\ll}$ is the set of contents absolutely continuous with respect to $\mathbb{P}$; see [6, Lemma 4.30 p. 172].

2.2 Minimal Penalty Functions

In the next sections we will show some of the difficulties that appear to prove the minimality of the penalty function when the probability space $(\Omega, \mathcal{F}, \mathbb{P})$ supports a Lévy process. We will also clarify the relevance of this property to get an optimal solution to the robust utility maximization problem in Sect. 6.

In order to establish the results of this section we only need to fix a probability space $(\Omega, \mathcal{F}, \mathbb{P})$. When we deal with a set of absolutely continuous probability measures $\mathcal{K} \subset \mathcal{Q}_{\ll}(\mathbb{P})$ it is necessary to make reference to some topological concepts, meaning that we are considering the corresponding set of densities and the strong topology in $L^1(\mathbb{P})$. Recall that within a locally convex space, a convex set $\mathcal{K}$ is weakly closed if and only if $\mathcal{K}$ is closed in the original topology [6, Thm A.59].

Lemma 2.1 *Let $\psi : \mathcal{K} \subset \mathcal{Q}_{\ll}(\mathbb{P}) \to \mathbb{R} \cup \{+\infty\}$ be a function with $\inf_{\mathbb{Q} \in \mathcal{K}} \psi(\mathbb{Q}) \in \mathbb{R}$, and define the extension $\psi(\mathbb{Q}) := \infty$ for each $\mathbb{Q} \in \mathcal{Q}_{cont} \setminus \mathcal{K}$, with $\mathcal{K}$ a convex closed set. Also, define the function Ψ, with domain in $L^1(\mathbb{P})$, as*

$$\Psi(D) := \begin{cases} \psi(\mathbb{Q}) \text{ if } D = d\mathbb{Q}/d\mathbb{P} \text{ for } \mathbb{Q} \in \mathcal{K} \\ \infty \text{ otherwise.} \end{cases}$$

Then, for the convex measure of risk $\rho(X) := \sup_{\mathbb{Q} \in \mathcal{Q}_{cont}} \{\mathbb{E}_{\mathbb{Q}}[-X] - \psi(\mathbb{Q})\}$ associated with ψ the following assertions hold:

(a) If ρ has as minimal penalty ψ_{ρ}^{*} the function ψ (i.e. $\psi = \psi_{\rho}^{*}$), then Ψ is a proper convex function and lower semicontinuous w.r.t. the (strong) L^1-topology or equivalently w.r.t. the weak topology $\sigma(L^1, L^{\infty})$.

(b) *If Ψ is convex and lower semicontinuous w.r.t. the (strong) L^1-topology or equivalently w.r.t. the weak topology $\sigma\left(L^1, L^\infty\right)$, then*

$$\psi \mathbf{1}_{\mathcal{Q}_{\ll}(\mathbb{P})} = \psi_\rho^* \mathbf{1}_{\mathcal{Q}_{\ll}(\mathbb{P})}. \tag{2.10}$$

Proof

(a) Recall that $\sigma\left(L^1, L^\infty\right)$ is the coarsest topology on $L^1(\mathbb{P})$ under which every linear operator is continuous, and hence $\Psi_0^X(Z) := \mathbb{E}_{\mathbb{P}}[Z(-X)]$, with $Z \in L^1$, is a continuous function for each $X \in \mathfrak{M}_b(\Omega, \mathcal{F})$ fixed. For $\delta(\mathcal{K}) := \{Z : Z = d\mathbb{Q}/d\mathbb{P} \text{ with } \mathbb{Q} \in \mathcal{K}\}$ we have that

$$\Psi_1^X(Z) := \Psi_0^X(Z)\mathbf{1}_{\delta(\mathcal{K})}(Z) + \infty \times \mathbf{1}_{L^1 \setminus \delta(\mathcal{K})}(Z)$$

is clearly lower semicontinuous on $\delta(\mathcal{K})$. For $Z' \in L^1(\mathbb{P}) \setminus \delta(\mathcal{K})$ arbitrary fixed we have from Hahn-Banach's Theorem that there is a continuous lineal functional $l(Z)$ with $l\left(Z'\right) < \inf_{Z \in \delta(\mathcal{K})} l(Z)$. Taking $\varepsilon := \frac{1}{2}\left\{\inf_{Z \in \delta(\mathcal{K})} l(Z) - l\left(Z'\right)\right\}$ we have that the weak open ball $B\left(Z', \varepsilon\right) := \left\{Z \in L^1(\mathbb{P}) : \left|l\left(Z'\right) - l(Z)\right| < \varepsilon\right\}$ satisfies $B\left(Z', \varepsilon\right) \cap \delta(\mathcal{K}) = \varnothing$. Therefore, $\Psi_1^X(Z)$ is weak lower semicontinuous on $L^1(\mathbb{P})$, as well as $\Psi_2^X(Z) := \Psi_1^X(Z) - \rho(X)$. If

$$\psi(\mathbb{Q}) = \psi_\rho^*(\mathbb{Q}) = \sup_{X \in \mathfrak{M}_b(\Omega, \mathcal{F})} \left\{\int Z(-X)\,d\mathbb{P} - \rho(X)\right\},$$

where $Z := d\mathbb{Q}/d\mathbb{P}$, we have that $\Psi(Z) = \sup_{X \in \mathfrak{M}_b(\Omega, \mathcal{F})}\left\{\Psi_2^X(Z)\right\}$ is the supremum of a family of convex lower semicontinuous functions with respect to the topology $\sigma\left(L^1, L^\infty\right)$, and $\Psi(Z)$ preserves both properties.

(b) For the Fenchel–Legendre transform (conjugate function) $\Psi^* : L^\infty(\mathbb{P}) \longrightarrow \mathbb{R}$ for each $U \in L^\infty(\mathbb{P})$

$$\Psi^*(U) = \sup_{Z \in \delta(\mathcal{K})} \left\{\int ZU\,d\mathbb{P} - \Psi(Z)\right\} = \sup_{\mathbb{Q} \in \mathcal{Q}_{cont}} \left\{\mathbb{E}_{\mathbb{Q}}[U] - \psi(\mathbb{Q})\right\} \equiv \rho(-U).$$

From the lower semicontinuity of Ψ w.r.t. the weak topology $\sigma\left(L^1, L^\infty\right)$ that $\Psi = \Psi^{**}$. Considering the weak*-topology $\sigma\left(L^\infty(\mathbb{P}), L^1(\mathbb{P})\right)$ for $Z = d\mathbb{Q}/d\mathbb{P}$ we have that

$$\psi(\mathbb{Q}) = \Psi(Z) = \Psi^{**}(Z) = \sup_{U \in L^\infty(\mathbb{P})} \left\{\int Z(-U)\,d\mathbb{P} - \Psi^*(-U)\right\} = \psi_\rho^*(\mathbb{Q}).$$

□

Remark 2.3

1. As it was pointed out in Remark 2.2, we have that

$$\mathbb{Q} \in \mathcal{Q}_{cont} \setminus \mathcal{Q}_{cont}^{\ll} \Longrightarrow \psi_{\rho}^{*}(\mathbb{Q}) = +\infty = \psi(\mathbb{Q}).$$

Therefore, under the conditions of Lemma 2.1 (b) the penalty function ψ might differ from ψ_{ρ}^{*} on $\mathcal{Q}_{cont}^{\ll} \setminus \mathcal{Q}_{\ll}$. For instance, the penalty function defined as $\psi(\mathbb{Q}) := \infty \times \mathbf{1}_{\mathcal{Q}_{cont} \setminus \mathcal{Q}_{\ll}(\mathbb{P})}(\mathbb{Q})$ leads to the worst case risk measure $\rho(X) := \sup_{\mathbb{Q} \in \mathcal{Q}_{\ll}(\mathbb{P})} \mathbb{E}_{\mathbb{Q}}[-X]$, which has as minimal penalty the function

$$\psi_{\rho}^{*}(\mathbb{Q}) = \infty \times \mathbf{1}_{\mathcal{Q}_{cont} \setminus \mathcal{Q}_{cont}^{\ll}}(\mathbb{Q}).$$

2. Note that the total variation distance $d_{TV}\left(\mathbb{Q}^1, \mathbb{Q}^2\right) := \sup_{A \in \mathcal{F}} \left|\mathbb{Q}^1[A] - \mathbb{Q}^2[A]\right|$, with $\mathbb{Q}^1, \mathbb{Q}^2 \in \mathcal{Q}_{\ll}$, fulfills that $d_{TV}\left(\mathbb{Q}^1, \mathbb{Q}^2\right) \leq \left\|d\mathbb{Q}^1/d\mathbb{P} - d\mathbb{Q}^2/d\mathbb{P}\right\|_{L^1}$. Therefore, the minimal penalty function is lower semicontinuous in the total variation topology; see Remark 4.16 (b) p. 163 in [6].

3 Fundamentals of Lévy and Semimartingales Processes

Let $(\Omega, \mathcal{F}, \mathbb{P})$ be a probability space. We say that $L := \{L_t\}_{t \in \mathbb{R}_+}$ is a Lévy process for this probability space if it is an adapted càdlàg process with independent stationary increments starting at zero. The filtration considered is $\mathbb{F} := \left\{\mathcal{F}_t^{\mathbb{P}}(L)\right\}_{t \in \mathbb{R}_+}$, the completion of its natural filtration, i.e. $\mathcal{F}_t^{\mathbb{P}}(L) := \sigma\{L_s : s \leq t\} \vee \mathcal{N}$ where $\mathcal{N}$ is the σ-algebra generated by all $\mathbb{P}$-null sets. The jump measure of L is denoted by $\mu : \Omega \times (\mathcal{B}(\mathbb{R}_+) \otimes \mathcal{B}(\mathbb{R}_0)) \to \mathbb{N}$ where $\mathbb{R}_0 := \mathbb{R} \setminus \{0\}$. The dual predictable projection of this measure, also known as its Lévy system, satisfies the relation $\mu^{\mathcal{P}}(dt, dx) = dt \times \nu(dx)$, where $\nu(\cdot) := \mathbb{E}[\mu([0,1] \times \cdot)]$ is the intensity or Lévy measure of L.

The Lévy-Itô decomposition of L is given by

$$L_t = bt + W_t + \int_{[0,t] \times \{0<|x|\leq 1\}} x d\left\{\mu - \mu^{\mathcal{P}}\right\} + \int_{[0,t] \times \{|x|>1\}} x\mu(ds, dx). \tag{3.1}$$

It implies that $L^c = W$ is the Wiener process, and hence $[L^c]_t = t$, where $(\cdot)^c$ and $[\cdot]$ denote the continuous martingale part and the process of quadratic variation of any semimartingale, respectively. For the predictable quadratic variation we use the notation $\langle \cdot \rangle$.

Denote by $\mathcal{V}$ the set of càdlàg, adapted processes with finite variation, and let $\mathcal{V}^+ \subset \mathcal{V}$ be the subset of non-decreasing processes in $\mathcal{V}$ starting at zero. Let $\mathcal{A} \subset \mathcal{V}$ be the class of processes with integrable variation, i.e. $A \in \mathcal{A}$ if and only if $\bigvee_0^{\infty} A \in$

$L^1(\mathbb{P})$, where $\bigvee_0^t A$ denotes the variation of A over the finite interval $[0,t]$. The subset $\mathcal{A}^+ = \mathcal{A} \cap \mathcal{V}^+$ represents those processes which are also increasing i.e. with non-negative right-continuous increasing trajectories. Furthermore, $\mathcal{A}_{loc}$ (resp. $\mathcal{A}_{loc}^+$) is the collection of adapted processes with locally integrable variation (resp. adapted locally integrable increasing processes). For a càdlàg process X we denote by $X_- := (X_{t-})$ the left hand limit process, where $X_{0-} := X_0$ by convention, and by $\triangle X = (\triangle X_t)$ the jump process $\triangle X_t := X_t - X_{t-}$.

Given an adapted càdlàg semimartingale U, the jump measure and its dual predictable projection (or compensator) are denoted by $\mu_U([0,t] \times A) := \sum_{s \le t} \mathbf{1}_A(\triangle U_s)$ and $\mu_U^{\mathcal{P}}$, respectively. Further, we denote by $\mathcal{P} \subset \mathcal{F} \otimes \mathcal{B}(\mathbb{R}_+)$ the predictable σ-algebra and by $\widetilde{\mathcal{P}} := \mathcal{P} \otimes \mathcal{B}(\mathbb{R}_0)$. With some abuse of notation, we write $\theta_1 \in \widetilde{\mathcal{P}}$ when the function $\theta_1 : \Omega \times \mathbb{R}_+ \times \mathbb{R}_0 \to \mathbb{R}$ is $\widetilde{\mathcal{P}}$-measurable and $\theta \in \mathcal{P}$ for predictable processes.

Let

$$\begin{aligned} \mathcal{L}(U^c) := \Big\{\theta \in \mathcal{P} : \exists \{\tau_n\}_{n \in \mathbb{N}} & \text{ sequence of stopping times with } \tau_n \uparrow \infty \\ & \text{and } \mathbb{E}\left[\int_0^{\tau_n} \theta^2 d[U^c]\right] < \infty \ \forall n \in \mathbb{N}\Big\} \end{aligned} \tag{3.2}$$

be the class of predictable processes $\theta \in \mathcal{P}$ integrable with respect to U^c in the sense of local martingale, and by

$$\Lambda(U^c) := \left\{\int \theta_0 dU^c : \theta_0 \in \mathcal{L}(U^c)\right\}$$

the linear space of processes which admits a representation as the stochastic integral with respect to U^c. For an integer valued random measure μ' we denote by $\mathcal{G}(\mu')$ the class of functions $\theta_1 : \Omega \times \mathbb{R}_+ \times \mathbb{R}_0 \to \mathbb{R}$ satisfying the following conditions:

(i) $\theta_1 \in \widetilde{\mathcal{P}}$,

(ii) $\int_{\mathbb{R}_0} |\theta_1(t,x)| (\mu')^{\mathcal{P}}(\{t\}, dx) < \infty \ \forall t > 0$,

(iii) The process

$$\left\{\sqrt{\sum_{s \le t}\left\{\int_{\mathbb{R}_0} \theta_1(s,x)\mu'(\{s\},dx) - \int_{\mathbb{R}_0} \theta_1(s,x)(\mu')^{\mathcal{P}}(\{s\},dx)\right\}^2}\right\}_{t \in \mathbb{R}_+} \in \mathcal{A}_{loc}^+.$$

The set $\mathcal{G}(\mu')$ represents the domain of the functional $\theta_1 \to \int \theta_1 d\left(\mu' - (\mu')^{\mathcal{P}}\right)$, which assign to θ_1 the unique purely discontinuous local martingale M with

$$\triangle M_t = \int_{\mathbb{R}_0} \theta_1(t,x)\mu'(\{t\},dx) - \int_{\mathbb{R}_0} \theta_1(t,x)\left(\mu'\right)^{\mathcal{P}}(\{t\},dx).$$

We use the notation $\int \theta_1 d\left(\mu' - (\mu')^{\mathcal{P}}\right)$ to write the value of this functional in θ_1. It is important to point out that this functional is not, in general, the integral with respect to the difference of two measures. For a detailed exposition on these topics see He et al. [9] or Jacod and Shiryaev [12], which are our basic references.

In particular, for the Lévy process L with jump measure μ,

$$\mathcal{G}(\mu) \equiv \left\{ \theta_1 \in \widetilde{\mathcal{P}} : \left\{ \sqrt{\sum_{s\leq t} \{\theta_1(s, \Delta L_s)\}^2 \mathbf{1}_{\mathbb{R}_0}(\Delta L_s)} \right\}_{t\in\mathbb{R}_+} \in \mathcal{A}^+_{loc} \right\}, \tag{3.3}$$

since $\mu^{\mathcal{P}}(\{t\} \times A) = 0$, for any Borel set A of $\mathbb{R}_0$.

We say that the semimartingale U has the *weak property of predictable representation* when

$$\mathcal{M}_{loc,0} = \Lambda\left(U^c\right) + \left\{ \int \theta_1 d\left(\mu_U - \mu_U^{\mathcal{P}}\right) : \theta_1 \in \mathcal{G}(\mu_U) \right\}, \tag{3.4}$$

where the previous sum is the linear sum of the vector spaces, and $\mathcal{M}_{loc,0}$ is the linear space of local martingales starting at zero.

Let $\mathcal{M}$ and $\mathcal{M}_\infty$ denote the class of càdlàg and càdlàg uniformly integrable martingale respectively. The following lemma is interesting by itself to understand the continuity properties of the quadratic variation for a given convergent sequence of uniformly integrable martingale . It will play a central role in the proof of the lower semicontinuity of the penalization function introduced in Sect. 4. Observe that the assertion of this lemma is valid in a general filtered probability space and not only for the completed natural filtration of the Lévy process introduced above.

Lemma 3.1 *For* $\left\{M^{(n)}\right\}_{n\in\mathbb{N}} \subset \mathcal{M}_\infty$ *and* $M \in \mathcal{M}_\infty$ *the following implication holds*

$$M_\infty^{(n)} \underset{n\to\infty}{\overset{L^1}{\longrightarrow}} M_\infty \Longrightarrow \left[M^{(n)} - M\right]_\infty \overset{\mathbb{P}}{\longrightarrow} 0.$$

Moreover,

$$M_\infty^{(n)} \underset{n\to\infty}{\overset{L^1}{\longrightarrow}} M_\infty \Longrightarrow \left[M^{(n)} - M\right]_t \underset{n\to\infty}{\overset{\mathbb{P}}{\longrightarrow}} 0 \ \forall t.$$

Proof From the L^1 convergence of $M_\infty^{(n)}$ to M_∞, we have that $\{M_\infty^{(n)}\}_{n\in\mathbb{N}} \cup \{M_\infty\}$ is uniformly integrable, which is equivalent to the existence of a convex and increasing function $G : [0, +\infty) \to [0, +\infty)$ such that

$$(i) \quad \lim_{x\to\infty} \frac{G(x)}{x} = \infty,$$

and

$$(ii) \quad \sup_{n\in\mathbb{N}} \mathbb{E}\left[G\left(\left|M_\infty^{(n)}\right|\right)\right] \vee \mathbb{E}\left[G\left(|M_\infty|\right)\right] < \infty.$$

Now, define the stopping times

$$\tau_k^n := \inf\left\{u > 0 : \sup_{t\leq u}\left|M_t^{(n)} - M_t\right| \geq k\right\}.$$

Observe that the estimation $\sup_{n\in\mathbb{N}} \mathbb{E}\left[G\left(\left|M_{\tau_k^n}^{(n)}\right|\right)\right] \leq \sup_{n\in\mathbb{N}} \mathbb{E}\left[G\left(\left|M_\infty^{(n)}\right|\right)\right]$ implies the uniformly integrability of $\left\{M_{\tau_k^n}^{(n)}\right\}_{n\in\mathbb{N}}$ for each k fixed. Since any uniformly integrable càdlàg martingale is of class $\mathcal{D}$, follows the uniform integrability of $\left\{M_{\tau_k^n}\right\}_{n\in\mathbb{N}}$ for all $k \in \mathbb{N}$, and hence $\left\{\sup_{t\leq\tau_k^n}\left|M_t^{(n)} - M_t\right|\right\}_{n\in\mathbb{N}}$ is uniformly integrable. This and the maximal inequality for supermartingales

$$\begin{aligned}\mathbb{P}\left[\sup_{t\in\mathbb{R}_+}\left|M_t^{(n)} - M_t\right| \geq \varepsilon\right] &\leq \frac{1}{\varepsilon}\left\{\sup_{t\in\mathbb{R}_+}\mathbb{E}\left[\left|M_t^{(n)} - M_t\right|\right]\right\} \\ &\leq \frac{1}{\varepsilon}\mathbb{E}\left[\left|M_\infty^{(n)} - M_\infty\right|\right] \longrightarrow 0,\end{aligned}$$

yields the convergence of $\left\{\sup_{t\leq\tau_k^n}\left|M_t^{(n)} - M_t\right|\right\}_{n\in\mathbb{N}}$ in L^1 to 0. The second Davis' inequality [9, Thm. 10.28] guarantees that, for some constant C,

$$\mathbb{E}\left[\sqrt{\left[M^{(n)} - M\right]_{\tau_k^n}}\right] \leq C\mathbb{E}\left[\sup_{t\leq\tau_k^n}\left|M_t^{(n)} - M_t\right|\right] \underset{n\to\infty}{\longrightarrow} 0 \quad \forall k \in \mathbb{N},$$

and hence $\left[M^{(n)} - M\right]_{\tau_k^n} \overset{\mathbb{P}}{\underset{n\to\infty}{\longrightarrow}} 0$ for all $k \in \mathbb{N}$.

Finally, to prove that $\left[M^{(n)} - M\right]_\infty \overset{\mathbb{P}}{\to} 0$ we assume that it is not true, and then $\left[M^{(n)} - M\right]_\infty \overset{\mathbb{P}}{\nrightarrow} 0$ implies that there exist $\varepsilon > 0$ and $\{n_k\}_{k\in\mathbb{N}} \subset \mathbb{N}$ with

$$d\left(\left[M^{(n_k)} - M\right]_\infty, 0\right) \geq \varepsilon$$

for all $k \in \mathbb{N}$,where $d(X, Y) := \inf\{\varepsilon > 0 : \mathbb{P}[|X - Y| > \varepsilon] \leq \varepsilon\}$ is the Ky Fan metric. We shall denote the subsequence as the original sequence, trying to keep the notation as simple as possible. Using a diagonal argument, a subsequence $\{n_i\}_{i\in\mathbb{N}} \subset \mathbb{N}$ can be chosen, with the property that $d\left(\left[M^{(n_i)} - M\right]_{\tau_k^{n_i}}, 0\right) < \frac{1}{k}$ for all $i \geq k$.

Since

$$\lim_{k\to\infty}\left[M^{(n_i)}-M\right]_{\tau_k^{n_i}}=\left[M^{(n_i)}-M\right]_\infty \quad \mathbb{P}-a.s.,$$

we can find some $k(n_i)\geq i$ such that

$$d\left(\left[M^{(n_i)}-M\right]_{\tau_{k(n_i)}^{n_i}},\left[M^{(n_i)}-M\right]_\infty\right)<\frac{1}{k}.$$

Then, using the estimation

$$\mathbb{P}\left[\left|\left[M^{(n_k)}-M\right]_{\tau_{k(n_k)}^{n_k}}-\left[M^{(n_k)}-M\right]_{\tau_k^{n_k}}\right|>\varepsilon\right]$$
$$\leq \mathbb{P}\left[\left\{\sup_{t\in\mathbb{R}_+}\left|M_t^{(n_k)}-M_t\right|\geq k\right\}\right],$$

it follows that

$$d\left(\left[M^{(n_k)}-M\right]_{\tau_{k(n_k)}^{n_k}},\left[M^{(n_k)}-M\right]_{\tau_k^{n_k}}\right)\underset{k\to\infty}{\longrightarrow}0,$$

which yields a contradiction with $\varepsilon\leq d\left(\left[M^{(n_k)}-M\right]_\infty,0\right)$. Thus, $\left[M^{(n)}-M\right]_\infty \overset{\mathbb{P}}{\to} 0$. The last part of the this lemma follows immediately from the first statement. □

Using the Doob's stopping theorem we can conclude that for $M\in\mathcal{M}_\infty$ and an stopping time τ, that $M^\tau\in\mathcal{M}_\infty$, and therefore it follows as a corollary the following result.

Corollary 3.1 *For $\left\{M^{(n)}\right\}_{n\in\mathbb{N}}\subset\mathcal{M}_\infty$, $M\in\mathcal{M}_\infty$ and τ any stopping time holds*

$$M_\tau^{(n)}\overset{L^1}{\to}M_\tau\Longrightarrow\left[M^{(n)}-M\right]_\tau\overset{\mathbb{P}}{\longrightarrow}0.$$

Proof $\left[\left(M^{(n)}\right)^\tau-M^\tau\right]_\infty=\left[M^{(n)}-M\right]_\infty^\tau=\left[M^{(n)}-M\right]_\tau\overset{\mathbb{P}}{\longrightarrow}0.$ □

3.1 Density Processes

Given an absolutely continuous probability measure $\mathbb{Q}\ll\mathbb{P}$ in a filtered probability space, where a semimartingale with the weak predictable representation property is defined, the structure of the density process has been studied extensively by several authors; see Theorem 14.41 in He et al. [9] or Theorem III.5.19 in Jacod and Shiryaev [12].

Denote by $D_t := \mathbb{E}\left[\frac{d\mathbb{Q}}{d\mathbb{P}}\middle|\mathcal{F}_t\right]$ the càdlàg version of the density process. For the increasing sequence of stopping times $\tau_n := \inf\left\{t \geq 0 : D_t < \frac{1}{n}\right\}$ $n \geq 1$ and $\tau_0 := \sup_n \tau_n$ we have $D_t(\omega) = 0\ \forall t \geq \tau_0(\omega)$ and $D_t(\omega) > 0\ \forall t < \tau_0(\omega)$, i.e.

$$D = D\mathbf{1}_{[\![0,\tau_0[\![}, \tag{3.5}$$

and the process

$$\frac{1}{D_{s-}}\mathbf{1}_{[\![D_- \neq 0]\!]} \text{ is integrable w.r.t. } D, \tag{3.6}$$

where we abuse of the notation by setting $[\![D_- \neq 0]\!] := \{(\omega, t) \in \Omega \times \mathbb{R}_+ : D_{t-}(\omega) \neq 0\}$. Both conditions (3.5) and (3.6) are necessary and sufficient in order that a semimartingale to be an *exponential semimartigale* [9, Thm. 9.41], i.e. $D = \mathcal{E}(Z)$ the Doléans-Dade exponential of another semimartingale Z. In that case we have

$$\tau_0 = \inf\{t > 0 : D_{t-} = 0 \text{ or } D_t = 0\} = \inf\{t > 0 : \Delta Z_t = -1\}. \tag{3.7}$$

It is well known that the Lévy-processes satisfy the weak property of predictable representation [9], when the completed natural filtration is considered. In the following lemma we present the characterization of the density processes for the case of these processes.

Lemma 3.2 *Given an absolutely continuous probability measure $\mathbb{Q} \ll \mathbb{P}$, there exist coefficients $\theta_0 \in \mathcal{L}(W)$ and $\theta_1 \in \mathcal{G}(\mu)$ such that*

$$\frac{d\mathbb{Q}_t}{d\mathbb{P}_t} = \frac{d\mathbb{Q}_t}{d\mathbb{P}_t}\mathbf{1}_{[\![0,\tau_0[\![} = \mathcal{E}\left(Z^\theta\right)(t), \tag{3.8}$$

where $Z_t^\theta \in \mathcal{M}_{loc}$ is the local martingale given by

$$Z_t^\theta := \int\limits_{]0,t]} \theta_0 dW + \int\limits_{]0,t]\times\mathbb{R}_0} \theta_1(s,x)\left(\mu(ds,dx) - ds\,\nu(dx)\right), \tag{3.9}$$

and $\mathcal{E}$ represents the Doleans-Dade exponential of a semimartingale. The coefficients θ_0 and θ_1 are dt-a.s and $\mu_{\mathbb{P}}^{\mathcal{P}}(ds,dx)$-a.s. unique on $[\![0,\tau_0]\!]$ and $[\![0,\tau_0]\!] \times \mathbb{R}_0$ respectively for $\mathbb{P}$-almost all ω. Furthermore, the coefficients can be chosen with $\theta_0 = 0$ on $]\!]\tau_0,\infty[\![$ and $\theta_1 = 0$ on $]\!]\tau_0,\infty[\![\times\mathbb{R}$.

Proof We only address the uniqueness of the coefficients θ_0 and θ_1, because the representation follows from (3.5) and (3.6). Let assume, that we have two possible vectors $\theta := (\theta_0, \theta_1)$ and $\theta' := \left(\theta_0', \theta_1'\right)$ satisfying the representation, i.e.

$$\begin{aligned} D_u \mathbf{1}_{[\![0,\tau_0[\![} &= \int D_{t-} d\{ \int\limits_{]0,t]} \theta_0(s)\, dW_s + \int\limits_{]0,t]\times\mathbb{R}_0} \theta_1(s,x)\,(\mu(ds,dx) - ds\, \nu(dx))\} \\ &= \int D_{t-} d\{ \int\limits_{]0,t]} \theta_0'(s)\, dW_s + \int\limits_{]0,t]\times\mathbb{R}_0} \theta_1'(s,x)\,(\mu(ds,dx) - ds\, \nu(dx))\}, \end{aligned}$$

and thus

$$\begin{aligned} \triangle D_t &= D_{t-} \triangle \left(\int\limits_{]0,t]\times\mathbb{R}_0} \theta_1(s,x)\,(\mu(ds,dx) - ds\, \nu(dx)) \right) \\ &= D_{t-} \triangle \left(\int\limits_{]0,t]\times\mathbb{R}_0} \theta_1'(s,x)\,(\mu(ds,dx) - ds\, \nu(dx)) \right). \end{aligned}$$

Since $D_{t-} > 0$ on $[\![0, \tau_0[\![$, it follows that

$$\triangle \left(\int\limits_{]0,t]\times\mathbb{R}_0} \theta_1(s,x)\,(\mu(ds,dx) - ds\, \nu(dx)) \right) = \triangle \left(\int\limits_{]0,t]\times\mathbb{R}_0} \theta_1'(s,x)\,(\mu(ds,dx) - ds\, \nu(dx)) \right).$$

Since two purely discontinuous local martingales with the same jumps are equal, it follows

$$\begin{aligned} &\int\limits_{]0,t]\times\mathbb{R}_0} \theta_1(s,x)\,(\mu(ds,dx) - ds\, \nu(dx)) \\ &\quad = \int\limits_{]0,t]\times\mathbb{R}_0} \widehat{\theta}_1(s,x)\,(\mu(ds,dx) - ds\, \nu(dx)) \end{aligned}$$

and thus

$$\int D_{t-} d\{ \int\limits_{]0,t]} \theta_0(s)\, dW_s\} = \int D_{t-} d\{ \int\limits_{]0,t]} \theta_0'(s)\, dW_s\}.$$

Then,

$$0 = \left[\int D_{s-} d \left\{ \int_{]0,s]} \left(\theta_0'(u) - \theta_0(u)\right) dW_u \right\} \right]_t = \int\limits_{]0,t]} (D_{s-})^2 \left\{\theta_0'(s) - \theta_0(s)\right\}^2 ds$$

and thus $\theta_0' = \theta_0\ dt\text{-}a.s$ on $[\![0, \tau_0]\!]$ for $\mathbb{P}$-almost all ω.

On the other hand,

$$
\begin{aligned}
0 &= \left\langle \int \left\{ \theta_1' (s, x) - \theta_1 (s, x) \right\} (\mu (ds, dx) - ds\, \nu (dx)) \right\rangle_t \\
&= \int\limits_{]0,t] \times \mathbb{R}_0} \left\{ \theta_1' (s, x) - \theta_1 (s, x) \right\}^2 \nu (dx)\, ds,
\end{aligned}
$$

implies that $\theta_1 (s, x) = \theta_1' (s, x) \quad \mu_{\mathbb{P}}^{\mathcal{P}} (ds, dx)$-a.s. on $[[0, \tau_0]] \times \mathbb{R}_0$ for $\mathbb{P}$-almost all ω. □

For $\mathbb{Q} \ll \mathbb{P}$ the function $\theta_1 (\omega, t, x)$ described in Lemma 3.2 determines the density of the predictable projection $\mu_{\mathbb{Q}}^{\mathcal{P}} (dt, dx)$ with respect to $\mu_{\mathbb{P}}^{\mathcal{P}} (dt, dx)$ (see He et al. [9] or Jacod and Shiryaev [12]). More precisely, for $B \in (\mathcal{B} (\mathbb{R}_+) \otimes \mathcal{B} (\mathbb{R}_0))$ we have

$$
\mu_{\mathbb{Q}}^{\mathcal{P}} (\omega, B) = \int_B (1 + \theta_1 (\omega, t, x))\, \mu_{\mathbb{P}}^{\mathcal{P}} (dt, dx) . \tag{3.10}
$$

In what follows we restrict ourself to the time interval $[0, T]$, for some $T > 0$ fixed, and we take $\mathcal{F} = \mathcal{F}_T$. The corresponding classes of density processes associated to $\mathcal{Q}_{\ll}(\mathbb{P})$ and $\mathcal{Q}_{\approx} (\mathbb{P})$ are denoted by $\mathcal{D}_{\ll} (\mathbb{P})$ and $\mathcal{D}_{\approx} (\mathbb{P})$, respectively. For instance, in the former case

$$
\mathcal{D}_{\ll} (\mathbb{P}) := \left\{ D = \{D_t\}_{t \in [0,T]} : \exists \mathbb{Q} \in \mathcal{Q}_{\ll} (\mathbb{P}) \text{ with } D_t = \left. \frac{d\mathbb{Q}}{d\mathbb{P}} \right|_{\mathcal{F}_t} \right\}, \tag{3.11}
$$

and the processes in this set are of the form

$$
\begin{aligned}
D_t = \exp &\left\{ \int\limits_{]0,t]} \theta_0 dW + \int\limits_{]0,t] \times \mathbb{R}_0} \theta_1 (s, x) (\mu (ds, dx) - \nu (dx)\, ds) \right. \\
&\left. - \frac{1}{2} \int\limits_{]0,t]} (\theta_0)^2 ds \right\} \times \\
\times \exp &\left\{ \int\limits_{]0,t] \times \mathbb{R}_0} \{ \ln (1 + \theta_1 (s, x)) - \theta_1 (s, x) \}\, \mu (ds, dx) \right\}
\end{aligned} \tag{3.12}
$$

for $\theta_0 \in \mathcal{L} (W)$ and $\theta_1 \in \mathcal{G} (\mu)$.

The set $\mathcal{D}_{\ll} (\mathbb{P})$ is characterized as follow.

Corollary 3.2 *The process D belongs to $\mathcal{D}_{\ll} (\mathbb{P})$ if and only if there are $\theta_0 \in \mathcal{L} (W)$ and $\theta_1 \in \mathcal{G} (\mu)$ with $\theta_1 \geq -1$ such that $D_t = \mathcal{E} \left(Z^{\theta} \right) (t)$ $\mathbb{P}$-a.s. $\forall t \in [0, T]$ and $\mathbb{E}_{\mathbb{P}} \left[\mathcal{E} \left(Z^{\theta} \right) (t) \right] = 1$ $\forall t \geq 0$, where $Z^{\theta} (t)$ is defined by (3.9).*

Proof The necessity follows from Lemma 3.2. Conversely, let $\theta_0 \in \mathcal{L}(W)$ and $\theta_1 \in \mathcal{G}(\mu)$ be arbitrarily chosen. Since $D_t = \int D_{s-} dZ_s^\theta \in \mathcal{M}_{loc}$ is a nonnegative local martingale, it is a supermartingale, with constant expectation from our assumptions. Therefore, it is a martingale, and hence the density process of an absolutely continuous probability measure. □

Since density processes are essentially uniformly integrable martingales, using Lemma 3.1 and Corollary 3.1 the following proposition follows immediately.

Proposition 3.1 *Let $\{\mathbb{Q}^{(n)}\}_{n\in\mathbb{N}}$ be a sequence in $\mathcal{Q}_{\ll}(\mathbb{P})$, with $D_T^{(n)} := \left.\frac{d\mathbb{Q}^{(n)}}{d\mathbb{P}}\right|_{\mathcal{F}_T}$ converging to $D_T := \left.\frac{d\mathbb{Q}}{d\mathbb{P}}\right|_{\mathcal{F}_T}$ in $L^1(\mathbb{P})$. For the corresponding density processes $D_t^{(n)} := \mathbb{E}_{\mathbb{P}}\left[D_T^{(n)} \,|\mathcal{F}_t\right]$ and $D_t := \mathbb{E}_{\mathbb{P}}\left[D_T \,|\mathcal{F}_t\right]$, for $t \in [0, T]$, we have*

$$\left[D^{(n)} - D\right]_T \overset{\mathbb{P}}{\to} 0.$$

4 Penalty Functions for Densities

Now, we shall introduce a family of penalty functions for the density processes described in Sect. 3.1, for the absolutely continuous measures $\mathbb{Q} \in \mathcal{Q}_{\ll}(\mathbb{P})$.

Let $h : \mathbb{R}_+ \to \mathbb{R}_+$ and $h_0, h_1 : \mathbb{R} \to \mathbb{R}_+$ be convex functions with $0 = h(0) = h_0(0) = h_1(0)$. Define the penalty function, with τ_0 as in (3.7), by

$$\begin{aligned}\vartheta(\mathbb{Q}) := \mathbb{E}_{\mathbb{Q}}\left[\int_0^{T\wedge\tau_0} h\left(h_0(\theta_0(t)) + \int_{\mathbb{R}_0} \delta(t,x)\, h_1(\theta_1(t,x))\, \nu(dx)\right) dt\right] \mathbf{1}_{\mathcal{Q}_{\ll}}(\mathbb{Q}) \\ +\infty \times \mathbf{1}_{\mathcal{Q}_{cont}\setminus\mathcal{Q}_{\ll}}(\mathbb{Q}),\end{aligned} \tag{4.1}$$

where θ_0, θ_1 are the processes associated to $\mathbb{Q}$ from Lemma 3.2 and $\delta(t,x) : \mathbb{R}_+ \times \mathbb{R}_0 \to \mathbb{R}_+$ is an arbitrary fixed nonnegative function $\delta(t,x) \in \mathcal{G}(\mu)$. Since $\theta_0 \equiv 0$ on $[\![\tau_0, \infty[\![$ and $\theta_1 \equiv 0$ on $[\![\tau_0, \infty[\![\times \mathbb{R}_0$ we have from the conditions imposed to h, h_0, and h_1

$$\begin{aligned}\vartheta(\mathbb{Q}) = \mathbb{E}_{\mathbb{Q}}\left[\int_0^{T} h\left(h_0(\theta_0(t)) + \int_{\mathbb{R}_0} \delta(t,x)\, h_1(\theta_1(t,x))\, \nu(dx)\right) dt\right] \mathbf{1}_{\mathcal{Q}_{\ll}}(\mathbb{Q}) \\ +\infty \times \mathbf{1}_{\mathcal{Q}_{cont}\setminus\mathcal{Q}_{\ll}}(\mathbb{Q}).\end{aligned} \tag{4.2}$$

Further, define the convex measure of risk

$$\rho(X) := \sup_{\mathbb{Q}\in\mathcal{Q}_{\ll}(\mathbb{P})} \left\{\mathbb{E}_{\mathbb{Q}}[-X] - \vartheta(\mathbb{Q})\right\}. \tag{4.3}$$

Notice that ρ is a normalized and sensitive measure of risk. For each class of probability measures introduced so far, the subclass of those measures with a finite penalization is considered. We will denote by $\mathcal{Q}^{\vartheta}$, $\mathcal{Q}^{\vartheta}_{\ll}(\mathbb{P})$ and $\mathcal{Q}^{\vartheta}_{\approx}(\mathbb{P})$ the corresponding subclasses, i.e.

$$\mathcal{Q}^{\vartheta} := \{\mathbb{Q} \in \mathcal{Q} : \vartheta(\mathbb{Q}) < \infty\}, \ \mathcal{Q}^{\vartheta}_{\ll}(\mathbb{P}) := \mathcal{Q}^{\vartheta} \cap \mathcal{Q}_{\ll}(\mathbb{P}) \text{ and}$$
$$\mathcal{Q}^{\vartheta}_{\approx}(\mathbb{P}) := \mathcal{Q}^{\vartheta} \cap \mathcal{Q}_{\approx}(\mathbb{P}). \tag{4.4}$$

Notice that $\mathcal{Q}^{\vartheta}_{\approx}(\mathbb{P}) \neq \varnothing$.

Next theorem establishes the minimality on $\mathcal{Q}_{\ll}(\mathbb{P})$ of the penalty function introduced above for the risk measure ρ, its proof is based on the sufficient conditions given in Theorem 2.1. This result is relevant to obtain one of the main results of this paper, namely Theorem 6.1.

Theorem 4.1 *The penalty function ϑ defined in (4.2) is equal to the minimal penalty function of the convex risk measure ρ, given by (4.3), on $\mathcal{Q}_{\ll}(\mathbb{P})$, i.e.*

$$\vartheta \mathbf{1}_{\mathcal{Q}_{\ll}(\mathbb{P})} = \psi_{\rho}^{*} \mathbf{1}_{\mathcal{Q}_{\ll}(\mathbb{P})}.$$

Proof From Lemma 2.1 (*b*), we need to show that the penalization ϑ is proper, convex and that the corresponding identification, defined as $\Theta(Z) := \vartheta(\mathbb{Q})$ if $Z \in \delta(\mathcal{Q}_{\ll}(\mathbb{P})) := \{Z \in L^1(\mathbb{P}) : Z = d\mathbb{Q}/d\mathbb{P} \text{ with } \mathbb{Q} \in \mathcal{Q}_{\ll}(\mathbb{P})\}$ and $\Theta(Z) := \infty$ on $L^1 \setminus \delta(\mathcal{Q}_{\ll}(\mathbb{P}))$, is lower semicontinuous with respect to the strong topology.

First, observe that the function ϑ is proper, since $\vartheta(\mathbb{P}) = 0$. To verify the convexity of ϑ, choose $\mathbb{Q}, \widetilde{\mathbb{Q}} \in \mathcal{Q}^{\vartheta}_{\ll}$ and define $\mathbb{Q}^{\lambda} := \lambda\mathbb{Q} + (1-\lambda)\widetilde{\mathbb{Q}}$, for $\lambda \in [0,1]$. Notice that the corresponding density process can be written as $D^{\lambda} := \dfrac{d\mathbb{Q}^{\lambda}}{d\mathbb{P}} = \lambda D + (1-\lambda)\widetilde{D}$ $\mathbb{P}$-a.s..

Now, from Lemma 3.2, let (θ_0, θ_1) and $(\widetilde{\theta}_0, \widetilde{\theta}_1)$ be the processes associated to $\mathbb{Q}$ and $\widetilde{\mathbb{Q}}$, respectively, and observe that from

$$D_t = 1 + \int_{[0,t]} D_{s-}\theta_0(s)\, dW_s + \int_{[0,t]\times\mathbb{R}_0} D_{s-}\theta_1(s,x)\, d(\mu(ds,dx) - ds\nu(dx)))$$

and the corresponding expression for $\widetilde{D}$ we have for $\tau_n^{\lambda} := \inf\left\{t \geq 0 : D_t^{\lambda} \leq \frac{1}{n}\right\}$

$$\int_0^{t\wedge\tau_n^{\lambda}} \left(D_{s-}^{\lambda}\right)^{-1} dD_s^{\lambda} = \int_0^{t\wedge\tau_n^{\lambda}} \frac{\lambda D_{s-}\theta_0(s) + (1-\lambda)\widetilde{D}_{s-}\widetilde{\theta}_0(s)}{(\lambda D_{s-} + (1-\lambda)\widetilde{D}_{s-})} dW_s +$$
$$\int_{[0,t\wedge\tau_n^{\lambda}]\times\mathbb{R}_0} \frac{\lambda D_{s-}\theta_1(s,x) + (1-\lambda)\widetilde{D}_{s-}\widetilde{\theta}_1(s,x)}{(\lambda D_{s-} + (1-\lambda)\widetilde{D}_{s-})} d\left(\mu - \mu_{\mathbb{P}}^{\mathcal{P}}\right).$$

The weak predictable representation property of the local martingale $\int_0^{t\wedge\tau_n^\lambda}\left(D_{s-}^\lambda\right)^{-1} dD_s^\lambda$, yield on the other hand

$$\int_0^{t\wedge\tau_n^\lambda}\left(D_{s-}^\lambda\right)^{-1} dD_s^\lambda = \int_0^{t\wedge\tau_n^\lambda}\theta_0^\lambda(s)\,dW_s + \int_{[0,t\wedge\tau_n^\lambda]\times\mathbb{R}_0}\theta_1^\lambda(s,x)\,d\left(\mu-\mu_{\mathbb{P}}^{\mathcal{P}}\right),$$

where identification

$$\theta_0^\lambda(s) = \frac{\lambda D_{s-}\theta_0(s) + (1-\lambda)\widetilde{D}_{s-}\widetilde{\theta}_0(s)}{\left(\lambda D_{s-} + (1-\lambda)\widetilde{D}_{s-}\right)},$$

and

$$\theta_1^\lambda(s,x) = \frac{\lambda D_{s-}\theta_1(s,x) + (1-\lambda)\widetilde{D}_{s-}\widetilde{\theta}_1(s,x)}{\left(\lambda D_{s-} + (1-\lambda)\widetilde{D}_{s-}\right)}.$$

This is possible thanks to the uniqueness of the representation in Lemma 3.2. The convexity follows now from the convexity of h, h_0 and h_1, using the fact that any convex function is continuous in the interior of its domain. More specifically,

$$\begin{aligned}
\vartheta\left(\mathbb{Q}^\lambda\right) \le\; & \mathbb{E}_{\mathbb{Q}^\lambda}\left[\int_{[0,T]}\frac{\lambda D_s}{(\lambda D_s+(1-\lambda)\widetilde{D}_s)}h\left(h_0(\theta_0(s))\right.\right.\\
&\left.\left.+\int_{\mathbb{R}_0}\delta(s,x)\,h_1(\theta_1(s,x))\,\nu(dx)\right)ds\right]\\
&+\mathbb{E}_{\mathbb{Q}^\lambda}\left[\int_{[0,T]}\frac{(1-\lambda)\widetilde{D}_s}{(\lambda D_s+(1-\lambda)\widetilde{D}_s)}h\left(h_0(\widetilde{\theta}_0(s))\right.\right.\\
&\left.\left.+\int_{\mathbb{R}_0}\delta(s,x)\,h_1(\widetilde{\theta}_1(s,x))\nu(dx)\right)ds\right]\\
=\; & \int_{[0,T]}\int_\Omega\frac{\lambda D_s}{\left(\lambda D_s+(1-\lambda)\widetilde{D}_s\right)}h\left(h_0(\theta_0(s))\right.\\
&\left.+\int_{\mathbb{R}_0}\delta(s,x)\,h_1(\theta_1(s,x))\,\nu(dx)\right)\\
&\times\left(\lambda D_s+(1-\lambda)\widetilde{D}_s\right)\mathbf{1}_{\{\lambda D_s+(1-\lambda)\widetilde{D}_s>0\}}d\mathbb{P}ds\\
&+\int_{[0,T]}\int_\Omega\frac{(1-\lambda)\widetilde{D}_s}{\left(\lambda D_s+(1-\lambda)\widetilde{D}_s\right)}h\left(h_0(\widetilde{\theta}_0(s))\right.
\end{aligned}$$

$$+ \int_{\mathbb{R}_0} \delta(s,x) h_1(\widetilde{\theta}_1(s,x)) \nu(dx) \Bigg)$$
$$\times \left(\lambda D_s + (1-\lambda)\widetilde{D}_s\right) \mathbf{1}_{\{\lambda D_s + (1-\lambda)\widetilde{D}_s > 0\}} d\mathbb{P} ds$$
$$= \lambda \vartheta(\mathbb{Q}) + (1-\lambda)\vartheta\left(\widetilde{\mathbb{Q}}\right),$$

where we used that

$$\left\{\int_{\mathbb{R}_0} \delta(t,x) h_1(\theta_1(t,x)) \nu(dx)\right\}_{t\in\mathbb{R}_+} \quad \text{and} \quad \left\{\int_{\mathbb{R}_0} \delta(t,x) h_1(\widetilde{\theta}_1(t,x)) \nu(dx)\right\}_{t\in\mathbb{R}_+}$$

are predictable processes.

It remains to prove the lower semicontinuity of Θ. As pointed out earlier, it is enough to consider a sequence of densities $Z^{(n)} := \frac{d\mathbb{Q}^{(n)}}{d\mathbb{P}} \in \delta\left(\mathcal{Q}_{\ll}(\mathbb{P})\right)$ converging in $L^1(\mathbb{P})$ to $Z := \frac{d\mathbb{Q}}{d\mathbb{P}}$. Denote the corresponding density processes by $D^{(n)}$ and D, respectively. In Proposition 3.1 it was verified the convergence in probability to zero of the quadratic variation process

$$\left[D^{(n)} - D\right]_T = \int_0^T \left\{D_{s-}^{(n)}\theta_0^{(n)}(s) - D_{s-}\theta_0(s)\right\}^2 ds$$
$$+ \int_{[0,T]\times\mathbb{R}_0} \left\{D_{s-}^{(n)}\theta_1^{(n)}(s,x) - D_{s-}\theta_1(s,x)\right\}^2 \mu(ds,dx).$$

This implies that

$$\left.\begin{array}{l} \int_0^T \left\{D_{s-}^{(n)}\theta_0^{(n)}(s) - D_{s-}\theta_0(s)\right\}^2 ds \xrightarrow{\mathbb{P}} 0, \\ \text{and} \\ \int_{[0,T]\times\mathbb{R}_0} \left\{D_{s-}^{(n)}\theta_1^{(n)}(s,x) - D_{s-}\theta_1(s,x)\right\}^2 \mu(ds,dx) \xrightarrow{\mathbb{P}} 0. \end{array}\right\} \tag{4.5}$$

Then, for an arbitrary but fixed subsequence, there exists a sub-subsequence such that $\mathbb{P}$-a.s.

$$\left\{D_{s-}^{(n)}\theta_0^{(n)}(s) - D_{s-}\theta_0(s)\right\}^2 \xrightarrow{L^1(\lambda)} 0$$

and

$$\left\{D_{s-}^{(n)}\theta_1^{(n)}(s,x) - D_{s-}\theta_1(s,x)\right\}^2 \xrightarrow{L^1(\mu)} 0,$$

where for simplicity we have denoted the sub-subsequence as the original sequence. Now, we claim that for the former sub-subsequence it also holds that

$$\begin{cases} D_{s-}^{(n)}\theta_0^{(n)}(s) \overset{\lambda\times\mathbb{P}\text{-a.s.}}{\longrightarrow} D_{s-}\theta_0(s), \\ D_{s-}^{(n)}\theta_1^{(n)}(s,x) \overset{\mu\times\mathbb{P}\text{-a.s.}}{\longrightarrow} D_{s-}\theta_1(s,x). \end{cases} \tag{4.6}$$

We present first the arguments for the proof of the second assertion in (4.6). Assuming the opposite, there exists $C \in \mathcal{B}([0,T])\otimes\mathcal{B}(\mathbb{R}_0)\otimes\mathcal{F}_T$, with $\mu\times\mathbb{P}[C] > 0$, and such that for each $(s,x,\omega)\in C$

$$\lim_{n\to\infty}\left\{D_{s-}^{(n)}\theta_1^{(n)}(s,x) - D_{s-}\theta_1(s,x)\right\}^2 = c \neq 0,$$

or the limit does not exist.

Let $C(\omega) := \{(t,x)\in[0,T]\times\mathbb{R}_0 : (t,x,\omega)\in C\}$ be the ω-section of C. Observe that $B := \{\omega\in\Omega : \mu[C(\omega)] > 0\}$ has positive probability: $\mathbb{P}[B] > 0$.

From (4.5), any arbitrary but fixed subsequence has a sub-subsequence converging $\mathbb{P}$-a.s. Denoting such a sub-subsequence simply by n, we can fix $\omega\in B$ with

$$\begin{aligned} &\int_{C(\omega)}\left\{D_{s-}^{(n)}\theta_1^{(n)}(s,x) - D_{s-}\theta_1(s,x)\right\}^2 d\mu(s,x) \\ &\leq \int_{[0,T]\times\mathbb{R}_0}\left\{D_{s-}^{(n)}\theta_1^{(n)}(s,x) - D_{s-}\theta_1(s,x)\right\}^2 d\mu(s,x) \underset{n\to\infty}{\longrightarrow} 0, \end{aligned}$$

and hence $\left\{D_{s-}^{(n)}\theta_1^{(n)}(s,x) - D_{s-}\theta_1(s,x)\right\}^2$ converges in μ-measure to 0 on $C(\omega)$. Again, for any subsequence there is a sub-subsequence converging μ-a.s. to 0. Furthermore, for an arbitrary but fixed $(s,x)\in C(\omega)$, when the limit does not exist

$$\begin{aligned} a &:= \liminf_{n\to\infty}\left\{D_{s-}^{(n)}\theta_1^{(n)}(s,x) - D_{s-}\theta_1(s,x)\right\}^2 \\ &\neq \limsup_{n\to\infty}\left\{D_{s-}^{(n)}\theta_1^{(n)}(s,x) - D_{s-}\theta_1(s,x)\right\}^2 =: b, \end{aligned}$$

and we can choose converging subsequences $n(i)$ and $n(j)$ with

$$\lim_{i\to\infty}\left\{D_{s-}^{n(i)}\theta_1^{n(i)}(s,x) - D_{s-}\theta_1(s,x)\right\}^2 = a$$

$$\lim_{j\to\infty}\left\{D_{s-}^{n(j)}\theta_1^{n(j)}(s,x) - D_{s-}\theta_1(s,x)\right\}^2 = b.$$

From the above argument, there are sub-subsequences $n(i(k))$ and $n(j(k))$ such that

$$a = \lim_{k\to\infty} \left\{D_{s-}^{n(i(k))}\theta_1^{n(i(k))}(s,x) - D_{s-}\theta_1(s,x)\right\}^2 = 0$$

$$b = \lim_{k\to\infty} \left\{D_{s-}^{n(j(k))}\theta_1^{n(j(k))}(s,x) - D_{s-}\theta_1(s,x)\right\}^2 = 0,$$

which is clearly a contradiction.

For the case when

$$\lim_{n\to\infty} \left\{D_{s-}^{(n)}\theta_1^{(n)}(s,x) - D_{s-}\theta_1(s,x)\right\}^2 = c \neq 0,$$

the same argument can be used, and get a subsequence converging to 0, having a contradiction again. Therefore, the second part of our claim in (4.6) holds.

Since $D_{s-}^{(n)}\theta_1^{(n)}(s,x)$, $D_{s-}\theta_1(s,x) \in \mathcal{G}(\mu)$, we have, in particular, that $D_{s-}^{(n)}\theta_1^{(n)}(s,x) \in \widetilde{\mathcal{P}}$ and $D_{s-}\theta_1(s,x) \in \widetilde{\mathcal{P}}$ and hence $C \in \widetilde{\mathcal{P}}$. From the definition of the predictable projection it follows that

$$0 = \mu \times \mathbb{P}[C] = \int_{\Omega}\int_{[0,T]\times\mathbb{R}_0} \mathbf{1}_C(s,\omega)\, d\mu d\mathbb{P} = \int_{\Omega}\int_{[0,T]\times\mathbb{R}_0} \mathbf{1}_C(s,\omega)\, d\mu_{\mathbb{P}}^{\mathcal{P}} d\mathbb{P}$$

$$= \int_{\Omega}\int_{\mathbb{R}_0}\int_{[0,T]} \mathbf{1}_C(s,\omega)\, ds d\nu d\mathbb{P} = \lambda \times \nu \times \mathbb{P}[C],$$

and thus

$$D_{s-}^{(n)}\theta_1^{(n)}(s,x) \overset{\lambda\times\nu\times\mathbb{P}\text{-a.s.}}{\longrightarrow} D_{s-}\theta_1(s,x).$$

Since

$$\int_{\Omega\times[0,T]} \left|D_{t-}^{(n)} - D_{t-}\right| d\mathbb{P}\times dt = \int_{\Omega\times[0,T]} \left|D_t^{(n)} - D_t\right| d\mathbb{P}\times dt \longrightarrow 0,$$

we have that

$$\left\{D_{t-}^{(n)}\right\}_{t\in[0,T]} \overset{L^1(\lambda\times\mathbb{P})}{\longrightarrow} \{D_{t-}\}_{t\in[0,T]} \quad \text{and} \quad \left\{D_t^{(n)}\right\}_{t\in[0,T]} \overset{L^1(\lambda\times\mathbb{P})}{\longrightarrow} \{D_t\}_{t\in[0,T]}.$$

Then, for an arbitrary but fixed subsequence $\{n_k\}_{k\in\mathbb{N}} \subset \mathbb{N}$, there is a sub-subsequence $\left\{n_{k_i}\right\}_{i\in\mathbb{N}} \subset \mathbb{N}$ such that

$$
\begin{array}{ccc}
D_{t-}^{(n_{k_i})}\theta_1^{(n_{k_i})}(t,x) & \overset{\lambda\times\nu\times\mathbb{P}\text{-a.s.}}{\longrightarrow} & D_{t-}\theta_1(t,x), \\
D_{t-}^{(n_{k_i})} & \overset{\lambda\times\mathbb{P}\text{-a.s.}}{\longrightarrow} & D_{t-}, \\
D_{t}^{(n_{k_i})} & \overset{\lambda\times\mathbb{P}\text{-a.s.}}{\longrightarrow} & D_{t}.
\end{array}
$$

Furthermore, $\mathbb{Q} \ll \mathbb{P}$ implies that $\lambda \times \nu \times \mathbb{Q} \ll \lambda \times \nu \times \mathbb{P}$, and then

$$
\begin{array}{ccc}
D_{t-}^{(n_{k_i})}\theta_1^{(n_{k_i})}(t,x) & \overset{\lambda\times\nu\times\mathbb{Q}\text{-a.s.}}{\longrightarrow} & D_{t-}\theta_1(t,x), \\
D_{t-}^{(n_{k_i})} & \overset{\lambda\times\nu\times\mathbb{Q}\text{-a.s.}}{\longrightarrow} & D_{t-},
\end{array}
$$

and

$$
D_{t}^{(n_{k_i})} \overset{\lambda\times\nu\times\mathbb{Q}\text{-a.s.}}{\longrightarrow} D_{t}. \tag{4.7}
$$

Finally, noting that $\inf D_t > 0$ $\mathbb{Q}$-a.s.

$$
\theta_1^{(n_{k_i})}(t,x) \overset{\lambda\times\nu\times\mathbb{Q}\text{-a.s.}}{\longrightarrow} \theta_1(t,x). \tag{4.8}
$$

The first assertion in (4.6) can be proved using essentially the same kind of ideas used above for the proof of the second part, concluding that for an arbitrary but fixed subsequence $\{n_k\}_{k\in\mathbb{N}} \subset \mathbb{N}$, there is a sub-subsequence $\left\{n_{k_i}\right\}_{i\in\mathbb{N}} \subset \mathbb{N}$ such that

$$
\left\{D_t^{(n_{k_i})}\right\}_{t\in[0,T]} \overset{\lambda\times\mathbb{Q}\text{-a.s.}}{\longrightarrow} \{D_t\}_{t\in[0,T]} \tag{4.9}
$$

and

$$
\left\{\theta_0^{(n_{k_i})}(t)\right\}_{t\in[0,T]} \overset{\lambda\times\mathbb{Q}\text{-a.s.}}{\longrightarrow} \{\theta_0(t)\}_{t\in[0,T]}. \tag{4.10}
$$

We are now ready to finish the proof of the theorem, observing that

$$
\begin{aligned}
\liminf_{n\to\infty}\vartheta\left(\mathbb{Q}^{(n)}\right) = \liminf_{n\to\infty} \int\limits_{\Omega\times[0,T]} &\left\{h\left(h_0\left(\theta_0^{(n)}(t)\right) + \int_{\mathbb{R}_0}\delta(t,x)\right.\right. \\
&\left.\left.\times h_1\left(\theta_1^{(n)}(t,x)\right)\nu(dx)\right)\right\}\frac{D_t^{(n)}}{D_t}d(\lambda\times\mathbb{Q}).
\end{aligned}
$$

Let $\{n_k\}_{k\in\mathbb{N}} \subset \mathbb{N}$ be a subsequence for which the limit inferior is realized. Using (4.7)–(4.10) we can pass to a sub-subsequence $\{n_{k_i}\}_{i\in\mathbb{N}} \subset \mathbb{N}$ and, from the continuity of h, h_0 and h_1, it follows

$$\begin{aligned}
&\liminf_{n\to\infty} \vartheta\left(\mathbb{Q}^{(n)}\right) \\
&\geq \int_{\Omega\times[0,T]} \liminf_{i\to\infty} \left(\left\{ h\left(h_0\left(\theta_0^{(n_{k_i})}(t)\right) + \int_{\mathbb{R}_0} \delta(t,x) h_1\left(\theta_1^{(n_{k_i})}(t,x)\right) \nu(dx) \right) \right\} \frac{D_t^{(n_{k_i})}}{D_t} \right) d(\lambda\times\mathbb{Q}) \\
&\geq \int_{\Omega\times[0,T]} h\left(h_0(\theta_0(t)) + \int_{\mathbb{R}_0} h_1(\theta_1(t,x))\, \nu(dx) \right) d(\lambda\times\mathbb{Q}) \\
&= \vartheta(\mathbb{Q}).
\end{aligned}$$

□

5 The Market Model: General Description and Martingale Measures

Let us now consider the stochastic process Y_t with dynamics given by

$$Y_t := \int_{]0,t]} \alpha_s ds + \int_{]0,t]} \beta_s dW_s + \int_{]0,t]\times\mathbb{R}_0} \gamma(s,x)\left(\mu(ds,dx) - \nu(dx)\, ds\right), \tag{5.1}$$

where α is an adapted process with left continuous paths (càg), β is càdlàg with $\beta \in \mathcal{L}(W)$, and $\gamma \in \mathcal{G}(\mu)$. Throughout we assume that the coefficients α, β and γ fulfill the following conditions:

$$\begin{aligned}
&(A\ 1)\ 0 < c \leq |\beta_t| \quad \forall t \in \mathbb{R}_+ \quad \mathbb{P}\text{-}a.s.. \\
&(A\ 2)\ \int_0^T \left(\tfrac{\alpha_u}{\beta_u}\right)^2 du \in \mathfrak{M}_b \text{ i.e. bounded.} \\
&(A\ 3)\ \gamma(t,\Delta L_t) \times \mathbf{1}_{\mathbb{R}_0}(\Delta L_t) \geq -1 \quad \forall t \in \mathbb{R}_+ \quad \mathbb{P}\text{-}a.s.. \\
&(A\ 4)\ \left\{\gamma(t,\Delta L_t)\mathbf{1}_{\mathbb{R}_0}(\Delta L_t)\right\}_{t\in\mathbb{R}_+} \text{ is a locally bounded process.}
\end{aligned} \tag{5.2}$$

The market model consists of two assets, one of them is the numéraire, having a strictly positive price. The dynamics of the other risky asset will be modeled as a function of the process Y_t defined above. More specifically, since we are interested in the analysis of problem of robust utility maximization, presented in the next section, the discounted capital process can be written in terms of the wealth invested in this asset, and hence the problem can be written using only the dynamics of the discounted price of this asset. For this reason, throughout we will be concentrated in the dynamics of this price.

The dynamic of the discounted price process S is determined by the process Y as its Doleans-Dade exponential

$$S_t = S_0 \mathcal{E}\left(Y_t\right), \tag{5.3}$$

where $\mathcal{E}$ represents the Doleans-Dade exponential of a semimartingale; condition $(A\ 3)$ ensures that the price process is non-negative. This process is an exponential semimartingale if and only if the following two conditions are fulfilled:

$$\begin{aligned} &(i)\ \ S = S\mathbf{1}_{[\![0,\tau[\![}, \text{ for } \tau := \inf\{t > 0 : S_t = 0 \text{ or } S_{t-} = 0\}, \\ &(ii)\ \ \tfrac{1}{S_{t-}}\mathbf{1}_{[\![S_- \neq 0]\!]} \text{ is integrable w.r.t. } S, \end{aligned} \tag{5.4}$$

where $[\![S_- \neq 0]\!] := \{(\omega, t) \in \Omega \times \mathbb{R}_+ : S_{t-}(\omega) \neq 0\}$. The first property in (5.4) is conceptually very appropriate when we are interested in modelling the dynamics of a price process. Recall that a stochastically continuous semimartingale has independent increments if and only if its predictable triplet is non-random. Therefore, in general, the price process S is not a Lévy exponential model, because $[Y^c]_t = \int_0^t (\beta_u)^2 du$ does not need to be deterministic. However, observe that the price dynamics (5.3) includes Lévy exponential models, for Lévy processes with $|\triangle L_t| \leq 1$.

For the model (5.3) the price process can be written explicitly as

$$\begin{aligned} S_t = S_0 \exp &\left\{ \int\limits_{]0,t]} \alpha_s ds + \int\limits_{]0,t]} \beta_s dW_s + \int\limits_{]0,t]\times\mathbb{R}_0} \gamma(s,x)\left(\mu(ds,dx) - \nu(dx)\, ds\right) \right. \\ &\left. -\frac{1}{2}\int\limits_{]0,t]} (\beta_s)^2 ds \right\} \\ &\times \exp\left\{ \int\limits_{]0,t]\times\mathbb{R}_0} \{\ln(1+\gamma(s,x)) - \gamma(s,x)\}\mu(ds,dx) \right\}. \end{aligned} \tag{5.5}$$

The predictable cádlág process $\{\pi_t\}_{t\in\mathbb{R}_+}$, satisfying the integrability condition $\int_0^t (\pi_s)^2 ds < \infty$ $\mathbb{P}$-a.s. for all $t \in \mathbb{R}_+$, shall denote the proportion of wealth at time t invested in the risky asset S. For an initial capital x, the discounted wealth $X_t^{x,\pi}$ associated with a self-financing investment strategy (x, π) fulfills the equation

$$X_t^{x,\pi} = x + \int_0^t \frac{X_{u-}^{x,\pi}\pi_u}{S_{u-}} \mathbf{1}_{[\![S_- \neq 0]\!]} dS_u. \tag{5.6}$$

We say that a self-financing strategy (x, π) is *admissible* if the wealth process $X_t^{x,\pi} > 0$ for all $t > 0$. The class of admissible wealth processes with initial wealth less than or equal to x is denoted by $\mathcal{X}(x)$. In what follows we restrict ourself to the time interval $[0, T]$, for some $T > 0$ fixed, and take $\mathcal{F} = \mathcal{F}_T$.

Let us recall briefly the notation introduced in Sect. 3.1. Denote by $\mathcal{Q}_{\ll}(\mathbb{P})$ the subclass of absolutely continuous probability measures with respect to $\mathbb{P}$ and by $\mathcal{Q}_{\approx}(\mathbb{P})$ the subclass of equivalent probability measures. The corresponding classes of density processes associated to $\mathcal{Q}_{\ll}(\mathbb{P})$ and $\mathcal{Q}_{\approx}(\mathbb{P})$ are denoted by $\mathcal{D}_{\ll}(\mathbb{P})$ and $\mathcal{D}_{\approx}(\mathbb{P})$, respectively. The processes in the class $\mathcal{D}_{\ll}(\mathbb{P})$ are of the form

$$\begin{aligned} D_t = \exp&\left\{ \int_{]0,t]} \theta_0 dW + \int_{]0,t]\times\mathbb{R}_0} \theta_1(s,x)\left(\mu(ds,dx) - \nu(dx)ds\right) \right. \\ &\left. -\frac{1}{2}\int_{]0,t]} (\theta_0)^2 ds \right\} \times \\ &\times \exp\left\{ \int_{]0,t]\times\mathbb{R}_0} \{\ln(1+\theta_1(s,x)) - \theta_1(s,x)\}\,\mu(ds,dx) \right\}, \end{aligned} \tag{5.7}$$

for $\theta_0 \in \mathcal{L}(W)$ and $\theta_1 \in \mathcal{G}(\mu)$. If $\int \theta_1(s,x)\,\mu(ds,dx) \in \mathcal{A}_{loc}(\mathbb{P})$ the previous formula can be written as

$$\begin{aligned} D_t = \exp&\left\{ \int_{]0,t]} \theta_0 dW - \frac{1}{2}\int_{]0,t]} (\theta_0(s))^2 ds \right. \\ &\left. + \int_{]0,t]\times\mathbb{R}_0} \ln(1+\theta_1(s,x))\,\mu(ds,dx) - \int_{]0,t]\times\mathbb{R}_0} \theta_1(s,x)\,\nu(dx)ds \right\}. \end{aligned} \tag{5.8}$$

Next result characterizes the class of *equivalent local martingale measures* defined as

$$\mathcal{Q}_{elmm} \equiv \{\mathbb{Q} \in \mathcal{Q}_{\approx}(\mathbb{P}) : \mathcal{X}(1) \subset \mathcal{M}_{loc}(\mathbb{Q})\} = \{\mathbb{Q} \in \mathcal{Q}_{\approx}(\mathbb{P}) : S \in \mathcal{M}_{loc}(\mathbb{Q})\}. \tag{5.9}$$

Observe that $(A\,4)$ is a necessary and sufficient condition for S to be a locally bounded process. This property is crucial in order to obtain the former equality in (5.9). The class of density processes associated with $\mathcal{Q}_{elmm}$ is denoted by $\mathcal{D}_{elmm}(\mathbb{P})$. Kunita [16] gave conditions on the parameters (θ_0, θ_1) of a measure $\mathbb{Q} \in \mathcal{Q}_{\approx}$ in order that it is a local martingale measure for a Lévy exponential model i.e. when $S = \mathcal{E}(L)$. Observe that in this case $\mathcal{Q}_{elmm}(S) = \mathcal{Q}_{elmm}(L)$. Next proposition extends this result, giving conditions on the parameters (θ_0, θ_1) under which an equivalent measure is a local martingale measure for the price model (5.3).

Proposition 5.1 *Given $\mathbb{Q} \in \mathcal{Q}_{\approx}$, let $\theta_0 \in \mathcal{L}(W)$ and $\theta_1 \in \mathcal{G}(\mu)$ be the corresponding processes describing the density processes found in Lemma 3.2.*

Then, the following equivalence holds:

$$\mathbb{Q} \in \mathcal{Q}_{elmm} \Longleftrightarrow \alpha_t + \beta_t \theta_0(t) + \int_{\mathbb{R}_0} \gamma(t,x)\theta_1(t,x)\nu(dx) = 0 \;\forall t \geq 0 \quad \mathbb{P}\text{-}a.s. \tag{5.10}$$

6 Robust Utility Maximization

The goal of the economic agent, with an initial capital $x > 0$, will be now to maximize the penalized expected utility from a terminal wealth in the worst case model. Given a penalty function ϑ, this means that the agent seeks to solve the associated robust expected utility problem with value function

$$u(x) := \sup_{X \in \mathcal{X}(x)} \inf_{\mathbb{Q} \in \mathcal{Q}^{\vartheta}_{\ll}(\mathbb{P})} \left\{ \mathbb{E}_{\mathbb{Q}}\left[U(X_T)\right] + \vartheta(\mathbb{Q}) \right\}, \tag{6.1}$$

where $\mathcal{Q}^{\vartheta}_{\ll} := \{\mathbb{Q} \ll \mathbb{P} : \vartheta(\mathbb{Q}) < \infty\}$ for a fixed reference measure $\mathbb{P}$; see (4.2). A utility function $U : (0,\infty) \longrightarrow \mathbb{R}$ will be hereafter a strictly increasing, strictly concave, continuously differentiable real function, which satisfies the Inada conditions, namely $U'(0+) = +\infty$ and $U'(\infty-) = 0$.

The Fenchel-Legendre transformation of the function $-U(-x)$ is defined by

$$V(y) = \sup_{x>0} \{U(x) - xy\}, \qquad y > 0. \tag{6.2}$$

This function V is continuously differentiable, decreasing, and strictly convex, satisfying: $V'(0+) = -\infty$, $V'(\infty) = 0$, $V(0+) = U(\infty)$, $V(\infty) = U(0+)$. Further, the biconjugate of U is again U itself, i.e.

$$U(x) = \inf_{y>0} \{V(y) + xy\}, \qquad x > 0.$$

For a fixed prior measure $\mathbb{Q}$, in Kramkov and Schachermayer [13] the dual problem was formulated in terms of the value function

$$v_{\mathbb{Q}}(y) := \inf_{Y \in \mathcal{Y}_{\mathbb{Q}}(y)} \left\{ \mathbb{E}_{\mathbb{Q}}\left[V(Y_T)\right] \right\}, \tag{6.3}$$

where

$$\mathcal{Y}_{\mathbb{Q}}(y) := \{Y \geq 0 : \; Y_0 = y, YX \; \mathbb{Q}\text{-supermartingale} \; \forall X \in \mathcal{X}(1)\}. \tag{6.4}$$

A similar problem was studied in [11] for diffusion processes and the logarithmic utility function.

Remark 6.1 To guarantee that the $\mathbb{Q}$-expectations in (6.1) and (6.3) are well defined, we extend the operator $\mathbb{E}_{\mathbb{Q}}[U(\cdot)]$ to $\mathcal{L}^0$, as in Schied [19, p. 111], in the following way

$$\mathbb{E}_{\mathbb{Q}}[X] := \sup_{n\in\mathbb{N}} \mathbb{E}_{\mathbb{Q}}[X \wedge n] = \lim_{n\to\infty} \mathbb{E}_{\mathbb{Q}}[X \wedge n] \quad X \in \mathcal{L}^0(\Omega, \mathcal{F}). \tag{6.5}$$

The corresponding dual value function, in the robust setting, is defined by

$$v(y) := \inf_{\mathbb{Q}\in\mathcal{Q}^{\vartheta}_{\ll}} \left\{ v_{\mathbb{Q}}(y) + \vartheta(\mathbb{Q}) \right\}. \tag{6.6}$$

In the rest of this section the connection between the penalty functions (4.1) and the existence of solutions to the penalized robust expected utility problem (6.1) is established. The first step in this direction is to notice that given Theorem 4.1, where the minimality of the penalty function was proved, it is possible to write the primal problem (6.1) as

$$u(x) = \inf_{\mathbb{Q}\in\mathcal{Q}^{\vartheta}_{\ll}(\mathbb{P})} \sup_{X\in\mathcal{X}(x)} \left\{ \mathbb{E}_{\mathbb{Q}}[U(X_T)] + \vartheta(\mathbb{Q}) \right\}.$$

See Schied [19, Theorem 2.3]. Then, based on the duality theory for solving the classical optimal investment problem, the dual problem (6.6) is solved using the analogous sufficient conditions introduced by Kramkov and Schachermayer [13]. More precisely, for the class of utility functions described at the beginning of this section, when

$$v_{\mathbb{Q}}(y) < \infty \quad \text{for all } \mathbb{Q} \in \mathcal{Q}^{\vartheta}_{\approx} \quad \text{and} \quad y > 0, \tag{6.7}$$

where $\mathcal{Q}^{\vartheta}_{\approx} := \{\mathbb{Q} \approx \mathbb{P} : \vartheta(\mathbb{Q}) < \infty\}$ and ϑ is the minimal penalty function of the associated convex measure of risk, we are able to conclude that there exists an optimal solution to the dual problem (6.6), from which we can obtain an optimal solution to (6.1), using Schied [19, Theorems 2.3 and 2.5]. For the proof of the main result of this section, namely Theorem 6.1, we shall verify that these sufficient conditions are satisfied.

6.1 Penalties and Solvability

Let us now introduce the class

$$\mathcal{C} := \left\{ \mathcal{E}\left(Z^{\xi}\right) : \begin{array}{c} \xi := \left(\xi^{(0)}, \xi^{(1)}\right),\ \xi^{(0)} \in \mathcal{L}(W),\ \xi^{(1)} \in \mathcal{G}(\mu),\ \text{with} \\ \alpha_t + \beta_t \xi^{(0)}_t + \int_{\mathbb{R}_0} \gamma(t,x)\xi^{(1)}(t,x)\,\nu(dx) = 0 \text{ Lebesgue } \forall t \end{array} \right\}, \tag{6.8}$$

where

$$Z_t^{\xi} := \int\limits_{]0,t]} \xi^{(0)} dW + \int\limits_{]0,t]\times\mathbb{R}_0} \xi^{(1)}(s,x)\left(\mu(ds,dx) - ds\,\nu(dx)\right).$$

Observe that $\mathcal{D}_{elmm}(\mathbb{P}) \subset \mathcal{C} \subset \mathcal{Y}_{\mathbb{P}}(1)$; see (6.4) for the definition of $\mathcal{Y}_{\mathbb{P}}(1)$. This relation between these three sets plays a crucial role in the formulation of the dual problem, even in the non-robust case.

Theorem 6.1 *For $q \in (-\infty,1)\setminus\{0\}$, let $U(x) := \frac{1}{q}x^q$ be the power utility function, and consider the functions h, h_0 and h_1 as in Sect. 4, satisfying the following conditions:*

$$h(x) \geq \exp\left(\kappa_1 x^2\right) - 1 \text{ where } \kappa_1 := 1 \vee 2\left(2p^2+p\right)T \text{ and } p := \tfrac{q}{1-q},$$
$$h_0(x) \geq |x|,$$
$$h_1(x) \geq \tfrac{|x|}{c}, \text{for } c \text{ as in assumption } (A\ 1).$$

Then, for the penalty function

$$\vartheta_{x^q}(\mathbb{Q}) := \mathbb{E}_{\mathbb{Q}}\left[\int\limits_0^T h\left(h_0(\theta_0(t)) + \int_{\mathbb{R}_0} |\gamma(t,x)|\, h_1(\theta_1(t,x))\,\nu(dx)\right)dt\right],$$

the penalized robust utility maximization problem (6.1) has a solution.

Proof The penalty function ϑ_{x^q} is bounded from below, and by Theorem 4.1 equals on $\mathcal{Q}_{\ll}(\mathbb{P})$ the minimal penalty function of the normalized and sensitive convex measure of risk defined in (4.3). Therefore, we only need to prove that condition (6.7) holds. In order to prove that, fix an arbitrary probability measure $\mathbb{Q} \in \mathcal{Q}_{\approx}^{\vartheta_{x^q}} = \{\mathbb{Q} \approx \mathbb{P} : \vartheta_{x^q}(\mathbb{Q}) < \infty\}$ and let $\theta = (\theta_0, \theta_1)$ be the corresponding coefficients obtained in Lemma 3.2.

(1) In Lemma 4.2, Schied [19] establishes that even for $\mathbb{Q} \in \mathcal{Q}_{\ll}$, with density process D, the next equivalence holds

$$Y \in \mathcal{Y}_{\mathbb{Q}}(y) \Leftrightarrow YD \in \mathcal{Y}_{\mathbb{P}}(y).$$

Therefore, for $\mathbb{Q} \in \mathcal{Q}_{\ll}^{\vartheta_{x^q}}$, with coefficient $\theta = (\theta_0, \theta_1)$, it follows that

$$v_{\mathbb{Q}}(y) = \inf_{Y\in\mathcal{Y}_{\mathbb{P}}(1)}\left\{\mathbb{E}_{\mathbb{Q}}\left[V\left(y\frac{Y_T}{D_T^{\mathbb{Q}}}\right)\right]\right\} \leq \inf_{\xi\in\mathcal{C}}\left\{\mathbb{E}_{\mathbb{Q}}\left[V\left(y\frac{\mathcal{E}\left(Z^{\xi}\right)_T}{\mathcal{E}\left(Z^{\theta}\right)_T}\right)\right]\right\}.$$

(2) Define

$$\varepsilon_t := \alpha_t + \beta_t \theta_0(t) + \int_{\mathbb{R}_0} \gamma(t,x)\,\theta_1(t,x)\,\nu(dx),$$

the process involved in the definition of the class $\mathcal{C}$ in (6.8).

When ε_t is identically zero for all $t > 0$, Proposition 5.1 implies that $\mathbb{Q} \in \mathcal{Q}_{elmm}$. However, for $\mathbb{Q} \in \mathcal{Q}_{elmm}$ the constant process $Y \equiv y$ belongs to $\mathcal{Y}_{\mathbb{Q}}(y)$, and it follows that $v_{\mathbb{Q}}(y) < \infty$, for all $y > 0$. In this case the proof is concluded.

If ε is not identically zero, consider $\xi_t^{(0)} := \theta_0(t) - \frac{\varepsilon_t}{\beta_t}$ and $\xi^{(1)} := \theta_1$. Since

$$\left\{ \frac{1}{\beta_t} \int_{\mathbb{R}_0} \gamma(t,x)\,\theta_1(t,x)\,\nu(dx) \right\}_{t\in[0,T]} \in \mathcal{P}$$

and

$$\infty > \vartheta_{x^q}(\mathbb{Q}) \geq \mathbb{E}_{\mathbb{Q}}\left[\int_0^T \left(\frac{1}{\beta_t} \int_{\mathbb{R}_0} \gamma(t,x)\,\theta_1(t,x)\,\nu(dx) \right)^2 dt \right] - T,$$

it follows that $\left\{ \frac{1}{\beta_t} \int_{\mathbb{R}_0} \gamma(t,x)\,\theta_1(t,x)\,\nu(dx) \right\}_{t\in[0,T]} \in \mathcal{L}(W')$ for W' a $\mathbb{Q}$-Wiener process and thus also $\xi^{(0)} \in \mathcal{L}(W')$. Moreover, for $\xi = \left(\xi^{(0)}, \xi^{(1)}\right)$ we have that $\mathcal{E}\left(Z^{\xi}\right) \in \mathcal{C}$.

Using Girsanov's theorem, we obtain further

$$\frac{\mathcal{E}\left(Z^{\xi}\right)_t}{\mathcal{E}\left(Z^{\theta}\right)_t} = \exp\left\{ \int_{]0,t]} \left(-\frac{\varepsilon_u}{\beta_u} \right) dW'_u - \frac{1}{2} \int_{]0,t]} \left(\frac{\varepsilon_u}{\beta_u} \right)^2 du \right\}.$$

(3) The Cauchy-Bunyakovsky-Schwarz inequality yields

$$\begin{aligned}
&\mathbb{E}_{\mathbb{Q}}\left[V\left(y \frac{\mathcal{E}(Z^{\xi})_T}{\mathcal{E}(Z^{\theta})_T} \right) \right] \\
&= \frac{1}{p} y^{-p} \mathbb{E}_{\mathbb{Q}}\left[\exp\left\{ p \int_{]0,T]} \left(\frac{\varepsilon_t}{\beta_t} \right) dW' + \frac{p}{2} \int_{]0,T]} \left(\frac{\varepsilon_t}{\beta_t} \right)^2 dt \right\} \right] \\
&\leq \frac{1}{p} y^{-p} \mathbb{E}_{\mathbb{Q}}\left[\exp\left\{ 2p \int_{]0,T]} \left(\frac{\varepsilon_t}{\beta_t} \right) dW' - \frac{4p^2}{2} \int_{]0,T]} \left(\frac{\varepsilon_t}{\beta_t} \right)^2 dt \right\} \right]^{\frac{1}{2}} \\
&\quad \times \mathbb{E}_{\mathbb{Q}}\left[\exp\left\{ \left(\frac{4p^2}{2} + p \right) \int_{]0,T]} \left(\frac{\varepsilon_t}{\beta_t} \right)^2 dt \right\} \right]^{\frac{1}{2}}.
\end{aligned} \tag{6.9}$$

On the other hand, the process

$$\exp\left\{2p\int_{]0,T]}\left(\frac{\varepsilon_t}{\beta_t}\right)dW' - \frac{4p^2}{2}\int_{]0,T]}\left(\frac{\varepsilon_t}{\beta_t}\right)^2 dt\right\} \in \mathcal{M}_{loc}(\mathbb{Q})$$

is a local $\mathbb{Q}$-martingale and, since it is positive, is a supermartingale. Hence,

$$\mathbb{E}_{\mathbb{Q}}\left[\exp\left\{2p\int_{]0,T]}\left(\frac{\varepsilon_t}{\beta_t}\right)dW' - \frac{4p^2}{2}\int_{]0,T]}\left(\frac{\varepsilon_t}{\beta_t}\right)^2 dt\right\}\right] \leq 1.$$

Therefore we need only to take care about $\mathbb{E}_{\mathbb{Q}}\left[\exp\left\{\left(\frac{4p^2}{2}+p\right)\int_{]0,T]}\left(\frac{\varepsilon_t}{\beta_t}\right)^2 dt\right\}\right]$ in order to have the desired integrability. From assumption $(A\ 2)$ we have

$$\mathbb{E}_{\mathbb{Q}}\left[\exp\left\{\left(2p^2+p\right)2\int_{]0,T]}\left(\left|\frac{\alpha_t}{\beta_t}\right|\right)^2 dt\right\}\right] < C,$$

and thus

$$\mathbb{E}_{\mathbb{Q}}\left[\exp\left\{\left(\frac{4p^2}{2}+p\right)\int_{]0,T]}\left(\frac{\varepsilon_t}{\beta_t}\right)^2 dt\right\}\right] \leq C\mathbb{E}_{\mathbb{Q}}\left[\exp\left\{2\left(2p^2+p\right)\right.\right.$$

$$\left.\left.\times\int_0^T\left(|\theta_0(t)| + \frac{1}{|\beta_t|}\left|\int_{\mathbb{R}_0}\gamma(t,x)\,\theta_1(t,x)\,\nu(dx)\right|\right)^2 dt\right\}\right].$$

Finally, observe that for $\mathbb{Q} \in \mathcal{Q}_{\approx}^{\vartheta_{x^q}}$, using that it has finite penalization $\vartheta_{x^q}(\mathbb{Q}) < \infty$ and Jensen's inequality, we have

$$\infty > \mathbb{E}_{\mathbb{Q}}\left[\exp\left\{\frac{\kappa_1}{T}\int_0^T\left(h_0(\theta_0(t)) + \int_{\mathbb{R}_0}|\gamma(t,x)|\,h_1(\theta_1(t,x))\,\nu(dx)\right)^2 dt\right\}\right]$$

$$\geq \mathbb{E}_{\mathbb{Q}}\left[\exp\left\{2\left(2p^2+p\right)\int_0^T\left(|\theta_0(t)| + \frac{1}{|\beta_t|}\left|\int_{\mathbb{R}_0}\gamma(t,x)\right.\right.\right.\right.$$

$$\left.\left.\left.\left.\theta_1(t,x)\,\nu(dx)\right|\right)^2 dt\right\}\right].$$

From the last two displays it follows that the r.h.s. of (6.9) is finite and the theorem follows. □

Next theorem establishes a sufficient condition for the existence of solution to the robust utility maximization problem (6.1) for an arbitrary utility function.

Theorem 6.2 *Suppose that the utility function $\widetilde{U}$ is bounded above by a power utility U, with penalty function ϑ_{x^q} associated to U as in Theorem 6.1. Then, the robust utility maximization problem (6.1) for $\widetilde{U}$ with penalty ϑ_{x^q} has an optimal solution.*

Proof Since $U(x) := \frac{1}{q}x^{-q} \geq \widetilde{U}(x)$ for all $x > 0$, for some $q \in (-\infty, 1) \setminus \{0\}$ the corresponding convex conjugate functions satisfy $V(y) \geq \widetilde{V}(y)$ for each $y > 0$. As it was pointed out in Remark 6.2, we can restrict ourself to the positive part $\widetilde{V}^{+}(y)$. From Proposition 6.1, we can fix some $Y \in \mathcal{Y}_{\mathbb{Q}}(y)$ such that $\mathbb{E}_{\mathbb{Q}}[V(Y_T)] < \infty$ for any $\mathbb{Q} \in \mathcal{Q}_{\approx}^{\vartheta_{x^q}}$ and $y > 0$, arbitrary, but fixed. Furthermore, the inequality $V(y) \geq \widetilde{V}(y)$ implies that their inverse functions satisfy $\left(V^{+}\right)^{(-1)}(n) \geq \left(\widetilde{V}^{+}\right)^{(-1)}(n)$ for all $n \in \mathbb{N}$, and hence

$$\sum_{n=1}^{\infty} \mathbb{Q}\left[Y_T \leq \left(\widetilde{V}^{+}\right)^{(-1)}(n)\right] \leq \sum_{n=1}^{\infty} \mathbb{Q}\left[Y_T \leq \left(V^{+}\right)^{(-1)}(n)\right] < \infty.$$

The Moments Lemma ($\mathbb{E}_{\mathbb{Q}}[|X|] < \infty \Leftrightarrow \sum_{n=1}^{\infty} \mathbb{Q}[|X| \geq n] < \infty$) yields $\mathbb{E}_{\mathbb{Q}}\left[\widetilde{V}^{+}(Y_T)\right] < \infty$, and the assertion follows. □

From the proof of Theorem 6.2 it is clear that the behavior of the convex conjugate function in a neighborhood of zero is fundamental. From this observation we conclude the following.

Corollary 6.1 *Let U be a utility function with convex conjugate V, and ϑ a penalization function such that the robust utility maximization problem (6.1) has a solution. For a utility function $\widetilde{U}$ such that their convex conjugate function $\widetilde{V}$ is majorized in an ε-neighborhood of zero by V, the corresponding utility maximization problem (6.1) has a solution.*

Remark 6.2 When the conjugate convex function V is bounded from above it follows immediately that the penalized robust utility maximization problem (6.1) has a solution for any proper penalty function ϑ. This is the case, for instance, of the power utility function $U(x) := \frac{1}{q}x^{q}$, for $q \in (-\infty, 0)$, where the convex conjugate function $V(x) = \frac{1}{p}x^{-p} \leq 0$, with $p := \frac{q}{1-q}$.

Next we give an alternative representation of the robust dual value function, introduced in (6.6), in terms of the family $\mathcal{C}$ of stochastic processes.

Theorem 6.3 *For a utility function U satisfying condition (6.7), the dual value function can be written as*

$$\begin{aligned} v(y) &= \inf_{\mathbb{Q}\in\mathcal{Q}_{\approx}^{\vartheta}} \left\{ \inf_{\xi\in\mathcal{C}} \left\{ \mathbb{E}_{\mathbb{Q}}\left[V\left(y\frac{\mathcal{E}(Z^{\xi})_T}{D_T^{\mathbb{Q}}} \right) \right] \right\} + \vartheta(\mathbb{Q}) \right\} \\ &= \inf_{\mathbb{Q}\in\mathcal{Q}_{\ll}} \left\{ \inf_{\xi\in\mathcal{C}} \left\{ \mathbb{E}_{\mathbb{Q}}\left[V\left(y\frac{\mathcal{E}(Z^{\xi})_T}{D_T^{\mathbb{Q}}} \right) \right] \right\} + \vartheta(\mathbb{Q}) \right\}. \end{aligned} \tag{6.10}$$

Proof Condition (6.7), together with Lemma 4.4 in [19] and Theorem 2 in [14] , imply the following identity

$$v(y) = \inf_{\mathbb{Q}\in\mathcal{Q}_{\approx}^{\vartheta}} \left\{ \inf_{\widetilde{\mathbb{Q}}\in\mathcal{Q}_{elmm}(\mathbb{Q})} \left\{ \mathbb{E}_{\mathbb{Q}}\left[V\left(yd\widetilde{\mathbb{Q}}/d\mathbb{Q} \right) \right] \right\} + \vartheta(\mathbb{Q}) \right\}.$$

Since $\mathcal{D}_{elmm}(\mathbb{P}) \subset \mathcal{C}$, we get

$$\begin{aligned} v(y) &\geq \inf_{\mathbb{Q}\in\mathcal{Q}_{\approx}^{\vartheta}} \left\{ \inf_{\xi\in\mathcal{C}} \left\{ \mathbb{E}_{\mathbb{Q}}\left[V\left(y\frac{\mathcal{E}(Z^{\xi})_T}{D_T^{\mathbb{Q}}} \right) \right] \right\} + \vartheta(\mathbb{Q}) \right\} \\ &\geq \inf_{\mathbb{Q}\in\mathcal{Q}_{\ll}} \left\{ \inf_{\xi\in\mathcal{C}} \left\{ \mathbb{E}_{\mathbb{Q}}\left[V\left(y\frac{\mathcal{E}(Z^{\xi})_T}{D_T^{\mathbb{Q}}} \right) \right] \right\} + \vartheta(\mathbb{Q}) \right\}. \end{aligned} \tag{6.11}$$

Finally, from Lemma 4.2 in Schied [19] and $\mathcal{C} \subset \mathcal{Y}_{\mathbb{P}}(1)$ follows

$$v_{\mathbb{Q}}(y) \leq \inf_{\xi\in\mathcal{C}} \left\{ \mathbb{E}_{\mathbb{Q}}\left[V\left(y\frac{\mathcal{E}(Z^{\xi})_T}{D_T^{\mathbb{Q}}} \right) \right] \right\},$$

and we have the inequalities (6.11) in the other direction, and the result follows. □

6.2 The Logarithmic Utility Case

The existence of solution to the robust problem for the logarithmic utility function $U(x) = \log(x)$ can be obtain using the relation between this utility function and the relative entropy function. Let h, h_0 and h_1 be as in Sect. 4, satisfying also the following growth conditions:

$$\begin{aligned} h(x) &\geq x, \\ h_0(x) &\geq \frac{1}{2}x^2, \\ h_1(x) &\geq \{|x| \vee x\ln(1+x)\}\,\mathbf{1}_{(-1,0)}(x) + x(1+x)\,\mathbf{1}_{\mathbb{R}_+}(x). \end{aligned}$$

Now, define the penalization function

$$\vartheta_{\log}(\mathbb{Q}) := \mathbb{E}_{\mathbb{Q}}\left[\int_0^T h\left(h_0(\theta_0(t)) + \int_{\mathbb{R}_0} h_1(\theta_1(t,x))\,\nu(dx)\right) dt\right] \mathbf{1}_{\mathcal{Q}_{\ll}}(\mathbb{Q})$$
$$+\infty \times \mathbf{1}_{\mathcal{Q}_{cont}\setminus\mathcal{Q}_{\ll}}(\mathbb{Q}). \tag{6.12}$$

Remark 6.3 Notice that when $\mathbb{Q} \in \mathcal{Q}_{\ll}^{\vartheta_{\log}}(\mathbb{P})$ with coefficient $\theta = (\theta_0, \theta_1)$ has a finite penalization, the following $\mathbb{Q}$-integrability properties hold:

(6.3.i) $\int_{[0,T]\times\mathbb{R}_0} \theta_1(t,x)\,\mu_{\mathbb{P}}^{\mathcal{P}}(dt,dx) \in \mathcal{L}^1(\mathbb{Q})$

(6.3.ii) $\int_{[0,T]\times\mathbb{R}_0} \{1+\theta_1(t,x)\}\ln(1+\theta_1(t,x))\,\mu_{\mathbb{P}}^{\mathcal{P}}(dt,dx) \in \mathcal{L}^1(\mathbb{Q})$

(6.3.iii) $\int_{[0,T]\times\mathbb{R}_0} \ln(1+\theta_1(s,x))\,\mu(ds,dx) \in \mathcal{L}^1(\mathbb{Q})$

(6.3.iv) $\mathbb{E}_{\mathbb{Q}}\left[\int_{]0,T]\times\mathbb{R}_0} \ln(1+\theta_1)\,d\mu\right] = \mathbb{E}_{\mathbb{Q}}\left[\int_{]0,T]\times\mathbb{R}_0} \{\ln(1+\theta_1)\}(1+\theta_1)\,d\mu_{\mathbb{P}}^{\mathcal{P}}\right]$

In addition, for $\mathbb{Q} \in \mathcal{Q}_{\approx}^{\vartheta_{\log}}(\mathbb{P})$ we have

(6.3.v) $\int_{[0,T]\times\mathbb{R}_0} \theta_1(s,x)\,\mu(ds,dx) < \infty$ $\mathbb{P}-a.s.$

For $\mathbb{Q} \in \mathcal{Q}_{\ll}(\mathbb{P})$, the relative entropy function is defined as

$$H(\mathbb{Q}|\mathbb{P}) = \mathbb{E}\left[D_T^{\mathbb{Q}} \log\left(D_T^{\mathbb{Q}}\right)\right].$$

Lemma 6.1 *Given* $\mathbb{Q} \in \mathcal{Q}_{\approx}^{\vartheta_{\log}}(\mathbb{P})$, *it follows that*

$$H(\mathbb{Q}\,|\mathbb{P}) \le \vartheta_{\log}(\mathbb{Q}).$$

Proof For $\mathbb{Q} \in \mathcal{Q}_{\approx}^{\vartheta_{\log}}(\mathbb{P})$ we have that θ_0 is integrable w.r.t. W' a $\mathbb{Q}$-Wiener process as an square integrable martingale. Further Remark 6.3 implies that

$$H(\mathbb{Q}\,|\mathbb{P}) = \mathbb{E}_{\mathbb{Q}}\left[\frac{1}{2}\int_0^T (\theta_0)^2\,ds + \int_{]0,T]\times\mathbb{R}_0} \ln(1+\theta_1(s,x))\,\mu(ds,dx)\right.$$
$$\left.-\int_0^T\int_{\mathbb{R}_0} \theta_1(s,x)\,\nu(dx)\,ds\right]$$

$$
\begin{aligned}
&\leq \mathbb{E}_{\mathbb{Q}}\left[\int_0^T \left\{\frac{1}{2}(\theta_0)^2 ds + \int_{\mathbb{R}_0} \{\ln(1+\theta_1(s,x))\}\theta_1(s,x)\,\nu(dx)\right\} ds\right] \\
&\leq \vartheta_{\log}(\mathbb{Q}).
\end{aligned}
$$

□

Using the previous result, the existence of solution to the primal problem (6.1) can be concluded.

Proposition 6.1 *Let* $U(x) = \log(x)$ *and* $\vartheta_{\log}$ *be as in (6.12). Then the robust utility maximization problem (6.1) has an optimal solution.*

Proof Again, we only need to verify that condition (6.7) holds. Observe that, for each $\mathbb{Q} \in \mathcal{Q}_{\approx}^{\vartheta_{\log}}(\mathbb{P})$, we have that

$$
v_{\mathbb{Q}}(y) \leq \inf_{\xi\in\mathcal{C}}\left\{\mathbb{E}\left[D_T^{\mathbb{Q}}\log\left(\frac{D_T^{\mathbb{Q}}}{\mathcal{E}(Z^{\xi})_T}\right) - \log(y) - 1\right]\right\}.
$$

Also, Proposition 5.1 and the Novikov condition yield for $\widetilde{\xi} \in \mathcal{C}$, with $\widetilde{\xi}^{(0)} := -\frac{\alpha_s}{\beta_s}$ and $\widetilde{\xi}^{(1)} := 0$, that $\widetilde{\mathbb{Q}} \in \mathcal{Q}_{elmm}$, where $d\widetilde{\mathbb{Q}}\backslash d\mathbb{P} = D_T^{\widetilde{\xi}} := \mathcal{E}\left(Z^{\widetilde{\xi}}\right)_T$. Further, from Lemma 6.1 we conclude for $\mathbb{Q} \in \mathcal{Q}_{\approx}^{\vartheta_{\log}}(\mathbb{P})$ that

$$
\mathbb{E}\left[D_T^{\mathbb{Q}}\log\left(\frac{D_T^{\mathbb{Q}}}{D_T^{\widetilde{\xi}}}\right)\right] = H(\mathbb{Q}\,|\mathbb{P}) + \mathbb{E}_{\mathbb{Q}}\left[\int_0^T \frac{\alpha_s}{\beta_s}\theta_s^{(0)} ds + \frac{1}{2}\int_0^T \left(\frac{\alpha_s}{\beta_s}\right)^2 ds\right] < \infty
$$

and the claim follows. □

References

1. P. Artzner, F. Delbaen, J.M. Eber, D. Heath, Thinking coherently. Risk Mag. **10**, 68–71 (1997)
2. P. Artzner, F. Delbaen, J.M. Eber, D. Heath, Coherent measures of risk. Math. Finance **9**, 203–228 (1999)
3. F. Delbaen, Coherent risk measures on general probability spaces, in *Advances in Finance and Stochastics, Essays in Honor of Dieter Sondermann*, ed. by K. Sandmann, Ph. Schönbucher (Springer, Berlin, 2002), pp. 1–37
4. H. Föllmer, A. Schied, Convex measures of risk and trading constraints. Finance Stoch. **6**, 429–447 (2002)
5. H. Föllmer, A. Schied, Robust preferences and convex risk measures, in *Advances in Finance and Stochastics, Essays in Honor of Dieter Sondermann* (Springer, Berlin, 2002), pp. 39–56
6. H. Föllmer, A. Schied, *Stochastic Finance. An Introduction in Discrete Time*, 2nd edn. de Gruyter Studies in Mathematics, vol. 27 (de Gruyter, Berlin, 2004)

7. M. Frittelli, E. Rosazza Gianin, Putting order in risk measures. J. Bank. Finance **26**, 1473–1486 (2002)
8. M. Frittelli, E. Rosazza Gianin, Dynamic convex risk measures, in *Risk Measures for the 21st Century*, ed. by G. Szegö (Wiley, New York, 2004), pp. 227–248
9. S.W. He, J.G. Wang, J.A. Yan, *Semimartingale Theory and Stochastic Calculus* (Science Press, Beijing, 1992)
10. D. Heath, Back to the future. Plenary Lecture at the First World Congress of the Bachelier Society, Paris (2000)
11. D. Hernández-Hernández, A. Schied, A control approach to robust utility maximization with logarithmic utility and time consistent penalties. Stoch. Process. Appl. **117**, 980–1000 (2007)
12. J. Jacod, A. Shiryaev, *Limit Theorems for Stochastic Processes*, 2nd edn. (Springer, Berlin, 2003)
13. D. Kramkov, W. Schachermayer, The asymptotic elasticity of utility functions and optimal investment in incomplete markets. Ann. Appl. Probab. **9**, 904–950 (1999)
14. D. Kramkov, W. Schachermayer, Necessary and sufficient conditions in the problem of optimal investment in incomplete markets. Ann. Appl. Probab. **13**, 1504–1516 (2003)
15. V. Krätschmer, Robust representation of convex risk measures by probability measures. Finance Stoch. **9**, 597–608 (2005)
16. H. Kunita, Representation of martingales with jumps and applications to mathematical finance, in *Stochastic Analysis and Related Topics in Kyoto*, ed. by H. Kunita et al. (Mathematical Society of Japan, Tokyo, 2004), pp. 209–232
17. F. Maccheroni, M. Marinacci, A. Rustichini, Ambiguity aversion, robustness and the variational representation of preferences. Econometrica **74**, 1447–1498 (2006)
18. M.-C. Quenez, Optimal portfolio in a multiple-priors model, in *Seminar on Stochastic Analysis, Random Fields and Applications IV*, ed. by R. Dalang, M. Dozzi, F. Russo. Progress in Probability, vol. 58 (Birkhäuser, Basel, 2004), pp. 291–321
19. A. Schied, Optimal investments for risk- and ambiguity-averse preferences: a duality approach. Finance Stoch. **11**, 107–129 (2007)

Blackwell-Nash Equilibria in Zero-Sum Stochastic Differential Games

Beatris Adriana Escobedo-Trujillo, Héctor Jasso-Fuentes, and José Daniel López-Barrientos

Abstract Advanced-type equilibria for a general class of zero-sum stochastic differential games have been studied in part by Escobedo-Trujillo et al. (J Optim Theory Appl 153:662–687, 2012), in which a comprehensive study of the so-named *bias and overtaking equilibria* was provided. On the other hand, a complete analysis of advanced optimality criteria in the context of optimal control theory such as bias, overtaking, sensitive discount, and Blackwell optimality was developed independently by Jasso-Fuentes and Hernández-Lerma (Appl Math Optim 57:349–369, 2008; J Appl Probab 46:372–391, 2009; Stoch Anal Appl 27:363–385, 2009). In this work we try to fill out the gap between the aforementioned references. Namely, the aim is to analyze Blackwell-Nash equilibria for a general class of zero-sum stochastic differential games. Our approach is based on the use of dynamic programming, the Laurent series and the study of sensitive discount optimality.

Keywords Zero-sum stochastic differential games · Average equilibrium · Bias equilibrium · Laurent series · Blackwell-Nash equilibrium

2010 Mathematics Subject Classification 91A10, 91A15, 91A25

B. A. Escobedo-Trujillo
Engineering Faculty, Universidad Veracruzana, Coatzacoalcos, Veracruz, México
e-mail: bescobedo@uv.mx

H. Jasso-Fuentes (✉)
Departamento de Matemáticas, CINVESTAV-IPN, México, D.F., México
e-mail: hjasso@math.cinvestav.mx

J. D. López-Barrientos
Facultad de Ciencias Actuariales, Universidad Anáhuac México, Huixquilucan, Estado de México, México
e-mail: daniel.lopez@anahuac.mx

© Springer International Publishing AG, part of Springer Nature 2018

D. Hernández-Hernández et al. (eds.), *XII Symposium of Probability and Stochastic Processes*, Progress in Probability 73,
https://doi.org/10.1007/978-3-319-77643-9_5

1 Introduction

Among the most common payoff functions existing in the literature in the general theory of dynamic games we can mention the (finite-horizon) Bolza-type payoff and the well-known (infinite-horizon) discounted and average payoffs. The key features of these two last criteria is that, whereas the discounted payoff only focuses on earlier revenues, the average reward ignores these and pays attention only to the asymptotic behavior of the utilities. A drawback of these points of view is that they do not consider what happens in the mid-run. For example, there can be N-tuples of strategies (N represents the number of players in the game) that might be "optimal" for all the players in the infinite-horizon average criterion, but in turn, they provide low profits (and/or high costs) to the players at *any* finite period of time. From another angle, there exist several applications in which the (infinite-horizon) discounted payoff criterion is used to model the real or present value (at the current time) of a company; the key ingredient is the use of a *discount factor*. However, in some other situations, this criterion might be used for some other purposes; one of them is to regard it as an estimate of criteria without discount.

To fix ideas, suppose we have a game such that $\bar{\pi} = (\pi_1 \cdots, \pi_N)$ represents an N-tuple associated to some choices of the players (i.e., π_i corresponds to the strategy of player i), and denote by r_i the associated payoff rate function of player i (for illustrative purposes let us assume for the moment that all players have the same reward rate; i.e., $r_i = r$ for all $i = 1, \cdots N$). The expected undiscounted and discounted payoffs of $\bar{\pi}$ for each player are defined, respectively, as

$$V(\bar{\pi}) = E\int_0^\infty r(x^{\bar{\pi}}(t))dt, \quad \text{and} \quad V_\alpha(\bar{\pi}) = E\int_0^\infty e^{-\alpha t} r(x^{\bar{\pi}}(t))dt,$$

where $x^{\bar{\pi}}(t)$ represents the state of the process under the policy $\bar{\pi}$ at time t, and $\alpha > 0$ is a given constant. A very important property of V_α is that, under mild assumptions, it is finite-valued; whereas the former requires very strong hypotheses to possess this feature. In this sense, if one is interested in studying optimality under the criterion V, one may regard such criterion as the limit of some sequence of V_α in the following sense:

$$V_{\alpha_n}(\bar{\pi}) \to V(\bar{\pi}) \quad \text{as} \quad \{\alpha_n\}_n \downarrow 0. \tag{1.1}$$

However, even when one can provide optimality results (Nash equilibria) to V_α for some *fixed* and of course *positive* and even *small* α, it turns out that this V_α, regarded as an estimate of V, is acceptable at *early* periods of times, but it is very imprecise in the long run.

An alternative approach that lies in the same direction of the limit (1.1) is the use of Blackwell-Nash equilibria. This consists essentially in seeking Nash equilibria that remain *optimal* for *all* the discounted payoffs V_α, $0 < \alpha < \alpha^*$, for some fixed $\alpha^* > 0$ (see Definition 8.1). Due to the nature of this class of equilibria, they turn out to be good "optimizers", when the payoff criterion under study is of type V.

The purpose of this work is to analyze Blackwell-Nash equilibria for a general class of zero-sum stochastic differential games; namely, we provide sufficient conditions for ensuring the existence and characterizations of these equilibria. This study is based on the analysis of the so-named sensitive discount equilibria introduced in Definition 8.2. It is worth noting that Blackwell-Nash equilibria have the property of being bias and overtaking equilibria too. In this sense, our present analysis is more general than [5], because we use the same set of assumptions. Finally, it is important to say that, due to the fact that our work studies only the zero-sum case, here and in the sequel, we consider only to the case $N = 2$ players.

Another interesting application concerning Blackwell games goes in the spirit of the so-named priority mean-payoff games, which are regarded as the limit of special multi-discounted games. In this type of games, Blackwell equilibria play an important role because of their stability property under small perturbations of the discount factor—see [7–9]. The study of Blackwell-Nash equilibria in zero-sum stochastic differential games also permits the extension to the theory of priority mean-payoff games in the stochastic differential games setting.

Bias and overtaking criteria have been studied in the context of zero-sum stochastic differential games; see, for example, [5, 17]. Nevertheless, to the best of our knowledge, the only works dealing with sensitive discount and Blackwell optimality, but in the context of controlled diffusions (i.e., the case of one player only) are [12, 13] and [22]. It is worth mentioning, however, that there are some works that are close to the present proposal. For instance, Arapostathis et al. [3] study a zero-sum stochastic differential game under a slightly different ergodicity assumption than ours. It states a parabolic Hamilton-Jacobi-Bellman (HJB) equation, and finds *risk-sensitive optimal selectors*, in the sense that the payoff form is "sensitive to higher moments of the running cost, and not merely its mean". This represents an alternative approach to ours, because while they deal with the concept of risk-sensitivity (as introduced in [25]), we rather choose the notion of *sensitive discount* in a Laurent series, as presented in [12] and [21]. Other works that are related to the selective criteria we study for stochastic diffusions are [5, 11–13, 17] and the references therein.

The rest of our work comprises eight short sections. In the next section we introduce the notation that we use, our game model, the main hypotheses, and the basic type of strategies we will deal with along our developments. Section 3 presents the long-run average optimality criterion, and a very well-known result on the existence of the corresponding Nash equilibria. Section 4 is devoted to the so called *bias criterion*. This is a first refinement of the criterion introduced in Sect. 3, and we profit from it by quoting the concepts introduced in that part in further sections. In Sect. 5 we extend the results from [21, Section 3] to the zero-sum case. There, we use an exponential ergodicity condition to characterize a discounted payoff in terms of a Laurent series. Sections 6–8 are extensions of the results from [12] and represent the main contribution of this paper. In Sect. 6 we define the so-called *Poisson system* and model its solution in terms of the criterion presented before in Sect. 3. Section 7 shows a connection between the Poisson system and the dynamic programming principle. There, we lay out the concept of *canonical equilibria* and represent it as the strategies for which certain HJB equations are met. In Sect. 8 we exhibit Blackwell-Nash and sensitive discount equilibria and relate them in some appropriated sense. We draw our conclusions in Sect. 9.

2 The Game Model and Main Assumptions

The Dynamic System Let us consider an n-dimensional diffusion process $x(\cdot)$ controlled by two players and evolving according to the stochastic differential equation

$$dx(t) = b(x(t), u_1(t), u_2(t))dt + \sigma(x(t))dW(t), \quad x(0) = x_0, \ t \geq 0, \tag{2.1}$$

where $b : \mathbb{R}^n \times U_1 \times U_2 \to \mathbb{R}^n$ and $\sigma : \mathbb{R}^n \to \mathbb{R}^{n\times d}$ are given functions, and $W(\cdot)$ is a d-dimensional standard Brownian motion. The sets $U_1 \subset \mathbb{R}^{m_1}$ and $U_2 \subset \mathbb{R}^{m_2}$ are given (Borel) sets. Moreover, for $i = 1, 2$, $u_i(\cdot)$ is a U_i-valued stochastic process representing the strategy of player i at each time $t \geq 0$.

Notation For vectors x and matrices A we consider the usual Euclidean norms

$$|x|^2 := \sum_k x_k^2 \ \text{ and } \ |A|^2 := \mathrm{Tr}(AA') = \sum_{i,j} A_{i,j}^2,$$

where A' and $\mathrm{Tr}(\cdot)$ denote the transpose and the trace of a matrix, respectively.

Assumption 2.1

(a) The action sets U_1 and U_2 are compact.

(b) $b(x, u_1, u_2)$ is continuous on $\mathbb{R}^n \times U_1 \times U_2$, and $x \mapsto b(x, u_1, u_2)$ satisfies a Lipschitz condition uniformly in $(u_1, u_2) \in U_1 \times U_2$; that is, there exists a positive constant K_1 such that

$$\sup_{(u_1,u_2)\in U_1\times U_2} |b(x, u_1, u_2) - b(y, u_1, u_2)| \leq K_1|x - y| \ \text{ for all } x, y \in \mathbb{R}^n.$$

(c) There exists a positive constant K_2 such that for all $x, y \in \mathbb{R}^n$,

$$|\sigma(x) - \sigma(y)| \leq K_2|x - y|.$$

(d) (Uniform ellipticity.) The matrix $a(x) := \sigma(x)\sigma'(x)$ satisfies that, for some constant $K_3 > 0$,

$$x'a(y)x \geq K_3|x|^2 \ \text{ for all } \ x, y \in \mathbb{R}^n.$$

For $(u_1, u_2) \in U_1 \times U_2$, and v in $C^2(\mathbb{R}^n)$, $p \geq 1$, let

$$\mathcal{L}^{u_1,u_2}v(x) := \sum_{i=1}^{n} b^i(x, u_1, u_2)\partial_i v(x) + \frac{1}{2}\sum_{i,j=1}^{n} a^{ij}(x)\partial_{ij}^2 v(x), \qquad (2.2)$$

where b^i is the i-th component of b, and a^{ij} is the (i, j)-component of the matrix $a(\cdot)$ defined in Assumption 2.1(d).

2.1 Strategies

Throughout this work, we will be interested in finding saddle points (see Theorem 3.5 below). To ensure that our search leads us to this result, we use the theory of relaxed controls—see for instance, [19, 24, 26]. The use of this class of controls, along with the semi-continuity properties of the cost/reward function (see Assumption 2.8(c) below) will give us the convex structure needed to guarantee the existence of non-cooperative Nash equilibria.

For each $k = 1, 2$, let $\mathcal{P}(U_k)$ be the space of probability measures on U_k endowed with the topology of weak convergence, and denote by $\mathcal{B}(U_k)$ the Borel σ-algebra of U_k.

Definition 2.2 A *randomized strategy* for player k is a family $\pi^k := \{\pi_t^k, t > 0\}$ of stochastic kernels on $\mathcal{B}(U_k) \times \mathbb{R}^n$ satisfying:

(a) for each $t \geq 0$ and $x \in \mathbb{R}^n$, $\pi_t^k(\cdot|x)$ is a probability measure on U_k such that $\pi_t^k(U_k|x) = 1$, and for each $D \in \mathcal{B}(U_k)$, $\pi_t^k(D|\cdot)$ is a Borel function on $\mathbb{R}^n$; and
(b) for each $D \in \mathcal{B}(U_k)$ and $x \in \mathbb{R}^n$, the mapping $t \longmapsto \pi_t^k(B|x)$ is Borel measurable.

We now introduce the notion of *stationary strategy*.

Definition 2.3 For each $k = 1, 2$, we say that a randomized strategy is *stationary* if and only if there is a probability measure $\pi^k(\cdot|x) \in \mathcal{P}(U_k)$ such that $\pi_t^k(\cdot|x) = \pi^k(\cdot|x)$ for all $x \in \mathbb{R}^n$ and $t \geq 0$.

The set of randomized stationary strategies for player $k = 1, 2$ is denoted by Π^k. It is important to state that we suppose the existence of a topology defined on Π^k, $k = 1, 2$, such that Π^k is compact—for more details see [14, Section 2].

For each pair of probability measures $(\phi, \psi) \in \mathcal{P}(U_1) \times \mathcal{P}(U_2)$ we write the drift coefficient b in (2.1) and the operator $\mathcal{L}$ in (2.2) in terms of these measures by means of the following expressions:

$$b(x,\phi,\psi) := \int_{U_2}\int_{U_1} b(x,u_1,u_2)\phi(du_1)\psi(du_2), \tag{2.3}$$

$$\mathcal{L}^{\phi,\psi}h(x) := \int_{U^2}\int_{U^1} \mathcal{L}^{u_1,u_2}h(x)\phi(du_1)\psi(du_2). \tag{2.4}$$

The notation above is valid also when the strategies $\pi^1 \in \Pi^1$ or/and $\pi^2 \in \Pi^2$ in (2.3)–(2.4) are interpreted as probability measures for each fixed $x \in \mathbb{R}^n$; that is, $\pi^k(\cdot|x) \in \mathcal{P}(U^k)$. In this case, unless the context requires further clarification, we shall simply write the "variable" π^k in the left-hand side of (2.3)–(2.4), rather than $\pi^k(\cdot|x)$.

Remark 2.4 Assumption 2.1 ensures that, for each pair of strategies $(\pi^1,\pi^2) \in \Pi^1 \times \Pi^2$ there exists an almost surely unique strong solution of (2.1) which is a Markov-Feller process. Furthermore, for each pair of strategies $(\pi^1,\pi^2) \in \Pi^1\times\Pi^2$, the operator $\mathcal{L}^{\pi^1,\pi^2}v$ in (2.4) becomes the infinitesimal generator of (2.1). (For more details, see the arguments of [2, Theorem 2.2.12] or [6, Theorem 2.1].)

Sometimes we write $x(\cdot)$ as $x^{\pi^1,\pi^2}(\cdot)$ to emphasize the dependence on $(\pi^1,\pi^2) \in \Pi^1 \times \Pi^2$. Also, we shall denote by $\mathbb{P}^{\pi^1,\pi^2}(t,x,\cdot)$ the corresponding transition probability of the process $x^{\pi^1,\pi^2}(\cdot)$, i.e., $\mathbb{P}^{\pi^1,\pi^2}(t,x,B) := \mathbb{P}(x^{\pi^1,\pi^2}(t) \in B|x(0) = x)$ for every Borel set $B \subset \mathbb{R}^n$ and $t \geq 0$. The symbol $\mathbb{E}_x^{\pi^1,\pi^2}(\cdot)$ stands for the associated conditional expectation.

Remark 2.5 In later sections, we will restrict ourselves to the space of stationary strategies within the class of randomized strategies. The reason is that the recurrence and ergodicity properties of the state system (2.1) can be easily verified through the use of such policies, but for a more general class of strategies (for instance, that of the so-called non-anticipative strategies), the corresponding state system might be time-inhomogeneous; which might present some technical difficulties. Thus even when it is possible to work with non-anticipative policies, our hypotheses ensure the existence of Nash equilibria in the class of stationary strategies for both players (see, [2, 15, 16]).

Definition 2.6 Let $\mathcal{O} \subset \mathbb{R}^n$ be an open set. We denote by $\mathcal{B}_w(\mathcal{O})$ the Banach space of real-valued measurable functions v on $\mathcal{O}$ with finite w-norm defined as follows:

$$\|v\|_w := \sup_{x\in\mathcal{O}} \frac{|v(x)|}{w(x)}.$$

2.2 Recurrence and Ergodicity

Assumption 2.7 *There exists a function* $w \in C^2(\mathbb{R}^n)$, *with* $w \geq 1$, *and constants* $d \geq c > 0$ *such that*

(i) $\lim_{|x|\to\infty} w(x) = +\infty$, *and*
(ii) $\mathcal{L}^{\pi^1,\pi^2} w(x) \leq -cw(x) + d$ *for each* $(\pi^1, \pi^2) \in \Pi^1 \times \Pi^2$ *and* $x \in \mathbb{R}^n$.

Assumption 2.7 ensures the existence of a unique invariant probability measure μ_{π^1,π^2} for the Markov process $x^{\pi^1,\pi^2}(\cdot)$, such that

$$\mu_{\pi^1,\pi^2}(w) := \int_{\mathbb{R}^n} w(x)\, \mu_{\pi^1,\pi^2}(dx) < \infty \quad \text{for all} \quad (\pi^1, \pi^2) \in \Pi^1 \times \Pi^2. \tag{2.5}$$

(See [2, 18] for details.) Moreover, for every $(\pi^1, \pi^2) \in \Pi^1 \times \Pi^2$, $x \in \mathbb{R}^n$, and $t \geq 0$, an application of Dynkin's formula to the function $v(t, x) := e^{ct} w(x)$, and Assumption 2.7(ii) yield

$$\mathbb{E}_x^{\pi^1,\pi^2} w(x(t)) \leq e^{-ct} w(x) + \frac{d}{c}(1 - e^{-ct}). \tag{2.6}$$

Hence, integrating both sides of (2.6) with respect to the invariant measure μ_{π^1,π^2} leads to

$$\mu_{\pi^1,\pi^2}(w) \leq \frac{d}{c}. \tag{2.7}$$

Assumption 2.8 *The process* $x^{\pi^1,\pi^2}(\cdot)$ *in (2.1) is uniformly* w-exponentially ergodic; *that is, there exist constants* $C > 0$ *and* $\delta > 0$ *such that*

$$\sup_{(\pi^1,\pi^2)\in\Pi^1\times\Pi^2} |\mathbb{E}_x^{\pi^1,\pi^2}[g(x(t))] - \mu_{\pi^1,\pi^2}(g)| \leq Ce^{-\delta t} \parallel g \parallel_w w(x) \tag{2.8}$$

for all $x \in \mathbb{R}^n$, $t \geq 0$, *and* $g \in \mathcal{B}_w(\mathbb{R}^n)$. *In this case,* $\mu_{\pi^1,\pi^1}(g)$ *equals the integral in (2.5) with* g *rather than* w.

Sufficient conditions for ensuring the w-exponential ergodicity of process $x^{\pi^1,\pi^2}(\cdot)$ are given in [11, Theorem 2.7].

2.3 The Payoff Rate

Let $r : \mathbb{R}^n \times U_1 \times U_2 \to \mathbb{R}$ be a measurable function, so-named the payoff (or reward/cost) rate, which satisfies the following conditions:

Assumption 2.9

(a) The function $r(x, u_1, u_2)$ *is continuous on* $\mathbb{R}^n \times U_1 \times U_2$ *and locally Lipschitz in* x *uniformly with respect to* $(u_1, u_2) \in U_1 \times U_2$*; that is, for each* $R > 0$*, there exists a constant* $K(R) > 0$ *such that*

$$\sup_{(u_1,u_2)\in U_1\times U_2} |r(x, u_1, u_2) - r(y, u_1, u_2)| \leq K(R)|x - y| \text{ for all } |x|, |y| \leq R.$$

(b) $r(\cdot, u_1, u_2)$ *is in* $\mathcal{B}_w(\mathbb{R}^n)$ *uniformly in* (u_1, u_2)*; that is, there exists* $M > 0$ *such that for all* $x \in \mathbb{R}^n$

$$\sup_{(u_1,u_2)\in U_1\times U_2} |r(x, u_1, u_2)| \leq Mw(x).$$

(c) $r(x, u_1, u_2)$ *is upper semicontinuous (u.s.c.) and concave in* $u_1 \in U_1$ *for every* $(x, u_2) \in \mathbb{R}^n \times U_2$*, and lower semicontinuous (l.s.c.) and convex in* $u_2 \in U_2$ *for every* $(x, u_1) \in \mathbb{R}^n \times U_1$*.*

Similar to (2.3)–(2.4), for each $(\phi, \psi) \in \mathcal{P}(U_1) \times \mathcal{P}(U_2)$ we write

$$r(x, \phi, \psi) := \int_{U_2}\int_{U_1} r(x, u_1, u_2)\phi(du_1)\psi(du_2), \quad x \in \mathbb{R}^n. \tag{2.9}$$

Note that this definition remains valid when the strategies $\pi^1 \in \Pi^1$ or/and $\pi^2 \in \Pi^2$ are applied in (2.9) as they are interpreted as probability measures, for each fixed $x \in \mathbb{R}^n$; that is, $\pi^k(\cdot|x) \in \mathcal{P}(U^k)$. As was agreed earlier, we shall simply write the "variable" π^k in the left-hand side of (2.9) rather than $\pi^k(\cdot|x)$.

Remark 2.10 Under Assumptions 2.1 and 2.9, the payoff rate $r(\cdot, \phi, \psi)$ and the infinitesimal generator $\mathcal{L}^{\phi,\psi}h(\cdot)$ (with $h \in C^2(\mathbb{R}^n) \bigcap B_w(\mathbb{R}^n)$) are u.s.c. in $\phi \in \mathcal{P}(U_1)$ and l.s.c. in $\psi \in \mathcal{P}(U_2)$. For further details see [5, Lemma 3.1].

3 Average Equilibria

We devote this section to the introduction of the basic optimality criterion we will use—and refine—along this study. We present the material in the spirit of [5, 11, 12, 17], and [20].

Definition 3.1 The *long-run average payoff* (also known as the *ergodic payoff*) when the players use the pair of strategies $(\pi^1, \pi^2) \in \Pi^1 \times \Pi^2$ given the initial state x is

$$J(x, \pi^1, \pi^2) := \limsup_{T\to\infty} \frac{1}{T}\mathbb{E}_x^{\pi^1,\pi^2}\Big[\int_0^T r(x(t), \pi^1, \pi^2)dt\Big]. \tag{3.1}$$

Given $(\pi^1, \pi^2) \in \Pi^1 \times \Pi^2$, let us define the constant

$$J(\pi^1, \pi^2) := \mu_{\pi^1,\pi^2}(r(\cdot, \pi^1, \pi^2)) = \int_{\mathbb{R}^n} r(x, \pi^1, \pi^2)\mu_{\pi^1,\pi^2}(dx). \tag{3.2}$$

with μ_{π^1,π^2} as in (2.5). Under our set of assumptions, it follows from (2.8) and (3.2) that the average payoff (3.1) coincides with the constant $J(\pi^1, \pi^2)$ for every $(\pi^1, \pi^2) \in \Pi^1 \times \Pi^2$—see [5, p. 669]. Moreover, by the definition (3.2) of $J(\pi^1, \pi^2)$, together with Assumption 2.9(b) and (2.7)

$$|J(\pi^1, \pi^2)| \le \int_{\mathbb{R}^n} | r(x, (\pi^1, \pi^2) | \mu_{\pi^1,\pi^2}(dx) \le M \cdot \frac{d}{c} \quad \forall (\pi^1, \pi^2) \in \Pi^1 \times \Pi^2, \tag{3.3}$$

so that the constant $J(\pi^1, \pi^2)$ is uniformly bounded on $\Pi^1 \times \Pi^2$.

Value of the Game Let

$$L := \sup_{\pi^1\in\Pi_1} \inf_{\pi^2\in\Pi_2} J(\pi^1, \pi^2) \quad \text{and} \quad U := \inf_{\pi^2\in\Pi_2} \sup_{\pi^1\in\Pi_1} J(\pi^1, \pi^2)$$

The function L is said to be the game's *lower value* whereas U is better known as the game's *upper value*. Clearly, we have $L \le U$. If the upper and lower values coincide, then the game is said to have a *value*, which we will denote by $\mathcal{V}$; in other words,

$$\mathcal{V} = L = U. \tag{3.4}$$

As a consequence of (3.3), L and U are finite; and hence, so is $\mathcal{V}$ if the second equality in (3.4) holds.

Definition 3.2 We say that a pair of stationary strategies $(\pi^{*1}, \pi^{*2}) \in \Pi^1 \times \Pi^2$ is an *average Nash equilibrium* (also known as an *average saddle point*) if

$$J(\pi^1, \pi^{*2}) \le J(\pi^{*1}, \pi^{*2}) \le J(\pi^{*1}, \pi^2) \quad \text{for every} \quad (\pi^1, \pi^2) \in \Pi^1 \times \Pi^2.$$

The set of average saddle points pairs is denoted by $(\Pi^1 \times \Pi^2)_{ao}$.

Remark 3.3 Note that if $(\pi^{*1}, \pi^{*2}) \in \Pi^1 \times \Pi^2$ is an average Nash equilibrium (in case it does exist), then the game has a value $J(\pi^{*1}, \pi^{*2}) =: \mathcal{V}$—see, for instance, [10, Proposition 4.2]. However, the converse is not necessarily true.

The following definition is crucial for our developments.

Definition 3.4 We say that a constant $J \in \mathbb{R}$, a function $h \in C^2(\mathbb{R}^n) \cap \mathcal{B}_w(\mathbb{R}^n)$, and a pair of strategies $(\pi^{*1}, \pi^{*2}) \in \Pi^1 \times \Pi^2$ verify the *average payoff optimality equations* if, for every $x \in \mathbb{R}^n$,

$$J = r(x, \pi^{*1}, \pi^{*2}) + \mathcal{L}^{\pi^{*1}, \pi^{*2}} h(x) \tag{3.5}$$

$$= \sup_{\phi \in \mathcal{P}(U_1)} \{r(x, \phi, \pi^{*2}) + \mathcal{L}^{\phi, \pi^{*2}} h(x)\} \tag{3.6}$$

$$= \inf_{\psi \in P(U_2)} \{r(x, \pi^{*1}, \psi) + \mathcal{L}^{\pi^{*1}, \psi} h(x)\} \quad \text{for all } x \in \mathbb{R}^n. \tag{3.7}$$

In this case, the pair of strategies $(\pi^{*1}, \pi^{*2}) \in \Pi^1 \times \Pi^2$ that satisfies (3.5)–(3.7) is called a pair of *canonical strategies*. We denote by $(\Pi^1 \times \Pi^2)_{ca}$ the family of canonical strategies.

Equation (3.5) is sometimes referred to as *Poisson equation*. This is the reason for which we call Eqs. (6.1)–(6.3) below, *Poisson system*.

The following result ensures the existence of solutions of Eqs. (3.5)–(3.7). It also states the existence of average saddle points, and provides us with their characterization. For a proof see [3, 5].

Theorem 3.5 *If Assumptions 2.1, 2.7, 2.8, and 2.9 hold, then:*

(i) There exist solutions $(J, h, (\pi^{*1}, \pi^{*2}))$ to the average payoff equations (3.5)–(3.7). Moreover, the constant J coincides with $\mathcal{V}$ defined in (3.4), and the function h is unique up to additive constants; in fact, h is unique under the additional condition that $h(0) = 0$.

(ii) A pair of strategies is an average saddle point if, and only if, it is canonical, that is, $(\Pi^1 \times \Pi^2)_{ao} = (\Pi^1 \times \Pi^2)_{ca}$.

Remark 3.6 One important aspect in the proof of the last result is that Remark 2.10 ensures that the mapping $\phi :\to r(x, \phi, \psi) + \mathcal{L}^{\phi,\psi} h(x)$ is u.s.c. on the compact set $\mathcal{P}(U_1)$, whereas $\psi :\to r(x, \phi, \psi) + \mathcal{L}^{\phi,\psi} h(x)$ is l.s.c. on the compact set $\mathcal{P}(U_2)$. Therefore, the existence of a canonical pair (π^{*1}, π^{*2}) as in (3.5)–(3.7) can be easily obtained from standard measurable selection theorems —see, for instance [23, Theorem 12.1].

4 Bias Equilibria

The first refinement of Definition 3.4 and Theorem 3.5 is presented in this section. Here, we will note that the set of bias equilibria is a subset of that of average equilibria. However, this section can be regarded as a list of some results that we have obtained in past works (see, for instance [5] and [17]).

Definition 4.1 Let $(\pi^1, \pi^2) \in \Pi^1 \times \Pi^2$. The *bias* of (π^1, π^2) is the function $h_{\pi^1,\pi^2} \in \mathcal{B}_w(\mathbb{R}^n)$ given by

$$h_{\pi^1,\pi^2}(x) := \int_0^\infty [\mathbb{E}_x^{\pi^1,\pi^2} r(x(t), \pi^1, \pi^2) - J(\pi^1, \pi^2)]dt \quad \text{for all } x \in \mathbb{R}^n. \tag{4.1}$$

Remark 4.2

(i) The w-exponential ergodicity of the process $x^{\pi^1,\pi^2}(\cdot)$ (see (2.8)) and the Assumption 2.9(b) ensure that the bias h_{π^1,π^2} is a finite-valued function and, in fact, it is in $\mathcal{B}_w(\mathbb{R}^n)$. Moreover, its w-norm is uniformly bounded in $(\pi^1, \pi^2) \in \Pi^1 \times \Pi^2$.

(ii) By Escobedo-Trujillo et al. [5, Proposition 5.2] we can prove that if $(\pi^1, \pi^2) \in \Pi^1 \times \Pi^2$ is average optimal, then its bias h_{π^1,π^2} and any function h satisfying the average optimality equations (3.5)–(3.7) coincide up to an additive constant; that is, for all $x \in \mathbb{R}^n$,

$$h_{\pi^1,\pi^2}(x) = h(x) - \mu_{\pi^1,\pi^2}(h).$$

Definition 4.3 (Bias Equilibrium) We say that an average saddle point $(\pi^{*1}, \pi^{*2}) \in (\Pi^1 \times \Pi^2)_{ao}$ is a *bias saddle point* if

$$h_{\pi^1,\pi^{*2}}(x) \le h_{\pi^{*1},\pi^{*2}}(x) \le h_{\pi^{*1},\pi^2}(x)$$

for every $x \in \mathbb{R}^n$ and every pair of strategies $(\pi^1, \pi^2) \in \Pi^1 \times \Pi^2$. The function $h_{\pi^{*1},\pi^{*2}}$ is called the *optimal bias function.*

We denote by $(\Pi^1 \times \Pi^2)_{bias}$ the set of bias saddle points. By Definition 4.3, $(\Pi^1 \times \Pi^2)_{bias} \subset (\Pi^1 \times \Pi^2)_{ao}$; that is,

$$\text{Bias equilibrium} \implies \text{Average equilibrium.}$$

Let (J, h) be a solution of the average payoff optimality equations (3.5)–(3.7). We define for each $x \in \mathbb{R}^n$ the sets

$$\Gamma_0^1(x) := \{\phi \in \mathcal{P}(U_1) | \ J = \inf_{\psi \in \mathcal{P}(U_2)} \{r(x, \phi, \psi) + \mathcal{L}^{\phi,\psi} h(x)\},$$

$$\Gamma_0^2(x) := \{\psi \in \mathcal{P}(U_2) | \ J = \sup_{\phi \in \mathcal{P}(U_1)} \{r(x, \phi, \psi) + \mathcal{L}^{\phi,\psi} h(x)\}.$$

Definition 4.4 We say that the constant $J \in \mathbb{R}$, the functions $h, \widetilde{h} \in C^2(\mathbb{R}^n) \cap \mathcal{B}_w(\mathbb{R}^n)$, and a pair $(\pi^{*1}, \pi^{*2}) \in \Pi^1 \times \Pi^2$ verify the bias optimality equations if and only if the triplet $(J, h, (\pi^{*1}, \pi^{*2}))$ satisfies the average optimality equations

(3.5)–(3.7) together with the following equations

$$h(x) = \mathcal{L}^{\pi^{*1},\pi^{*2}}\widetilde{h}(x) \tag{4.2}$$

$$= \sup_{\phi\in\Gamma_0^1(x)} \{\mathcal{L}^{\phi,\pi^{*2}}\widetilde{h}(x)\} \tag{4.3}$$

$$= \inf_{\psi\in\Gamma_0^2(x)} \{\mathcal{L}^{\pi^{*1},\psi}\widetilde{h}(x)\}. \tag{4.4}$$

The next result summarizes important results on the existence of bias equilibria. For further details, see [5, Section 5] or [17, Theorem 7.7].

Proposition 4.5 *Under Assumptions 2.1, 2.7, 2.8, and 2.9, the following holds:*

(i) $(\Pi^1 \times \Pi^2)_{bias}$ *is nonempty.*
(ii) $\Gamma_0^1(x)$ *and* $\Gamma_0^2(x)$ *are convex compact sets.*
(iii) The triplet $(J, h_{\pi^{*1},\pi^{*2}}, \tilde{h})$ *consisting of the constant* J *in Definition 3.4, the optimal bias function* $h_{\pi^{*1},\pi^{*2}}$ *in Definition 4.3 and some other function* $\tilde{h} \in C^2(\mathbb{R}^n \bigcap B_w(\mathbb{R}^n)$*, form the unique solution satisfying the bias optimality equations (3.5)–(3.7) and (4.2)–(4.4).*
(iv) $(\pi^1, \pi^2) \in \Pi^1 \times \Pi^2$ *is a bias saddle point if and only if it verifies the bias optimality equations (4.2)–(4.4).*

5 The Laurent Series

This section presents an extension of the results shown in [12, Section 3] or in [21, Section 3] to the zero-sum case. Here, we use the exponential ergodicity condition from Assumption 2.8 to characterize a discounted payoff in terms of a Laurent series. This will be very useful in our later developments. This is the essence of Theorem 5.5, which is the main result of this part.

Recall the definition of w in Assumption 2.7 and let μ_{π^1,π^2} be the invariant measure whose existence is ensured by Assumption 2.7.

Definition 5.1 Let $\mathcal{B}_w(\mathbb{R}^n \times U_1 \times U_2)$ be the space of measurable functions $v : \mathbb{R}^n \times U_1 \times U_2 \to \mathbb{R}$ such that

$$\sup_{(u_1,u_2)\in U_1\times U_2} |v(x, u_1, u_2)| \le M_v w(x) \quad \forall x \in \mathbb{R}^n, \tag{5.1}$$

where M_v is a positive constant depending of v.

As in (2.9) for $v \in \mathcal{B}_w(\mathbb{R}^n \times U_1 \times U_2)$ and $(\phi, \psi) \in \mathcal{P}(U_1) \times \mathcal{P}(U_2)$, we write

$$v(x, \phi, \psi) := \int_{U_2} \int_{U_1} v(x, u_1, u_2)\phi(du_1)\psi(du_2) \quad \forall x \in \mathbb{R}^n.$$

Now use $(\pi^1, \pi^2) \in \Pi^1 \times \Pi^2$ in lieu of $(\phi, \psi) \in \mathcal{P}(U_1) \times \mathcal{P}(U_2)$. Let us define

$$\overline{v}(\pi^1, \pi^2) := \int_{\mathbb{R}^n} v(x, \pi^1, \pi^2)\mu_{\pi^1,\pi^2}(dx), \quad \text{and}$$

$$Z_t^{\pi^1,\pi^2} v(x) := \mathbb{E}_x^{\pi^1,\pi^2} v(x(t), \pi^1, \pi^2) - \overline{v}(\pi^1, \pi^2).$$

With these ingredients, we define the *v-bias operator* $G_{\pi^1,\pi^2} : \mathcal{B}_w(\mathbb{R}^n \times U_1 \times U_2) \to \mathcal{B}_w(\mathbb{R}^n)$ as follows

$$G_{\pi^1,\pi^2} v(x) := \int_0^\infty [\mathbb{E}_x^{\pi^1,\pi^2} v(x(t), \pi^1, \pi^2) - \overline{v}(\pi^1, \pi^2)]dt. \tag{5.2}$$

Remark 5.2 Note that the w-exponential ergodicity of the process $x^{\pi^1,\pi^2}(\cdot)$ established in (2.8), and (5.1) yield that

$$|Z_t^{\pi^1,\pi^2} v(x)| \leq C M_v e^{-\delta t} w(x),$$

and thus,

$$|G_{\pi^1,\pi^2} v(x)| \leq \delta^{-1} C M_v w(x) \quad \text{or equivalently} \quad \|G_{\pi^1,\pi^2} v(x)\|_w \leq \delta^{-1} C M_v. \tag{5.3}$$

The following result shows some properties of both, the operator G_{π^1,π^2}, and the operators that result from its compositions with itself. Its proof delves into the discussion that led from (3.10) to (3.11) in [12].

Lemma 5.3 *For $j \geq 0$, let $G^{j+1}_{\pi^1,\pi^2}$ be the $j+1$-composition of G_{π^1,π^2} with itself. Then*

$$G^{j+1}_{\pi^1,\pi^2} v \quad \textit{is in } \mathcal{B}_w(\mathbb{R}^n), \quad \textit{and} \quad \mu_{\pi^1,\pi^2}\left(G^{j+1}_{\pi^1,\pi^2} v\right) = 0.$$

Proof By (5.3), G_{π^1,π^2} is in $B_w(\mathbb{R}^n)$. Now, the fact that $\mu_{\pi^1,\pi^2}(G_{\pi^1,\pi^2}) = 0$ is straightforward from (3.2) and (5.2). The rest of the proof easily follows by applying mathematical induction on j. □

Definition 5.4 Given a discount factor $\alpha > 0$. The expected α-discounted v-payoff when the players use $(\pi^1, \pi^2) \in \Pi^1 \times \Pi^2$, given the initial state $x \in \mathbb{R}^n$, is

$$V_\alpha(x, \pi^1, \pi^2, v) := \mathbb{E}_x^{\pi^1,\pi^2}\left[\int_0^\infty e^{-\alpha t} v(x(t), \pi^1, \pi^2)dt\right]. \tag{5.4}$$

The following result provides a useful characterization of the α-discounted υ-payoff in terms of a Laurent series (see, for instance [4, Chapter 6]). The proof uses essentially the same steps of the proof of Theorem 3.1 and Proposition 3.2 in [12], so we shall omit it.

Theorem 5.5

(a) Let $\delta > 0$ be the constant in Assumption 2.8. If (π^1, π^2) an arbitrary pair of strategies in $\Pi^1 \times \Pi^2$ and υ is a function in $\mathcal{B}_w(\mathbb{R}^n \times U_1 \times U_2)$, then, for $\alpha \in (0, \delta)$, the α-discounted υ-payoff (5.4) can be written as

$$V_\alpha(x, \pi^1, \pi^2, \upsilon) = \frac{1}{\alpha}\overline{\upsilon}(\pi^1, \pi^2) + \sum_{j=0}^{\infty}(-\alpha)^j G^{j+1}_{\pi^1,\pi^2}\upsilon(x). \tag{5.5}$$

Moreover the above series converges in w-norm.

(b) Let $\theta \in \mathbb{R}$ be such that $0 < \theta < \delta$, where δ is the constant in Assumption 2.8. For each $\upsilon \in \mathcal{B}_w(\mathbb{R}^n \times U_1 \times U_2)$, $(\pi^1, \pi^2) \in \Pi^1 \times \Pi^2$, and $i = 0, 1, \ldots$ define the i-residual of the Laurent series (5.5) as

$$R_i(\pi^1, \pi^2, \upsilon, \alpha) := \sum_{j=i}^{\infty}(-\alpha)^j G^{j+1}_{\pi^1,\pi^2}\upsilon.$$

Then, for all $|\alpha| \le \theta$ and $i = 0, 1, \ldots,$

$$\sup_{(\pi^1,\pi^2)\in\Pi^1\times\Pi^2} \left|\left| R_i((\pi^1, \pi^2), \upsilon, \alpha)\right|\right|_w \le \frac{CM_\upsilon}{\delta^i(\delta - \theta)}|\alpha|^k. \tag{5.6}$$

For each $\upsilon \in \mathcal{B}_w(\mathbb{R}^n \times U_1 \times U_2)$, (π^1, π^2) in $\Pi^1 \times \Pi^2$, and $i = 0, 1, \ldots,$ define $h^i_{\pi^1,\pi^2}\upsilon$ as

$$h^i_{\pi^1,\pi^2}\upsilon(x) := (-1)^i G^{i+1}_{\pi^1,\pi^2}\upsilon(x) \quad \text{for all } x \in \mathbb{R}^n \text{ and } i = 1, 2. \tag{5.7}$$

It is obvious that, for each $\upsilon \in \mathcal{B}_w(\mathbb{R}^n \times U_1 \times U_2)$, $h^i_{\pi^1,\pi^2}\upsilon$ belongs to $B_w(\mathbb{R}^n)$ because $G^{i+1}_{\pi^1,\pi^2}\upsilon$ does.

Notation For $\upsilon = r$, with r as in Assumption 2.9, we simply write the operator in (5.7) as $h^i_{\pi^1,\pi^2}$; that is,

$$h^i_{\pi^1,\pi^2}r := h^i_{\pi^1,\pi^2}.$$

Note that for $i = 0$, $h^0_{\pi^1,\pi^2}$ equals to the bias function defined in (4.1), i.e.,

$$h^0_{\pi^1,\pi^2}(x) = G_{\pi^1,\pi^2}r(x) = h_{\pi^1,\pi^2}(x) \quad \text{for all } \; x \in \mathbb{R}^n.$$

Moreover,

$$h^1_{\pi^1,\pi^2} = -G^2_{\pi^1,\pi^2} r(x) = G_{\pi^1,\pi^2}(-h^0_{\pi^1,\pi^2}),$$

is the bias of (π^1, π^2) when the payoff is $-h^0_{\pi^1,\pi^2}$. In general, using mathematical induction, we can obtain that

$$h^i_{\pi^1,\pi^2} = G_{\pi^1,\pi^2}(-h^{i-1}_{\pi^1,\pi^2}) \quad i = 1, 2, \ldots$$

By Theorem 5.5(a) and the expression (5.2), the α-discounted payoff (5.4)—with r in lieu v—can be written in terms of operator $h^i_{\pi^1,\pi^2}$ as follows

$$V_\alpha(x, \pi^1, \pi^2, r) = \frac{1}{\alpha} J(\pi^1, \pi^2) + \sum_{i=0}^{\infty} \alpha^i h^i_{\pi^1,\pi^2}(x), \tag{5.8}$$

and, by Lemma 5.3,

$$\mu_{\pi^1,\pi^2}(h^i_{\pi^1,\pi^2}) = 0 \quad \text{for all} \quad i = 0, 1, 2, \ldots \tag{5.9}$$

6 The Poisson System

We now define the so-called *Poisson system* and characterize its solution in terms of the basic average optimality criterion, and the recursive operator G_{π^1,π^2} introduced in Sect. 5.

For the following definition, recall that Eq. (3.5) is sometimes dubbed *Poisson equation*.

Definition 6.1 Let $(\pi^1, \pi^2) \in \Pi^1 \times \Pi^2$ fixed. We say that a constant $J \in \mathbb{R}$ and the functions $h^0, h^1, \ldots, h^{m+1} \in C^2(\mathbb{R}^n) \cap B_w(\mathbb{R}^n)$ verify the *Poisson system* for $(\pi^1, \pi^2) \in \Pi^1 \times \Pi^2$ if

$$J = r(x, \pi^1, \pi^2) + \mathcal{L}^{\pi^1,\pi^2} h^0(x), \tag{6.1}$$

$$h^0(x) = \mathcal{L}^{\pi^1,\pi^2} h^1(x), \tag{6.2}$$

$$\cdots$$

$$h^m(x) = \mathcal{L}^{\pi^1,\pi^2} h^{m+1}(x). \tag{6.3}$$

Theorem 6.2 *Let $m \geq -1$ be fixed. The constant $J \in \mathbb{R}$ and the functions $h^0, h^1, \ldots, h^{m+1} \in C^2(\mathbb{R}^n) \cap B_w(\mathbb{R}^n)$ are solutions to the Poisson system (6.1)–(6.3) if and only if $J = J(\pi^1, \pi^2)$, $h^i = h^i_{\pi^1,\pi^2}$ for $0 \leq i \leq m$, and $h^{m+1} =$*

$h^{m+1}_{\pi^1,\pi^2} + z$ *for* $z \in \mathbb{R}$*, where* J *and* $h^i_{\pi^1,\pi^2}$*,* $0 \le i \le m+1$*, are the functions in (3.2) and (5.7), respectively.*

Proof We will use mathematical induction over Eqs. (6.1)–(6.3).

1. Case $m = -1$ follows from Lemma 3.2 and Proposition 5.1 in [5].
2. Now, suppose the result is valid for some $m \ge -1$.
3. Case $m + 1$:

   The "if" part: Suppose that $J = J(\pi^1, \pi^2)$, $h^i = h^i_{\pi^1,\pi^2}$ for $0 \le i \le m$, and $h^{m+1} = h^{m+1}_{\pi^1,\pi^2} + z$ for $z \in \mathbb{R}$. Then, we need to prove that $h^{m+1}_{\pi^1,\pi^2}$ verifies the $(m+1)$-th Poisson equation. To this end, observe that $h^{m+2}_{\pi^1,\pi^2}$ is the bias function of (π^1, π^2) when we consider as reward rate $-h^{m+1}_{\pi^1,\pi^2}(x)$. It is easy to verify through a mathematical induction procedure that $-h^{m+1}_{\pi^1,\pi^2}$ satisfies Assumption 2.9, then we can invoke Theorem 4.1 in [5], to ensure the existence of a function $h^{m+2} \in C^2(\mathbb{R}^n) \cap \mathcal{B}_w(\mathbb{R}^n)$, a constant $\overline{J}$ and a pair of strategies (π^1, π^2) that satisfy the average optimality equation

$$\begin{aligned}\overline{J} &= -h^{m+1}(x) + \mathcal{L}^{\pi^1,\pi^2} h^{m+2}(x)\\ &= \sup_{\phi \in \mathcal{P}(U_1)} \{-h^{m+1}(x) + \mathcal{L}^{\phi,\pi^2} h^{m+2}(x)\},\\ &= \inf_{\psi \in \mathcal{P}(U_2)} \{-h^{m+1}(x) + \mathcal{L}^{\pi^1,\psi} h^{m+2}(x)\},\end{aligned}$$

   with $\overline{J} = \mu_{\pi^1,\pi^2}(-h^{m+1}) = \mu_{\pi^1,\pi^2}(-h^{m+1}_{\pi^1,\pi^2})$. Now, Proposition 5.1 in [5] gives that the bias function with reward rate $-h^{m+1}(x) = -h^{m+1}_{\pi^1,\pi^2}(x)$ satisfies the following Poisson equation

$$\mu_{\pi^1,\pi^2}(-h^{m+1}_{\pi^1,\pi^2}) = -h^{m+1}_{\pi^1,\pi^2}(x) + \mathcal{L}^{\pi^1,\pi^2} h^{m+2}(x),$$

   which implies that

$$h^{m+1}_{\pi^1,\pi^2}(x) = \mathcal{L}^{\pi^1,\pi^2} h^{m+2}(x), \tag{6.4}$$

   since that (5.9) gives $\mu_{\pi^1,\pi^2}(-h^{m+1}_{\pi^1,\pi^2}) = 0$. Thus, (6.4) implies that $h^{m+1}_{\pi^1,\pi^2}$, satisfies the $(m+1)$-th Poisson equation.

   The "only if" part: Suppose that $J \in \mathbb{R}$ and $h^0, h^1, \ldots, h^{m+1} \in C^2(\mathbb{R}^n) \cap \mathcal{B}_w(\mathbb{R}^n)$ are solutions to (6.1)–(6.3). By the induction hypothesis the result holds for some $m \ge 0$, i.e.,

$$h^m_{\pi^1,\pi^2}(x) = h^m(x) = \mathcal{L}^{\pi^1,\pi^2} h^{m+1}(x). \tag{6.5}$$

Therefore, we only need to prove that $h^{m+1} = h^{m+1}_{\pi^1,\pi^2}$. Namely, the bias function $h^{m+1}_{\pi^1,\pi^2}(x)$ when the payoff rate is $-h^m_{\pi^1,\pi^2}$, verifies the following Poisson equation

$$\mu_{\pi^1,\pi^2}(-h^m_{\pi^1,\pi^2}) = -h^m_{\pi^1,\pi^2}(x) + \mathcal{L}^{\pi^1,\pi^2} h^{m+1}_{\pi^1,\pi^2}(x),$$

then, by (5.9) we obtain

$$h^m_{\pi^1,\pi^2}(x) = \mathcal{L}^{\pi^1,\pi^2} h^{m+1}_{\pi^1,\pi^2}(x). \tag{6.6}$$

Thus, subtracting equation (6.5) to (6.6) we obtain

$$0 = \mathcal{L}^{\pi^1,\pi^2}(h^{m+1}_{\pi^1,\pi^2}(x) - h^{m+1}(x)).$$

Therefore, $h^{m+1}_{\pi^1,\pi^2} - h^{m+1}$ is a harmonic function and as a consequence, Lemma 2.1 in [5], yields

$$h^{m+1}_{\pi^1,\pi^2}(x) = h^{m+1}(x) + \mu_{\pi^1,\pi^2}(h^{m+1}). \tag{6.7}$$

Since μ_{π^1,π^2} is an invariant probability measure, and $h^{m+1} \in C^2(\mathbb{R}^n) \cap \mathcal{B}_w(\mathbb{R}^n)$ satisfies the $(m+1)$-th Poisson equation, we have

$$\mu_{\pi^1,\pi^2}(h^{m+1}) = \int_{\mathbb{R}^n} \mathcal{L}^{\pi^1,\pi^2} h^{m+2}(y)\mu_{\pi^1,\pi^2}(dy) = 0 \quad \text{for all}$$
$$h^{m+2} \in C^2(\mathbb{R}^n) \cap \mathcal{B}_w(\mathbb{R}^n), \tag{6.8}$$

where the last equality follows from a well-known result of invariant probability measures—see, for example [2]. Therefore, $h^{m+1} = h^{m+1}_{\pi^1,\pi^2}$ follows from (6.7) and (6.8).

□

7 The Average Payoff Optimality System

We devote this section to link the Poisson system (6.1)–(6.3) from Sect. 6 with the optimization problem we are trying to solve (see Definitions 8.1 and 8.2 below). We do this by means of a system of average optimality equations, and the characterization of their solutions as a sequence of *canonical equilibria* of a collection of average payoff games. This is the purpose of the main result of this part, namely, Theorem 7.4.

Definition 7.1 We say that a constant $J \in \mathbb{R}$ and functions $h^0, h^1, \ldots, h^{m+1} \in C^2(\mathbb{R}^n) \cap \mathcal{B}_w(\mathbb{R}^n)$ verify the -1-th, 0-th,..., m-th *average payoff optimality system* for $(\pi^{*1}, \pi^{*2}) \in \Pi^1 \times \Pi^2$ and $x \in \mathbb{R}$ if

$$J = r(x, \pi^{*1}, \pi^{*2}) + \mathcal{L}^{\pi^{*1}, \pi^{*2}} h^0(x), \tag{7.1}$$

$$= \sup_{\phi \in \mathcal{P}(U_1)} r(x, \phi, \pi^{*2}) + \mathcal{L}^{\phi, \pi^{*2}} h^0(x), \tag{7.2}$$

$$= \inf_{\psi \in \mathcal{P}(U_2)} r(x, \pi^{*1}, \psi) + \mathcal{L}^{\pi^{*1}, \psi} h^0(x) \tag{7.3}$$

$$h^0(x) = \mathcal{L}^{\pi^{*1}, \pi^{*2}} h^1(x) \tag{7.4}$$

$$= \sup_{\phi \in \Gamma_0^1(x)} \mathcal{L}^{\phi, \pi^{*2}} h^1(x) \tag{7.5}$$

$$= \inf_{\psi \in \Gamma_0^2(x)} \mathcal{L}^{\pi^{*1}, \psi} h^1(x) \tag{7.6}$$

$$\ldots$$

$$h^m(x) = \mathcal{L}^{\pi^{*1}, \pi^{*2}} h^{m+1}(x) \tag{7.7}$$

$$= \sup_{\phi \in \Gamma_m^1(x)} \mathcal{L}^{\phi, \pi^{*2}} h^{m+1}(x) \tag{7.8}$$

$$= \inf_{\psi \in \Gamma_m^2(x)} \mathcal{L}^{\pi^{*1}, \psi} h^{m+1}(x) \tag{7.9}$$

where letting $\Gamma_{-1}^1(x) := \mathcal{P}(U_1)$ and $\Gamma_{-1}^2(x) := \mathcal{P}(U_2)$ for all $x \in \mathbb{R}^n$, then the sets $\Gamma_j^k(x)$, for $0 \le j \le m$ and $k = 1, 2$, consist of probability measures $\phi \in \Gamma_{j-1}^1(x)$ and $\psi \in \Gamma_{j-1}^2(x)$ attaining the maximum and minimum in the $(j-1)$-th average payoff optimality equation, respectively; that is, for each $x \in \mathbb{R}^n$,

$$\Gamma_0^1(x) := \left\{ \phi \in \mathcal{P}(U_1) \mid J = \inf_{\psi \in \mathcal{P}(U_2)} \left[r(x, \phi, \psi) + \mathcal{L}^{\phi, \psi} h^0(x) \right] \right\},$$

$$\Gamma_0^2(x) := \left\{ \psi \in \mathcal{P}(U_2) \mid J = \sup_{\phi \in \mathcal{P}(U_1)} \left[r(x, \phi, \psi) + \mathcal{L}^{\phi, \psi} h^0(x) \right] \right\}$$

and, for $1 \leq j \leq m$,

$$\Gamma_j^1(x) := \left\{ \phi \in \Gamma_{j-1}^1(x) \mid h^{j-1}(x) = \inf_{\psi \in \Gamma_{j-1}^2(x)} \mathcal{L}^{\phi,\psi} h^j(x) \right\},$$

$$\Gamma_j^2(x) := \left\{ \psi \in \Gamma_{j-1}^2(x) \mid h^{j-1}(x) = \sup_{\phi \in \Gamma_{j-1}^1} \mathcal{L}^{\phi,\psi} h^j(x) \right\}.$$

Proposition 7.2 *For each $k = 1, 2$, and $-1 \leq j \leq m$, the sets $\{\Gamma_j^k(x)\}_{j \geq 0}$ are convex compact sets.*

Proof We use mathematical induction on j:

1. Case $j = -1, 0$. Since $\mathcal{P}(U_1)$ and $\mathcal{P}(U_2)$ are compact and convex sets (see, for instance, [1, Theorem 15.11]), Lemma 5.1 in [5], gives that $\Gamma_0^1(x)$ and $\Gamma_0^2(x)$ are also convex and compact sets.
2. Suppose now that for some $0 \leq j \leq m$, $\Gamma_j^1(x)$ and $\Gamma_j^2(x)$, are convex compact sets.
3. Let us prove the result for $m = j + 1$. To this end, note that

$$\Gamma_{j+1}^1(x) := \left\{ \phi \in \Gamma_j^1(x) | \, h^j(x) = \inf_{\psi \in \Gamma_j^2(x)} \mathcal{L}^{\phi,\psi} h^{j+1}(x) \right\},$$

   and

$$\Gamma_{j+1}^2(x) := \left\{ \psi \in \Gamma_j^2(x) | \, h^j(x) = \sup_{\phi \in \Gamma_j^1(x)} \mathcal{L}^{\phi,\psi} h^{j+1}(x) \right\},$$

   and by induction hypothesis $\Gamma_j^1(x)$ and $\Gamma_j^2(x)$ are convex compact sets. Then, to verify if $\Gamma_{j+1}^1(x)$ and $\Gamma_{j+1}^2(x)$ are compact sets it is sufficient to prove that they are closed, but this property follows due to the compactness of $\Gamma_j^1(x)$ and $\Gamma_j^2(x)$ (induction hypothesis) and the u.s.c in ϕ (l.s.c. ψ) of $\mathcal{L}^{\phi,\psi}$ established in the Remark 2.10.

   The proof that $\Gamma_{j+1}^1(x)$ and $\Gamma_{j+1}^2(x)$ are convex sets mimicks that of Lemma 4.6 in [20].

□

Since $\{\Gamma_j^k(x)\}_{j \geq 0}$, $k = 1, 2$, is a nonincreasing sequence of nonempty compact sets, the set

$$\Gamma_\infty^k(x) := \bigcap_{m \geq -1} \Gamma_m^k(x) \tag{7.10}$$

is nonempty and compact as well.

The following definition concerns the pair of strategies $(\pi^1, \pi^2) \in \Pi^1 \times \Pi^2$ that attain the maximum and minimum respectively in Eqs. (7.1)–(7.9).

Definition 7.3 We define

$$\Pi^1_m \times \Pi^2_m := \{(\pi^1, \pi^2) \in \Pi^1 \times \Pi^2 \mid (\pi^1(\cdot|x), \pi^2(\cdot|x)) \in \Gamma^1_{m+1}(x) \times \Gamma^2_{m+1}(x), \forall x \in \mathbb{R}^n\}.$$

A pair $(\pi^1, \pi^2) \in \Pi^1_m \times \Pi^2_m$ will be referred to as a *canonical equilibrium* for the -1-th, 0-th, ..., m-th average payoff optimality system (7.1)–(7.9).

From Definition 7.3, it is clear that $\Pi^1_{m+1} \times \Pi^2_{m+1} \subseteq \Pi^1_m \times \Pi^2_m$, for all $m = -1, 0, 1, \cdots$.

Theorem 7.4 *The -1-th, 0-th,..., m-th average reward HJB system (7.1)–(7.9) admits a unique solution $J \in \mathbb{R}$, $h^0, h^1, \ldots, h^{m+1} \in C^2(\mathbb{R}^n) \cap \mathcal{B}_w(\mathbb{R}^n)$, where $J, h^0, h^1, \ldots, h^m$ are unique, and h^{m+1} is unique up to an additive constant. Moreover, the set $\Pi^1_m \times \Pi^2_m$ is nonempty.*

Proof We will use mathematical induction on m.

1. Case $m = 0$. It follows from Theorems 4.1, 5.1 and 5.2 in [5].
2. Suppose that the result holds for some $m = j$.
3. Now, we prove that the result holds for $m = j + 1$.

   The induction hypothesis ensures the existence of $J \in \mathbb{R}$, $h^0, h^1, \ldots, h^j \in C^2(\mathbb{R}^n) \cap \mathcal{B}_w(\mathbb{R}^n)$ which are unique solutions of the -1-th, 0-th,..., j-th average payoff optimality system and that that Π^1_j and Π^2_j are nonempty.

   Let us consider now a new game, so-named j-bias game, consisting in:

   - The dynamic system (2.1).
   - The payoff function $-h^j$. (7.11)
   - The set of control actions $\Gamma^1_j(x)$ and $\Gamma^2_j(x)$.

   It is easy to verify that this new game satisfies all of our hypotheses. Then, Theorem 3.5(i)–(ii) ensures the existence of solutions $(\overline{J}, h^{j+1}, (\pi^{*1}, \pi^{*2}))$ to the following average optimality equations

$$\begin{aligned}
\overline{J} &= -h^j(x) + \mathcal{L}^{\pi^{*1},\pi^{*2}} h^{j+1}(x) \\
&= \sup_{\phi \in \Gamma^1_j(x)} \{-h^j(x) + \mathcal{L}^{\phi,\pi^{*2}} h^{j+1}(x)\} \\
&= \inf_{\psi \in \Gamma^2_j(x)} \{-h^j(x) + \mathcal{L}^{\pi^{*1},\psi} h^{j+1}(x)\}.
\end{aligned}$$

The existence of a function $h^{j+2} \in C^2(\mathbb{R}^n) \cap \mathcal{B}_w(\mathbb{R}^n)$ satisfying

$$\begin{aligned} h^{j+1}(x) &= \mathcal{L}^{\pi^{*1},\pi^{*2}} h^{j+2}(x) \\ &= \sup_{\phi \in \Gamma_j^1(x)} \mathcal{L}^{\phi,\pi^{*2}} h^{j+2}(x) \\ &= \inf_{\psi \in \Gamma_j^2(x)} \mathcal{L}^{\pi^{*1},\psi} h^{j+2}(x). \end{aligned}$$

is ensured by Proposition 4.5, and the fact that $\overline{J} = \mu_{\pi^{*1},\pi^{*2}}(-h^j)$. In this case h^{j+1} is unique, and h^{j+2} is unique up to additive constants. Thus, h^{j+1} satisfies the $(j+1)$-th average reward HJB equations.

It remains to prove that $\Pi_m^1 \times \Pi_m^2$ is nonempty. To this end, we proceed again by mathematical induction on m. Namely, for the case $m = 0$, the result follows by Theorems 5.1 and 5.2 in [5]. Now assume that $\Pi_j^1 \times \Pi_j^2$ is nonempty for some $j = 0, 1, \ldots$; that is, there is at least an element $(\pi_j^1, \pi_j^2) \in \Pi_j^1 \times \Pi_j^2$ or equivalently, $(\pi_{j+1}^1(\cdot|x), \pi_{j+1}^2(\cdot|x)) \in \Gamma_{j+1}^1(x) \times \Gamma_{j+1}^2(x)$ for all $x \in \mathbb{R}^n$. We want to prove that $\Pi_{j+1}^1 \times \Pi_{j+1}^2$ is nonempty. For this, we consider again the j-bias game (7.11). Since this game satisfies all of our hypotheses, we can invoke Proposition 4.5(i) to ensure the existence of a bias equilibrium (π^1, π^2) associated to the j-bias game. Hence, Proposition 4.5(iii) yields that, in fact, such equilibrium satisfies both, the j-th and the $(j+1)$-th average payoff equations. This completes the proof. □

Remark 7.5 It is worth noting the relation of the pairs $(\pi^{*1}, \pi^{*2}) \in \Pi_m^1 \times \Pi_m^2$ with the m-bias game in (7.11); namely, if we apply iteratively Proposition 4.5, we can easily verify that $(\pi^{*1}, \pi^{*2}) \in \Pi_m^1 \times \Pi_m^2$ if and only if such a pair is an average Nash equilibrium for the j-bias game (7.11) for $j = -1, \cdots, m$.

We define

$$\Pi_\infty^1 \times \Pi_\infty^2 := \bigcap_{m=-1}^{\infty} (\Pi_m^1 \times \Pi_m^2). \tag{7.12}$$

As a consequence of (7.10) and Theorem 7.4, we deduce the following result.

Corollary 7.6 *There exists a strategy* $(\pi^1, \pi^2) \in \Pi^1 \times \Pi^2$ *that satisfies the m-th average reward HJB equation for all* $m = -1, 0, \ldots$. *In other words,* $\Pi_\infty^1 \times \Pi_\infty^2$ *is nonempty.*

8 Blackwell-Nash Equilibria

In this section we present a zero-sum type of Nash equilibrium so-named *Blackwell-Nash equilibrium*; we will also introduce a sensitive discount concept related to a family of optimality criteria so-named m-discount equilibria, for $m \geq -1$. We will see that a Blackwell-Nash equilibrium becomes the limit, as $m \to \infty$, of a sequence of m-discount equilibria and prove the existence of each element of this sequence based on the results given in previous sections. To begin with this analysis, we first define the aforementioned concepts as follows.

Definition 8.1 (Blackwell-Nash Equilibrium) A pair $(\pi^{*1}, \pi^{*2}) \in \Pi^1 \times \Pi^2$ is called Blackwell-Nash equilibrium if for each $(\pi^1, \pi^2) \in \Pi^1 \times \Pi^2$ and each state $x \in \mathbb{R}^n$, there exists a discount factor $\alpha^* = \alpha^*(x, \pi^1, \pi^2)$ such that

$$V_\alpha(x, \pi^1, \pi^{*2}) \leq V_\alpha(x, \pi^{*1}, \pi^{*2}) \leq V_\alpha(x, \pi^{*1}, \pi^2) \tag{8.1}$$

for all $0 < \alpha < \alpha^*$.

Definition 8.2 (Sensitive Discount Equilibrium)

(a) Let $m \geq -1$ be an integer. A pair $(\pi^{*1}, \pi^{*2}) \in \Pi^1 \times \Pi^2$ is called an *m-discount equilibrium* if

$$\liminf_{\alpha \to 0} \alpha^{-m} [V_\alpha(x, \pi^{*1}, \pi^{*2}) - V_\alpha(x, \pi^1, \pi^{*2})] \geq 0 \quad \text{for all } \pi^1 \in \Pi^1,$$

and

$$\limsup_{\alpha \to 0} \alpha^{-m} [V_\alpha(x, \pi^{*1}, \pi^{*2}) - V_\alpha(x, \pi^{*1}, \pi^2)] \leq 0 \quad \text{for all } \pi^2 \in \Pi^2.$$

(b) We call *sensitive discount equilibria* to the family $\{(\pi_m^{*1}, \pi_m^{*2}) \mid m \geq -1\}$ of all the m-discount equilibria ($m \geq -1$).

We denote by $\Pi_m^{1,d}$ and $\Pi_m^{2,d}$ the sets of strategies m-discount optimal for player 1 and 2, respectively.

Theorem 8.3

(i) Let $m \geq -1$ be an integer, then $\Pi_m^1 \times \Pi_m^2 \subseteq \Pi_m^{1,d} \times \Pi_m^{2,d}$.
(ii) If $(\pi^{*1}, \pi^{*2}) \in \Pi_\infty^1 \times \Pi_\infty^2$, then it is a Blackwell-Nash equilibrium.

Proof

(i) Consider the pair $(\pi^{*1}, \pi^{*2}) \in \Pi_m^1 \times \Pi_m^2$, and use the series (5.8) to deduce the following

$$\begin{aligned}
&\frac{1}{\alpha^m}[V_\alpha(x, \pi^{*1}, \pi^{*2}) - V_\alpha(x, \pi^1, \pi^{*2})] \\
&\quad = \frac{1}{\alpha}\Big[\frac{1}{\alpha^m}\left(J(\pi^{*1}, \pi^{*2}) - J(\pi^1, \pi^{*2})\right) + \frac{1}{\alpha^{m-1}}\left(h^0_{\pi^{*1},\pi^{*2}}(x) - h^0_{\pi^1,\pi^{*2}}(x)\right) \\
&\quad + \cdots + \left(h^{m-1}_{\pi^{*1},\pi^{*2}}(x) - h^{m-1}_{\pi^1,\pi^{*2}}(x)\right)\Big] + \left(h^m_{\pi^{*1},\pi^{*2}}(x) - h^m_{\pi^1,\pi^{*2}}(x)\right) + \\
&\quad + \frac{1}{\alpha^m}\sum_{i=m+1}^{\infty} \alpha^i \left(h^i_{\pi^{*1},\pi^{*2}}(x) - h^i_{\pi^1,\pi^{*2}}(x)\right),
\end{aligned} \tag{8.2}$$

for all $\pi^1 \in \Pi^1$. By virtue of Remark 7.5, (π^{*1}, π^{*2}) is a Nash equilibrium for the -1-th, 0-th, ..., m-th bias game (7.11). Then, the first $m + 2$ elements in equality (8.2) are greater or equal to zero. Finally, letting $\alpha \to 0$ in both sides of (8.2) and using Theorem 5.5(b), we get

$$\frac{1}{\alpha^m}[V_\alpha(x, \pi^{*1}, \pi^{*2}) - V_\alpha(x, \pi^1, \pi^{*2})] \geq 0.$$

Similar arguments yield

$$\frac{1}{\alpha^m}[V_\alpha(x, \pi^{*1}, \pi^{*2}) - V_\alpha(x, \pi^{*1}, \pi^2)] \leq 0 \quad \text{for all } \pi^2 \in \Pi^2.$$

Therefore, $\Pi_m^1 \times \Pi_m^2 \subset \Pi_m^{1,d} \times \Pi_m^{2,d}$, which proves (i).

(ii) Let $\pi^1 \in \Pi^1$ and $x \in \mathbb{R}^n$ arbitrary and suppose that $(\pi^{*1}, \pi^{*2}) \in \Pi_\infty^1 \times \Pi_\infty^2$, then using again (5.8) we can write

$$\begin{aligned}
V_\alpha(x, \pi^{*1}, \pi^{*2}) - V_\alpha(x, \pi^1, \pi^{*2}) &= \frac{1}{\alpha}[J(\pi^{*1}, \pi^{*2}) - J(\pi^1, \pi^{*2})] \\
&\quad + \sum_{i=0}^{\infty} \alpha^i [h^i_{\pi^{*1},\pi^{*2}}(x) - h^i_{\pi^1,\pi^{*2}}(x)].
\end{aligned} \tag{8.3}$$

By virtue of (7.12), $(\pi^{*1}, \pi^{*2}) \in \Pi_m^1 \times \Pi_m^2$ for $-1 \leq m \leq \infty$. So, (π^{*1}, π^{*2}) is a Nash equilibrium for the m-bias game (7.11) for all $m = -1, 0, 1, \cdots$. Therefore, the equality in (8.3) is nonnegative for every $\alpha > \alpha^*$, where α^* depends on the residual term (5.6), which yields the first inequality in (8.1). We can also mimic the same arguments but now for arbitrary $\pi^2 \in \Pi^2$ and thus to obtain the second inequality in (8.1), yielding that (π^{*1}, π^{*2}) is a Blackwell-Nash equilibrium.

□

We use Theorems 7.4, 8.3, and Corollary 7.6, to state our final claim.

Corollary 8.4 *Under Assumptions 2.1, 2.7, 2.8, and 2.9,*

(i) *For each* $m \geq -1$, *the set* $\Pi_m^{1,d} \times \Pi_m^{2,d}$ *of* m*-discount optimal strategies is nonempty.*
(ii) *There exist Blackwell optimal strategies in* $\Pi^1 \times \Pi^2$.

9 Final Remarks

In this paper we have shown the existence and provide some characterizations of the sensitive discount equilibria in a class of zero-sum stochastic differential games with a uniform ellipticity assumption. This yields a Blackwell-Nash equilibrium in the limit as $m \to \infty$. To this end, we truncated the Laurent series of the expected discounted reward/cost, and thus stated the so-called *Poisson system*, which allowed us to characterize the equilibria as the collection of strategies that meet it.

It is worth pointing out the fact that Theorem 8.3 and Corollary 8.4 show that, for a zero-sum stochastic differential game, an m-discount equilibrium is equivalent to a Blackwell-Nash equilibrium only when $m \to \infty$. This agrees with the controlled diffusion scheme (see [12, 22]).

Some possible extensions of our work are, for example, to do this same analysis but considering a more general dynamics type, such is the case of stochastic differential equations with jumps (in the context of Lévy processes) or using the same dynamic than ours but under weaker assumptions than those considered here, such is the case of *degenerate* diffusions.

Acknowledgement This research was supported in part by CONACyT grant 238045.

References

1. C. Aliprantis, K. Border, *Infinite Dimensional Analysis*, 3rd edn. (Springer, Berlin, 2006)
2. A. Arapostathis, M. Ghosh, V. Borkar, *Ergodic Control of Diffusion Processes*, vol. 143. Encyclopedia of Mathematics and Its Applications (Cambridge University Press, Cambridge, 2012)
3. A. Arapostathis, V. Borkar, K. Suresh, Relative value iteration for stochastic differential games, in *Advances in Dynamic Games: Theory, Applications, and Numerical Methods* (Springer International Publishing, Cham, 2013), pp. 3–27
4. J.B. Conway, *A Course in Functional Analysis* (Springer, New York, 1994)
5. B.A. Escobedo-Trujillo, J.D. López-Barrientos, O. Hernández-Lerma, Bias and overtaking equilibria for zero-sum stochastic differential games. J. Optim. Theory Appl. **153**, 662–687 (2012)
6. M. Ghosh, A. Arapostathis, S. Marcus, Optimal control of switching diffusions with applications to flexible manufacturing systems. SIAM J. Control Optim. **30**, 1–23 (1992)

7. H. Gimbert, W. Zielonka, Deterministic priority mean-payoff games as limits of discounted games, in *Automata, Languages and Programming*, Part II. Lecture Notes in Computer Science, vol. 4052 (Springer, Berlin, 2006), pp. 312–323
8. H. Gimbert, W. Zielonka, Applying Blackwell optimality: priority mean-payoff games as limits of multi-discounted games, in *Logic and Automata*. Texts in Logic and Games, vol. 2 (Amsterdam University Press, Amsterdam, 2008), pp. 331–355
9. H. Gimbert, W. Zielonka, Blackwell optimal strategies in priority mean-payoff games. Int. J. Found. Comput. Sci. **23**, 687–711 (2012)
10. H. Jasso-Fuentes, Noncooperative continuous-time Markov games. Morfismos **9**, 39–54 (2005)
11. H. Jasso-Fuentes, O. Hernández-Lerma, Characterizations of overtaking optimality for controlled diffusion processes. Appl. Math. Optim. **57**, 349–369 (2008)
12. H. Jasso-Fuentes, O. Hernández-Lerma, Blackwell optimality for controlled diffusion processes. J. Appl. Probab. **46**, 372–391 (2009)
13. H. Jasso-Fuentes, O. Hernández-Lerma, Ergodic control, bias and sensitive discount optimality for Markov diffusion processes. Stoch. Anal. Appl. **27**, 363–385 (2009)
14. H. Jasso-Fuentes, J.D. López-Barrientos, B.A. Escobedo-Trujilo, Infinite horizon nonzero-sum stochastic differential games with additive structure. IMA J. Math. Control Inform. **34**, 283–309 (2017)
15. A. Leizarowitz, Controlled diffusion process on infinite horizon with the overtaking criterion. Appl. Math. Optim. **17**, 61–78 (1988)
16. A. Leizarowitz, Optimal controls for diffusion in $\mathbb{R}^d$-min-max max-min formula for the minimal cost growth rate. J. Math. Anal. Appl. **149**, 180–209 (1990)
17. J.D. López-Barrientos, Basic and advanced optimality criteria for zero-sum stochastic differential games. Ph.D. dissertation. Department of Mathematics, CINVESTAV-IPN, 2012. Available at http://www.math.cinvestav.mx/sites/default/files/tesis-daniel-2012.pdf
18. S. Meyn, R. Tweedie, Stability of Markovian processes III. Foster-Lyapunov criteria for continuous-time processes. Adv. Appl. Probab. **25**, 518–548 (1993)
19. A.S. Nowak, Correlated relaxed equilibria in nonzero-sum linear differential games. J. Math. Anal. Appl. **163**, 104–112 (1992)
20. T. Prieto-Rumeau, O. Hernández-Lerma, Bias and overtaking equilibria for zero-sum continuous- time Markov games. Math. Meth. Oper. Res. **61**, 437–454 (2005)
21. T. Prieto-Rumeau, O. Hernández-Lerma, The Laurent series, sensitive discount and Blackwell optimality for continuous-time controlled Markov chains. Math. Meth. Oper. Res. **61**, 123–145 (2005)
22. M.L. Puterman, Sensitive discount optimality in controlled one-dimensional diffusions. Ann. Probab. **2**, 408–419 (1974)
23. M. Schäl, Conditions for optimality and for the limit of n-stage optimal policies to be optimal. Z. Wahrs. Verw. Gerb. **32**, 179–196 (1975)
24. J. Warga, *Optimal Control of Differential and Functional Equations* (Academic, New York, 1972)
25. P. Whittle, *Risk-Sensitive Optimal Control*. Wiley-Interscience Series in Systems and Optimization (Wiley, Chichester, 1990)
26. L.C. Young, Generalized curves and the existence of an attained absolute minimum in the calculus of variations. Comptes Rendus de la Société des Sciences et des Lettres de Varsovie, Classe III **30**, 212–234 (1937)

A Note on Γ-Convergence of Monotone Functionals

Erick Treviño Aguilar

Abstract In this note we present a criterion under which a functional defined on vectors of non-decreasing functions is the Γ-limit of a functional defined on vectors of continuous non-decreasing functions. To this end, we present a separation principle in which a weakly converging sequence of continuous non-decreasing functions is decomposed in two parts, one converging to a non-decreasing function with a finite number of jumps and the other to the complementary jumps.

Keywords Γ-Convergence · Monotone functionals · Singular control · Skorokhod representation

Mathematics Subject Classification 60B10, 60B05, 49J45, 90C30

1 Introduction

For $\mathbb{T} > 0$ fixed, we denote by C the class of right-continuous with left-limits functions defined on the interval $[0, \mathbb{T}]$, which are non-negative and non-decreasing. We denote by C_{finite} the elements of C with a finite number of jumps and by C^0 the elements of C with no jumps. For $\mathsf{c} \in \mathsf{C}$, the jump at time $t \in [0, \mathbb{T}]$ is denoted by $\Delta\mathsf{c}(t)$ and is defined as the difference $\mathsf{c}(t) - \mathsf{c}(t-)$. If $\mathsf{c}(0) > 0$ then we consider a jump of size $\mathsf{c}(0)$ at time $t = 0$. Thus $\Delta\mathsf{c}(0) := \mathsf{c}(0)$. An element of C defines a unique positive measure in the interval $[0, \mathbb{T}]$ and we will consider the topology of weak convergence on C. Recall that a sequence of measures $\{\mu_n\}_{n\in\mathbf{N}}$ converges weakly to a measure μ if for each continuous bounded function $f : [0, \mathbb{T}] \to \mathbf{R}$ we have $\lim_{n\to\infty} \int f d\mu_n = \int f d\mu$. An equivalent property to weak convergence is formulated in terms of the elements of C (which can be seen as "distribution functions"). A sequence $\{\mathsf{c}(n)\}_{n\in\mathbf{N}} \subset \mathsf{C}$ converges pointwise to an element $\mathsf{c} \in \mathsf{C}$

E. Treviño Aguilar (✉)
Department of Economics and Finance, University of Guanajuato, Guanajuato, Mexico

© Springer International Publishing AG, part of Springer Nature 2018

D. Hernández-Hernández et al. (eds.), *XII Symposium of Probability and Stochastic Processes*, Progress in Probability 73,
https://doi.org/10.1007/978-3-319-77643-9_6

for each continuity point of c if and only if the corresponding measures converges weakly. By a slight abuse of language we will say that the sequence $\{\mathsf{c}(n)\}_{n\in\mathbf{N}}$ converges weakly to c.

An important property of weak convergence is that it is metrizable on separable spaces. Indeed, the well-known Prokhorov distance is a metric which on separable spaces characterizes weak convergence; see e.g., Ethier and Kurtz [6, Section 3.1]. This property will be crucial for our results here.

Now consider a functional $\mathsf{J} : \mathsf{C}^0 \times \mathsf{C}^0 \to \mathbf{R}$ and suppose we are required to consider the functional in all of the space $\mathsf{C} \times \mathsf{C}$. One reason why we might need the functional in an enlarged space is related to the problem of minimizing the functional. Indeed, a minimizer may fail to exist in the class of continuous elements and we might need to consider an enlarged space. A classical method to construct a functional in an enlarged space is by density and approximation. In this method, we take a point $(\mathsf{c}^1, \mathsf{c}^2) \in \mathsf{C} \times \mathsf{C}$ and a sequence $\{(\mathsf{c}^1(n), \mathsf{c}^2(n))\}_{n\in\mathbf{N}} \subset \mathsf{C}^0 \times \mathsf{C}^0$ which componentwise converges weakly. We might define a functional J^* in the point $(\mathsf{c}^1, \mathsf{c}^2)$ by the limit:

$$\mathsf{J}^*(\mathsf{c}^1, \mathsf{c}^2) = \lim_{n\to\infty} \mathsf{J}(\mathsf{c}^1(n), \mathsf{c}^2(n)).$$

The method requires that the limit always exists and to be independent of the particular sequence. However, we will illustrate in Sect. 3 that for weak convergence, with a very simple functional one gets different limits and even oscillatory behaviors. Note also that even for elements of $\mathsf{C}^0 \times \mathsf{C}^0$ the functionals J^* and J does not necessarily coincide and J^* is not necessarily an extension of J. Thus, the method does not work in general and we might need to consider "envelopes" instead of extensions. A convenient solution still keeping in mind problems of minimization is that of Γ-convergence. The concept was introduced in the study of variational problems by De Giorgi [5]. It is systematically presented by Dal Maso [4] and its relevance in optimal control, which is our main motivation here, is presented by e.g., Buttazzo and Dal Maso [3]. The Γ-convergence is a far reaching concept providing a powerful framework covering a wide range of applications; see e.g., Braides [2] and its references. In Sect. 2 below, we give more detail on this concept for our specific setting. Let us at this point formulate on the relevance of Γ-convergence in optimal control. Consider two topological spaces U (the space of controls) and Y (the space of state variables), and a function $\mathsf{J} : U \times Y \to [0, +\infty]$. Given a set of "admissible control-states" $\mathcal{A} \subset U \times Y$, consider the minimization problem:

$$\min_{(u,y)\in\mathcal{A}} \mathsf{J}(u, y).$$

This general problem may be difficult to study directly and instead, it might be convenient to study related problems formulated with other sets $\mathcal{A}^h \subset U \times Y$ and other functions J^h for $h \in \mathbf{N}$. In principle, the minimization problem formulated in terms of the pair $(\mathcal{A}^h, \mathsf{J}^h)$ should be easier and provide information about the original minimization problem formulated in terms of $\mathcal{A}$ and J. A way in which the sequence of auxiliary minimization problems help to understand the original problem is that of convergence of minimal values and convergence of optimal controls for the auxiliary problems, possibly along a subsequence, to an optimal control of the original problem. This is one of the main properties of Γ-Convergence; see e.g., Buttazzo and Dal Maso [3, Theorem 2.1].

The construction of Γ-limits is a highly non trivial task and in this paper we obtain a substantial reduction based on an assumption of monotonicity.

Definition 1.1 A functional J is monotone if for each $\mathsf{c}^1, \mathsf{c}^2 \in \mathsf{C}^0$ and $\nu^1, \nu^2 \in \mathsf{C}^0$ we have

$$\mathsf{J}(\mathsf{c}^1 + \nu^1, \mathsf{c}^2 + \nu^2) \geq \mathsf{J}(\mathsf{c}^1, \mathsf{c}^2). \tag{1.1}$$

In this note, we prove the Γ-convergence in $\mathsf{C} \times \mathsf{C}$ for monotone functionals as a consequence of the property for elements of $\mathsf{C}_{finite} \times \mathsf{C}_{finite}$ which have a finite number of atoms. This is a non trivial reduction that makes use of Skorokhod's representation of weak convergence and depends strongly on the property of monotonicity of the functional.

After this introduction, the note is organized as follows. In Sect. 2, we elaborate on the concept of Γ-convergence in our specific setting. In Sect. 3, we illustrate the phenomenon of oscillatory behavior. In Sect. 4 we prove a separation principle for sequences of continuous distributions by making use of Skorokhod's representation of weak convergence. In Sect. 5 we prove the sufficient condition for Γ-convergence.

2 Γ-Convergence

The next definition can be seen as a special case of the concept systematically presented by Dal Maso [4].

Definition 2.1 For a functional $\mathsf{J} : \mathsf{C}^0 \times \mathsf{C}^0 \to \mathbf{R}$ we say that the functional $\mathsf{J}^* : \mathsf{C} \times \mathsf{C} \to \mathbf{R}$ is the Γ-limit of J if the following conditions are satisfied:

1. for each point $(\mathsf{c}^1, \mathsf{c}^2) \in \mathsf{C} \times \mathsf{C}$ and sequence $\{(\mathsf{c}^1(n), \mathsf{c}^2(n))\}_{n \in \mathbf{N}} \subset \mathsf{C}^0 \times \mathsf{C}^0$ which componentwise weakly-converges to $(\mathsf{c}^1, \mathsf{c}^2)$ we have:

$$\mathsf{J}^*(\mathsf{c}^1, \mathsf{c}^2) \leq \liminf_{n \to \infty} \mathsf{J}(\mathsf{c}^1(n), \mathsf{c}^2(n)),$$

2. there exists a sequence $\{(\mathsf{c}^{1*}(n), \mathsf{c}^{2*}(n))\}_{n \in \mathbf{N}} \subset \mathsf{C}^0 \times \mathsf{C}^0$ which weakly-converges component by component to $(\mathsf{c}^1, \mathsf{c}^2)$ with the property

$$\mathsf{J}^*(\mathsf{c}^1, \mathsf{c}^2) = \lim_{n \to \infty} \mathsf{J}(\mathsf{c}^{1*}(n), \mathsf{c}^{2*}(n)).$$

3 An Example of Oscillatory Behavior

In this section we illustrate the phenomenon of oscillatory behavior with the very simple functional J defined by

$$\mathsf{J}(\mathsf{c}^1, \mathsf{c}^2) := \int_0^{\mathbb{T}} d\mathsf{c}_s^2 \int_0^s d\mathsf{c}_z^1.$$

In particular this example illustrates the convenience of considering the concept of Γ-convergence.

For $\tau \in (0, \mathbb{T})$, let

$$\mathsf{c}_t^2 := \mathsf{c}_t^1 := 1_{[\tau,\mathbb{T}]}(t). \tag{3.1}$$

Now we define continuous approximations. Take $\epsilon > 0$ with $\tau + \epsilon < \mathbb{T}$ and for $\alpha \in (0, 1)$ let $\epsilon' := \alpha\epsilon$. Let

$$\begin{aligned} u_t^1(\epsilon, \alpha) &:= \int_0^{t\wedge(\tau+\epsilon)} m_\epsilon^1 1_{[\tau,\tau+\epsilon]}(s)ds \\ u_t^2(\epsilon, \alpha) &:= \int_0^{t\wedge(\tau+\epsilon)} m_{\epsilon,\alpha}^2 1_{[\tau+\epsilon',\tau+\epsilon]}(s)ds. \end{aligned} \tag{3.2}$$

where

$$\begin{aligned} m_\epsilon^1 &:= \frac{\mathsf{c}_{\tau+\epsilon}^1 - \mathsf{c}_{\tau-}^1}{\epsilon} = \frac{1}{\epsilon} \\ m_{\epsilon,\alpha}^2 &:= \frac{\mathsf{c}_{\tau+\epsilon}^2 - \mathsf{c}_{\tau-}^2}{\epsilon - \epsilon'} = \frac{1}{(1-\alpha)\epsilon}. \end{aligned}$$

The functions u_1 and u_2 are illustrated in Fig. 1.

Proposition 3.1 *The functions* $u^1(\epsilon, \alpha)$ *and* $u^2(\epsilon, \alpha)$ *defined in* (3.2) *converge weakly as* $\epsilon \searrow 0$ *to* c^1 *and* c^2 *respectively, and*

$$\mathsf{J}(u^1(\epsilon, \alpha), u^2(\epsilon, \alpha)) = \frac{1+\alpha}{2}.$$

Proof Note that $u^1(\epsilon, \alpha), u^2(\epsilon, \alpha)$ converge pointwise as $\epsilon \searrow 0$ in $[0, \mathbb{T}]/\{\tau\}$ to $\mathsf{c}^1, \mathsf{c}^2$, respectively, and therefore converge weakly. For the second part of the

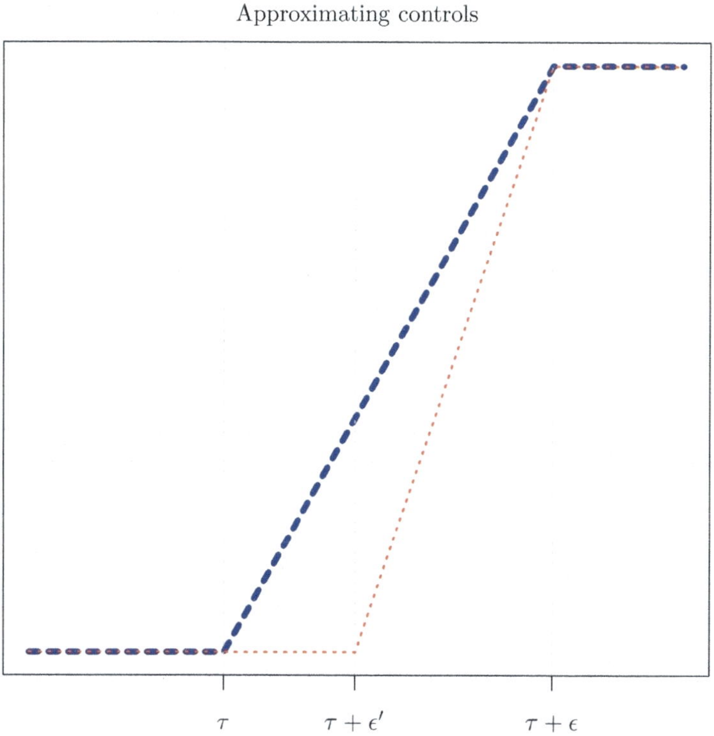

Fig. 1 The two functions u^1 and u^2, defined in Eq. (3.2)

proposition, we have

$$\begin{aligned}\int_0^{\mathbb{T}} du_s^2(\epsilon,\alpha)\int_0^s du_z^1(\epsilon,\alpha) &= \int_{\tau+\epsilon'}^{\tau+\epsilon} m_{\epsilon,\alpha}^2 1_{[\tau+\epsilon',\tau+\epsilon]}(s)ds \int_0^s m_\epsilon^1 1_{[\tau,\tau+\epsilon]}(z)dz\\ &= m_\epsilon^1 m_{\epsilon,\alpha}^2 \int_{\tau+\epsilon'}^{\tau+\epsilon} ds \int_\tau^s dz\\ &= m_\epsilon^1 m_{\epsilon,\alpha}^2 \int_{\tau+\epsilon'}^{\tau+\epsilon}(s-\tau)ds\\ &= m_\epsilon^1 m_{\epsilon,\alpha}^2 \frac{1}{2}\epsilon^2(1-\alpha^2)\\ &= \frac{1+\alpha}{2}.\end{aligned}$$

□

Remark 3.2 Note that it is possible to select a sequence $\{\alpha_m\}_{m\in\mathbf{N}}$ in such a way that the sequence $\left\{\mathsf{J}(u^1(\epsilon,\alpha_m),u^2(\epsilon,\alpha_m))\right\}_{m\in\mathbf{N}}$ generates a dense subset of the interval $[\frac{1}{2},1]$, due to Proposition 3.1.

4 A Separation Principle of Sequences

We start this section with Skorokhod's representation of weak convergence in the following form. Let $\{\mu_n\}_{n\in\mathbf{N}}$ be a sequence of probability measures in the interval $[0,\mathbb{T}]$ converging weakly to the measure μ. Then, there exists a probability space $(\Omega,\mathcal{F},\mathbb{P})$, a random variable Y and a sequence of random variables $\{Y_n\}_{n\in\mathbf{N}}$ defined in this common space, such that Y_n has distribution μ_n and Y has distribution μ and the sequence converges to Y everywhere in Ω. See e.g., Billingsley [1, Theorem 25.6] for the proof.

Lemma 4.1 *Let $\{F_m\}_{m\in\mathbf{N}}$ be a sequence of elements of C^0. Assume the sequence converges weakly to an element F of C. Thus, the sequence converges pointwise to F except, possibly, for the points $\{\tau_k\}_{k=0}^{\infty}$ where F jumps.*

Then, for $k_0\in\mathbf{N}$ fixed, there exist sequences of non-negative, non-decreasing continuous functions $\{G_m\}_{m\in\mathbf{N}}$ and $\{H_m\}_{m\in\mathbf{N}}$ such that

1. *$F_m=G_m+H_m$ for $m\in\mathbf{N}$.*
2. *The sequence $\{H_m\}_{m\in\mathbf{N}}$ converges pointwise to the function*

$$H(t):=\sum_{k=k_0+1}^{\infty}\Delta F(\tau_k)1_{\{\tau_k\le t\}},\tag{4.1}$$

   for $t\in[0,\mathbb{T}]$.
3. *The sequence $\{G_m\}_{m\in\mathbf{N}}$ converges pointwise to the function*

$$G(t):=F(t)-H(t),\tag{4.2}$$

   for $t\in[0,\mathbb{T}]$.

Proof We will do the proof only in the case that $F_m(\mathbb{T})=F(\mathbb{T})=1$, the general case following by normalization.

There exist a probability space $(\Omega,\mathcal{F},P)$ and a sequence of random variables $\{X_n\}_{n\in\mathbf{N}}$ converging to a random variable X, with $X_n\sim F_n$ and $X\sim F$, due to Skorokhod's representation theorem; see e.g., Billingsley [1, Theorem 25.6].

Let

$$A := X^{-1}([0, \mathbb{T}]/\{\tau_{k_0+1}, \tau_{k_0+2} \dots\}),$$
$$B := X^{-1}(\tau_{k_0+1}, \tau_{k_0+2} \dots).$$

Let us verify that the function G satisfies

$$G(t) = P\left[\{X \le t\} \cap A\right]. \tag{4.3}$$

Note that

$$\begin{aligned} F(t) - P\left[\{X \le t\} \cap A\right] &= P\left[\{X \le t\}\right] - P\left[\{X \le t\} \cap A\right] \\ &= P\left[\{X \le t\} \cap B\right] \\ &= \sum_{k=k_0+1}^{\infty} P\left[\{X \le t\} \cap X^{-1}(\tau_k)\right] \\ &= \sum_{k=k_0+1}^{\infty} \Delta F(\tau_k) 1_{\{\tau_k \le t\}} \\ &= H(t), \end{aligned}$$

and the equality (4.3) follows. Let

$$G_m(t) := P\left[\{X_m \le t\} \cap A\right], \text{ for } t \in [0, \mathbb{T}] \text{ and } m \in \mathbf{N}.$$

The function G_m has the following properties:

1. The function is clearly non-negative and non-decreasing.
2. G_m is a continuous function. Suppose by way of contradiction that G_m has a jump in $t_0 \in [0, \mathbb{T}]$. Take $\epsilon > 0$ smaller than the size of the jump

$$0 < \epsilon \le \Delta G_m(t_0).$$

   Then

$$\epsilon \le P\left[\{X_m = t_0\} \cap A\right] \le P\left[\{X_m = t_0\}\right],$$

   a contradiction with the fact that the function F_m is continuous. Thus, it was false to assume that G_m has a jump.
3. For $t \in [0, \mathbb{T}]$ we claim

$$\lim_{m \to \infty} G_m(t) = G(t). \tag{4.4}$$

Indeed, we have

$$\begin{aligned}\lim_{m\to\infty} P\left[\{X_m \le t\} \cap A\right] &= \lim_{m\to\infty} E_P\left[1_{(-\infty,t]}(X_m)1_A\right] \\ &= E_P\left[1_{(-\infty,t]}(X)1_A\right] \\ &= P\left[\{X \le t\} \cap A\right] \\ &= G(t),\end{aligned}$$

where the second equality holds true due to Lebesgue dominated convergence and the last equality is just (4.3).

Let

$$H_m := P\left[\{X_m \le t\} \cap B\right].$$

Analogously to the sequence $\{G_m\}_{m\in\mathbf{N}}$ we can prove that H_m

1. is a non-decreasing non-negative function,
2. is a continuous function
3. and $\lim_{m\to\infty} H_m(t) = H(t)$, for $t \in [0, \mathbb{T}]$.

The proof concludes with the equalities

$$\begin{aligned}G_m(t) + H_m(t) &= P\left[\{X_m \le t\} \cap A\right] + P\left[\{X_m \le t\} \cap B\right] \\ &= P\left[\{X_m \le t\}\right] \\ &= F_m(t).\end{aligned}$$

□

5 The Γ-Limit Under Monotonicity

Theorem 5.1 *Let* $\mathsf{J} : \mathsf{C}^0 \times \mathsf{C}^0 \to \mathbf{R}$ *be a monotone functional. Assume* $\mathsf{J}^* : \mathsf{C} \times \mathsf{C} \to \mathbf{R}$ *is the* Γ*-limit of* J *for elements in* $\mathsf{C}_{finite} \times \mathsf{C}_{finite}$ *of distributions with a finite number of jumps. Then, the* Γ*-limit of* J *in* $\mathsf{C} \times \mathsf{C}$ *is given as follows. For a pair* $(\mathsf{c}^1, \mathsf{c}^2) \in \mathsf{C} \times \mathsf{C}$ *with a countable number of jumps* $\{\tau_0, \tau_1, \ldots\}$ *we have*

$$\mathsf{J}^*(\mathsf{c}^1, \mathsf{c}^2) = \lim_{k\to\infty} \mathsf{J}^*(\widetilde{\mathsf{c}}^1(k), \widetilde{\mathsf{c}}^2(k)),$$

where

$$\widetilde{\mathsf{c}}^i_t(k) := \widehat{\mathsf{c}}^i_t + \sum_{j=0}^{k} \Delta\mathsf{c}^i_{\tau_j} 1_{\{\tau_j \le t\}}$$

and $\widehat{\mathsf{c}}^i$ *is the continuous part of* c^i, *for* $i = 1, 2$.

Proof

1. Let $\{(\mathsf{s}^1(m), \mathsf{s}^2(m))\}_{m\in\mathbf{N}} \subset \mathsf{C}^0 \times \mathsf{C}^0$ be a sequence componentwise weakly-converging to $(\mathsf{c}^1, \mathsf{c}^2) \in \mathsf{C} \times \mathsf{C}$. We first prove that

$$\liminf_{m\to\infty} \mathsf{J}(\mathsf{s}^1(m), \mathsf{s}^2(m)) \geq \lim_{k\to\infty} \mathsf{J}^*(\widetilde{\mathsf{c}}^1(k), \widetilde{\mathsf{c}}^2(k)). \tag{5.1}$$

For $k \in \mathbf{N}$ fixed and arbitrary $m \in \mathbf{N}$, take a decomposition $\mathsf{s}^i(m) = G^i(m) + H^i(m)$ as in Lemma 4.1 with $G^i(m)$ converging to $\widetilde{\mathsf{c}}^i(k)$ as $m \to \infty$. The functional J is monotone and therefore

$$\mathsf{J}(\mathsf{s}^1(m), \mathsf{s}^2(m)) \geq \mathsf{J}(G^1(m), G^2(m)).$$

As a consequence

$$\liminf_{m\to\infty} \mathsf{J}(\mathsf{s}^1(m), \mathsf{s}^2(m)) \geq \liminf_{m\to\infty} \mathsf{J}(G^1(m), G^2(m)) \geq \mathsf{J}^*(\widetilde{\mathsf{c}}^1(k), \widetilde{\mathsf{c}}^2(k)),$$

where the last inequality holds true since $G^i(m)$ weakly converges to $\widetilde{\mathsf{c}}^i(k)$. The sequence $\{\mathsf{J}^*(\widetilde{\mathsf{c}}^1(k), \widetilde{\mathsf{c}}^2(k))\}_{k\in\mathbf{N}}$ is non decreasing and we obtain the inequality (5.1).

2. Now we construct a sequence where the inequality (5.1) is satisfied with equality.

   Let $\{\mathsf{k}^i(k, j)\}_{j\in\mathbf{N}}$ be a sequence of continuous functions weakly converging to $\widetilde{\mathsf{c}}^i(k)$ for $i = 1, 2$ and

$$\mathsf{J}^*(\widetilde{\mathsf{c}}^1(k), \widetilde{\mathsf{c}}^2(k)) = \lim_{j\to\infty} \mathsf{J}(\mathsf{k}^1(k, j), \mathsf{k}^2(k, j)),$$

such a sequence exists since J^* is the Γ-limit of J in $\mathsf{C}_{finite} \times \mathsf{C}_{finite}$. Let ρ denote the Prokhorov metric on the space of probability measures defined on the interval $[0, \mathbb{T}]$. Next, identify distributions with probability measures. For $k \in \mathbf{N}$ let $j_k \in \mathbf{N}$ be such that $j_k > j_{k-1}$ and for $j \geq j_k$ and $i = 1, 2$

$$\rho(\widetilde{\mathsf{c}}^i(k), \mathsf{k}^i(k, j)) < \frac{1}{2k}$$

$$\rho(\widetilde{\mathsf{c}}^i(k), \mathsf{c}^i) < \frac{1}{2k}$$

$$\left|\mathsf{J}^*(\widetilde{\mathsf{c}}^1(k), \widetilde{\mathsf{c}}^2(k)) - \mathsf{J}(\mathsf{k}^1(k, j), \mathsf{k}^2(k, j))\right| < \frac{1}{k}.$$

Then, the sequence $\{(\mathsf{k}^1(k, j_k), \mathsf{k}^2(k, j_k))\}_{k\in\mathbf{N}}$ satisfies (5.1) with equality, since it has the properties

$$\rho(\mathsf{c}^i, \mathsf{k}^i(k, j_k)) < \frac{1}{k},$$

$$\left|\mathsf{J}^*(\widetilde{\mathsf{c}}^1(k), \widetilde{\mathsf{c}}^2(k)) - \mathsf{J}(\mathsf{k}^1(k, j_k), \mathsf{k}^2(k, j_k))\right| < \frac{1}{k}.$$

□

Let us give an application of Theorem 5.1. To this end, take a non-negative Radon measure η with support in the interval $[0, \mathbb{T}]$. Consider a functional of the form

$$\mathsf{J}(\mathsf{c}^1, \mathsf{c}^2) = \int_{[0,\mathbb{T}]} f(t, \mathsf{c}^1_t, \mathsf{c}^2_t)d\eta_t, \text{ for } (\mathsf{c}^1, \mathsf{c}^2) \in \mathsf{C}^0 \times \mathsf{C}^0,$$

where f is a normal integrand. That is, the correspondence

$$t \in [0, \mathbb{T}] \to \{(c^1, c^2, \alpha) \in \mathbf{R}^2_+ \times \mathbf{R} \mid f(t, c^1, c^2) \leq \alpha\},$$

is closed-valued and measurable. Recall that a set valued mapping (or correspondence) $S : \Xi \mapsto \mathbf{R} \cup \{\infty\}$ defined in a measurable space (Ξ, σ) is measurable if the inverse image $S^{-1}(O) := \{\xi \in \Xi \mid S(\xi) \cap O \neq \emptyset\}$ of every open set O is measurable. We will assume that $f(t, \cdot, \cdot)$ is a continuous non decreasing function for each $t \in [0, \mathbb{T}]$ and it is dominated by an η-integrable function. It is clear that J is a monotone functional. The Γ-limit of J is given in the next result.

Proposition 5.2 *For* $(\mathsf{c}^1, \mathsf{c}^2) \in \mathsf{C} \times \mathsf{C}$ *let* $\mathcal{D}$ *be the set of points where* c^1 *or* c^2 *jumps and let* A *be the set of atoms of the Radon measure* η*. Let* $(A \cap \mathcal{D})^c$ *be the complement of* $A \cap \mathcal{D}$ *in the interval* $[0, \mathbb{T}]$*. The* Γ*-limit of* J *in* $(\mathsf{c}^1, \mathsf{c}^2)$ *is given by*

$$\mathsf{J}^*(\mathsf{c}^1, \mathsf{c}^2) = \int_{(A\cap\mathcal{D})^c} f(t, \mathsf{c}^1_t, \mathsf{c}^2_t)d\eta_t + \sum_{t\in A\cap\mathcal{D}} \eta(\{t\}) f(t, \mathsf{c}^1_{t-}, \mathsf{c}^2_{t-}). \tag{5.2}$$

Proof Take $(\mathsf{c}^1, \mathsf{c}^2) \in \mathsf{C}_{finite} \times \mathsf{C}_{finite}$. For $i = 1, 2$, take a sequence $\{w^i(n)\}_{n\in\mathbf{N}} \subset \mathsf{C}^0$ converging weakly to c^i. We clearly have that

$$\liminf_{n\to\infty} \int_{[0,\mathbb{T}]} f(t, w^1_t(n), w^2_t(n))d\eta_t$$

$$= \int_{(A\cap\mathcal{D})^c} f(t, \mathsf{c}^1_t, \mathsf{c}^2_t)d\eta_t + \liminf_{n\to\infty} \int_{A\cap\mathcal{D}} f(t, w^1_t(n), w^2_t(n))d\eta_t,$$

due to the weak convergence, since f is a continuous function.

Take $t \in A \cap \mathcal{D}$. We will do the proof for $t \in (0, \mathbb{T})$, the other cases being more simple. For $\epsilon > 0$ and $\delta > 0$ with $t-\delta, t+\delta \in (0,\mathbb{T})/A \cup \mathcal{D}$ let $N \in \mathbf{N}$ be such that $\left|w^i_{t-\delta}(n) - \mathsf{c}^i_{t-\delta}\right| \le \epsilon$ and $\left|w^i_{t+\delta}(n) - \mathsf{c}^i_{t+\delta}\right| \le \epsilon$, for $n \ge N$. Then

$$-\epsilon + \mathsf{c}^i_{t-\delta} \le w^i_t(n) \le \epsilon + \mathsf{c}^i_{t+\delta}.$$

As a consequence

$$\mathsf{c}^i_{t-} \le \liminf_{n\to\infty} w^i_t(n) \le \limsup_{n\to\infty} w^i_t(n) \le \mathsf{c}^i_t.$$

The monotonicity and continuity of f implies now that

$$\liminf_{n\to\infty} \int_{A\cap\mathcal{D}} f(t, w^1_t(n), w^2_t(n))d\eta_t \ge \int_{A\cap\mathcal{D}} f(t, \mathsf{c}^1_{t-}, \mathsf{c}^2_{t-})d\eta_t.$$

Thus, we have proved that $\mathsf{J}^*(\mathsf{c}^1, \mathsf{c}^2) \le \liminf_{n\to\infty} \mathsf{J}(w^1(n), w^1(n))$.

Now we are going to construct a sequence $\{(v^1(n), v^2(n))\}_{n\in\mathbf{N}}$ converging weakly to $(\mathsf{c}^1, \mathsf{c}^2)$ with $\mathsf{J}^*(\mathsf{c}^1, \mathsf{c}^2) = \lim_{n\to\infty} \mathsf{J}(v^1(n), v^2(n))$. For $t \in \mathcal{D} \cap (0, \mathbb{T})$ let $B_t(\delta) := (t, t+\delta]$ where $\delta > 0$ is small enough so that $t+\delta \in (0,\mathbb{T})/(A \cup \mathcal{D})$ and the sets $B_t(\delta)$ are pairwise disjoint. For $i = 1, 2$, let l^i_t be the linear function defined by

$$l^i_t(z) = (z-t)\frac{\mathsf{c}^i(t+\delta) - \mathsf{c}^i(t-)}{\delta} + \mathsf{c}^i(t-).$$

We define

$$v^i_z(\delta) := \begin{cases} \mathsf{c}^i_{z-} & \text{for } \; z \notin \bigcup_{t\in\mathcal{D}\cap(0,\mathbb{T})} B_t(\delta), \\ l^i_t(z) & \text{for } \; z \in B_t(\delta). \end{cases}$$

Let $\{\delta_n\}_{n\in\mathbf{N}}$ be a sequence with $\delta_n \le \frac{1}{n}$ and satisfying the requirements that $t+\delta_n \in (0,\mathbb{T})/(A\cup\mathcal{D})$ and the sets $B_t(\delta_n)$ are pairwise disjoint. It is clear that the sequence $\{(v^1(\delta_n), v^2(\delta_n))\}_{n\in\mathbf{N}}$ converges weakly to $(\mathsf{c}^1, \mathsf{c}^2)$. Indeed, $(v^1(\delta_n), v^2(\delta_n)) = (\mathsf{c}^1, \mathsf{c}^2)$ outside the set $\bigcup_{t\in\mathcal{D}\cap(0,\mathbb{T})} B_t(\delta_n)$. Moreover

$$\int_{A\cap\mathcal{D}} f(t, v^1_t(\delta_n), v^2_t(\delta_n))d\eta_t = \int_{A\cap\mathcal{D}} f(t, \mathsf{c}^1_{t-}, \mathsf{c}^2_{t-})d\eta_t,$$

due to the definition of $(v^1_t(\delta_n), v^2_t(\delta_n))$.

We have proved that J^* as defined in (5.2), is the Γ-limit of J for elements in $\mathsf{C}_{finite} \times \mathsf{C}_{finite}$ of distributions with a finite number of jumps. Then, after Theorem 5.1, the Γ-limit of J in $\mathsf{C} \times \mathsf{C}$ is given as follows. For a pair $(\mathsf{c}^1, \mathsf{c}^2) \in$

$\mathsf{C} \times \mathsf{C}$ with a countable number of jumps $\{\tau_0, \tau_1, \ldots\}$ we have

$$\mathsf{J}^*(\mathsf{c}^1, \mathsf{c}^2) = \lim_{k\to\infty} \mathsf{J}^*(\widetilde{\mathsf{c}}^1(k), \widetilde{\mathsf{c}}^2(k)),$$

with the notation of Theorem 5.1. Note that

$$\mathsf{J}^*(\widetilde{\mathsf{c}}^1(k), \widetilde{\mathsf{c}}^2(k)) = \int_{[0,\mathbb{T}]} f(t, \widetilde{\mathsf{c}}^1_{t-}(k), \widetilde{\mathsf{c}}^2_{t-}(k)) d\eta_t.$$

Moreover, $\lim_{k\to\infty} \widetilde{\mathsf{c}}^i_{t-}(k) = \mathsf{c}^i_{t-}$ uniformly in $t \in [0, \mathbb{T}]$ and $i = 1, 2$. As a consequence

$$\lim_{k\to\infty} \mathsf{J}^*(\widetilde{\mathsf{c}}^1(k), \widetilde{\mathsf{c}}^2(k)) = \int_{[0,\mathbb{T}]} f(t, \mathsf{c}^1_{t-}, \mathsf{c}^2_{t-}) d\eta_t,$$

due to the continuity of the function f. The right-hand side of the last equation coincides with the right-hand side of (5.2). This proves the proposition. □

References

1. P. Billingsley, *Probability and Measure*. Wiley Series in Probability and Mathematical Statistics, Anniversary edition (Wiley, New York, 2012)
2. A. Braides, A handbook of Γ-convergence, in *Handbook of Differential Equations, Stationary Partial Differential Equations*, ed. by M. Chipot, P. Quittner, vol. 2 (North-Holland, Amsterdam, 2006), pp. 101–213
3. G. Buttazzo, G. Dal Maso, Gamma-convergence and optimal control problems. J. Optim. Theory Appl. **38**(3), 385–407 (1982)
4. G. Dal Maso, *An Introduction to Gamma-Convergence* (Birkhäuser, Boston, 1993)
5. D. De Giorgi, Γ-convergenza e G-convergenza. Bollettino dell' Unione Matematica Italiana **14**, 213–220 (1977)
6. S.N. Ethier, T.G. Kurtz, *Markov Processes, Characterization and Convergence*. Wiley Series in Probability and Mathematical Statistics (Wiley, New York, 1986)

A Criterion for Blow Up in Finite Time of a System of 1-Dimensional Reaction-Diffusion Equations

Eugenio Guerrero and José Alfredo López-Mimbela

Abstract We give a criterion for blow up in finite time of the system of semilinear partial differential equations $\frac{\partial u_i(t,x)}{\partial t} = \frac{1}{2}\frac{\partial^2 u_i(t,x)}{\partial x^2} + \frac{\varphi_i'(x)}{\varphi_i(x)}\frac{\partial u_i(t,x)}{\partial x} + u_j^{1+\beta_i}(t,x), t > 0$, $x \in \mathbb{R}$, with initial values of the form $u_i(0,x) = h_i(x)/\varphi_i(x)$, where $0 < \varphi_i \in L^2(\mathbb{R}, \mathrm{d}x) \cap C^2(\mathbb{R}), 0 \le h_i \in L^2(\mathbb{R}, \mathrm{d}x), \beta_i > 0$ and $i = 1, 2$, $j = 3-i$. Moreover, we find an upper bound T^* for the blowup time of such system which depends both on the initial values f_1, f_2, and the measures $\mu_i(\mathrm{d}x) = \varphi_i^2(x)\,\mathrm{d}x, i = 1, 2$.

Keywords Semilinear system of PDEs · Local mild solution · Finite time blow up

2000 Mathematics Subject Classification Primary 60H30, 35K57, 35B35, 60J57

1 Introduction

Consider the semilinear partial differential equation

$$\frac{\partial u(t,x)}{\partial t} = \frac{1}{2}\frac{\partial^2 u(t,x)}{\partial x^2} + \frac{\varphi'(x)}{\varphi(x)}\frac{\partial u(t,x)}{\partial x} + u^{1+\beta}(t,x), \quad t > 0, \quad x \in \mathbb{R}, \tag{1}$$

where $\beta > 0$, $\varphi \in C^2(\mathbb{R})$ is a square-integrable, strictly positive function, and the initial value is of the form $u(0,x) = h(x)/\varphi(x)$ with $h \in L^2(\mathbb{R}, \mathrm{d}x)$ and $\varphi\check{S}(x) = d\varphi(x)/dx$. Setting $\varphi(x) = e^{-x^2/2}$ in (1) it becomes

$$\frac{\partial u(t,x)}{\partial t} = L^{\varphi}u(t,x) + u^{1+\beta}(t,x), \quad t > 0, \quad x \in \mathbb{R},$$

E. Guerrero (✉) · J. A. López-Mimbela
Área de Probabilidad y Estadística, Centro de Investigación en Matemáticas, Guanajuato, Mexico
e-mail: eguerrero@cimat.mx

© Springer International Publishing AG, part of Springer Nature 2018

D. Hernández-Hernández et al. (eds.), *XII Symposium of Probability and Stochastic Processes*, Progress in Probability 73,
https://doi.org/10.1007/978-3-319-77643-9_7

where $L^{\varphi} := \frac{1}{2}\frac{\partial^2}{\partial x^2} - x\frac{\partial}{\partial x}$ is the infinitesimal generator of the Ornstein-Uhlenbeck semigroup $\{T_t,\ t \geq 0\}$. Using essentially Jensen's inequality and the fact that the measure $\mu(\mathrm{d}x) = \varphi^2(x)\,\mathrm{d}x$ is invariant for $\{T_t,\ t \geq 0\}$, in [8] we were able to prove that Eq. (1) exhibits blow up in finite time for any nontrivial initial value of the form $u(0,x) = h(x)/\varphi(x)$, $x \in \mathbb{R}$.

Motivated by this example, in this note we provide a criterion for explosion in finite time of positive mild solutions of the 1-dimensional semilinear system

$$
\begin{aligned}
\frac{\partial u_1(t,x)}{\partial t} &= \frac{1}{2}\frac{\partial^2 u_1(t,x)}{\partial x^2} + \frac{\varphi_1'(x)}{\varphi_1(x)}\frac{\partial u_1(t,x)}{\partial x} + u_2^{1+\beta_1}(t,x), \quad t>0, \quad x\in\mathbb{R},\\
\frac{\partial u_2(t,x)}{\partial t} &= \frac{1}{2}\frac{\partial^2 u_2(t,x)}{\partial x^2} + \frac{\varphi_2'(x)}{\varphi_2(x)}\frac{\partial u_2(t,x)}{\partial x} + u_1^{1+\beta_2}(t,x), \quad t>0, \quad x\in\mathbb{R}, \quad (2)\\
u_i(0,x) &= f_i(x), \quad x\in\mathbb{R}, \quad i=1,2,
\end{aligned}
$$

where $\beta_1, \beta_2 > 0$ are constants, f_1, f_2 are nonnegative functions and $\varphi_1, \varphi_2 \in C^2(\mathbb{R}) \cap L^2(\mathbb{R}, \mathrm{d}x)$ are strictly positive. Semilinear systems of this type have been investigated intensively in last years, starting with the pioneering work of Galaktionov et al. [4] (see also [2, 3, 5, 7, 9] and the review papers [1, 6]). This kind of systems arise as simplified models of the process of diffusion of heat and burning in a two-component continuous media, where u_1 and u_2 represent the temperatures of the two reactant components.

Recall that a pair (u_1, u_2) of measurable functions is termed *mild solution* of system (2) if it solves the system of integral equations

$$
u_i(t,x) = T_t^i\left(f_i(x)\right) + \int_0^t T_{t-s}^i\left(u_j^{1+\beta_i}(s,x)\right)\mathrm{d}s, \quad t\geq 0, \quad x\in\mathbb{R}, \qquad (3)
$$

where $i = 1, 2,\ j = 3 - i$ and $\{T_t^i, t \geq 0\}$ is the semigroup of continuous linear operators on $L^\infty(\mathbb{R}, \mathrm{d}x)$ having infinitesimal generator

$$
L^{\varphi_i} = \frac{1}{2}\frac{\partial^2}{\partial x^2} + \frac{\varphi_i'}{\varphi_i}\frac{\partial}{\partial x}; \quad i=1,2.
$$

If there exists $T \in (0,\infty)$ such that $\|u_1(t,\cdot)\|_{L^\infty(\mathbb{R},\mathrm{d}x)} = \infty$ or $\|u_2(t,\cdot)\|_{L^\infty(\mathbb{R},\mathrm{d}x)} = \infty$ for all $t \geq T$, then it is said that (u_1, u_2) *blows up (or explodes) in finite time*, and in this case the infimum of such T's is called the *blow up time* (or the *explosion time*) of (u_1, u_2).

Notice that for any $g \in L^{\infty}(\mathbb{R}, \mathrm{d}x)$ and $i = 1, 2$,

$$T_t^i(g(x)) = \mathbb{E}\left[g\left(X_t^{x,i}\right)\right], \quad t \geq 0, \quad x \in \mathbb{R},$$

where $\{X_t^{x,i}, t \geq 0\}$ is the unique strong solution of the stochastic differential equation

$$Y_t = x + B_t + \int_0^t \frac{\varphi_i'}{\varphi_i}(Y_s)\,\mathrm{d}s, \quad t \geq 0, \quad x \in \mathbb{R};$$

here $\{B_t, t \geq 0\}$ is a standard 1-dimensional Brownian motion. It turns out that under our assumptions both processes $\{X_t^{x,i}, t \geq 0\}$, $i = 1, 2$, are recurrent and, moreover, possess corresponding invariant measures

$$\mu_i(\mathrm{d}x) = \varphi_i^2(x)\,\mathrm{d}x, \quad i = 1, 2. \tag{4}$$

The intuitive explanation of the blow up phenomenon in non-linear heat equations of the archetype

$$\frac{\partial u}{\partial t} = \mathcal{A}u + u^{1+\beta}; \quad u(0) = f \geq 0,$$

where $\beta > 0$ and $\mathcal{A}$ is the generator of a strong Markov process on a locally compact space, is that if the initial value f is "small" then the tendency of the solution to blow up (which it would do if $u^{1+\beta}$ were the only term in the left-hand side of the equation) can be inhibited by the dissipative effect of the migration with generator $\mathcal{A}$; see e.g. [6, 9] or [10]. In view of the ergodicity of the processes $\{X_t^{x,i}, t \geq 0\}$, $i = 1, 2$, the mild solution of (2) should therefore blow up in finite time, at least for certain non-trivial positive initial values f_i, $i = 1, 2$.

In this work we give conditions which imply blow up in finite time of system (2) under the assumption that φ_1/φ_2 is a strictly positive bounded function such that $\inf_{x\in\mathbb{R}}\{\varphi_1(x)/\varphi_2(x)\} > 0$, and the initial values are of the form $f_i = h_i/\varphi_i$, where $h_i \in L^2(\mathbb{R}, \mathrm{d}x)$, $i = 1, 2$. We distinguish two cases: if $\beta_1 = \beta_2$ we show that any non-trivial positive mild solution of (2) blows up in finite time. If $\beta_1 \neq \beta_2$ we prove that a condition on the "sizes" of f_1 and f_2 and on the measures μ_1, μ_2 of the form

$$\int f_1\,\mathrm{d}\mu_1 + \int f_2\,\mathrm{d}\mu_2 > c_0,$$

(where the constant $c_0 > 0$ is determined by the system parameters) already implies finite time explosion of (2); see Theorem 2 below. Moreover, we find an upper bound T^* for the blowup time of system (2) which depends both on the initial values f_1, f_2, and the invariant measures (4). Our setting allows us to consider a wide range of choices for φ_1 and φ_2, for instance

$$\varphi_1(x) = (\sin(x) + 2)\,\varphi_2(x) \text{ with } \varphi_2(x) = e^{-x^2/2},$$

or else

$$\varphi_1(x) = \left(e^{-x^2/2} + 1\right)\varphi_2(x) \text{ with } \varphi_2(x) = 1/(1+x^2).$$

In these two cases the functions h_i, $i = 1, 2$, can be chosen of the form $h_i(x) = P_i(|x|)/Q_i(|x|)$, where P_i, Q_i are polynomial functions with non-negative coefficients such that their degrees satisfy $2 \leq \deg(Q_i) - \deg(P_i)$, and $Q_i(0) > 0$.

In the next section we prove existence and uniqueness of local mild solutions of (2) using the classical fixed-point argument, adapted to our context. Our main result, Theorem 2, is stated and proved in Sect. 3.

2 Local Existence and Uniqueness of Mild Solutions

Our proof of existence, uniqueness and positiveness of mild solutions of system (2) is based on [14, Theorem 2.1], (see also [12, Theorem 2.1], [15, Theorem 3], [7, Theorem 2] or [11, Theorem 1]).

For each $\tau \in (0, \infty)$ we define the set

$$E_\tau := \left\{(u_1, u_2)\,|u_1, u_2 : [0, \tau] \to L^\infty(\mathbb{R}, \mathrm{d}x), |||(u_1, u_2)||| < \infty\right\},$$

where

$$|||(u_1, u_2)||| := \sup_{t\in[0,\tau]} \left\{\|u_1(t, \cdot)\|_{L^\infty(\mathbb{R},\mathrm{d}x)} + \|u_2(t, \cdot)\|_{L^\infty(\mathbb{R},\mathrm{d}x)}\right\}.$$

Then $(E_\tau, |||\cdot|||)$ is a Banach space and the sets

$$P_\tau := \{(u_1, u_2) \in E_\tau : u_1 \geq 0, u_2 \geq 0\} \quad \text{and}$$
$$B_R := \{(u_1, u_2) \in E_\tau : |||(u_1, u_2)||| \leq R\}$$

are closed subsets of E_τ for any $R \in (0, \infty)$. Therefore $(P_\tau \cap B_R, |||\cdot|||)$ is a Banach space for all $\tau, R \in (0, \infty)$.

Theorem 1 *There exist $\tau, R \in (0, \infty)$ such that system (2) has a unique positive mild solution in $P_\tau \cap B_R$.*

Proof We will prove that the operator $\Psi : P_\tau \cap B_R \to P_\tau \cap B_R$ defined by

$$\Psi\left(\left(u_1(t,x), u_2(t,x)\right)\right) = \left(T_t^1\left(f_1(x)\right) + \int_0^t T_{t-s}^1\left(u_2^{1+\beta_1}(s,x)\right) \mathrm{d}s,\right.$$
$$\left. T_t^2\left(f_2(x)\right) + \int_0^t T_{t-s}^2\left(u_1^{1+\beta_2}(s,x)\right) \mathrm{d}s\right),$$

is a contraction for certain $\tau, R \in (0,\infty)$. We start by verifying that Ψ is in fact an operator from $P_\tau \cap B_R$ onto $P_\tau \cap B_R$ for suitably chosen $\tau, R \in (0,\infty)$. Let $\tau_0, R_0 \in (0,\infty)$ be such that

$$R_0 > \left(\|f_1\|_{L^\infty(\mathbb{R},\mathrm{d}x)} + \|f_2\|_{L^\infty(\mathbb{R},\mathrm{d}x)}\right) \text{ and}$$
$$\tau_0 \leq \frac{R_0 - \left(\|f_1\|_{L^\infty(\mathbb{R},\mathrm{d}x)} + \|f_2\|_{L^\infty(\mathbb{R},\mathrm{d}x)}\right)}{R_0^{1+\beta_1} + R_0^{1+\beta_2}}.$$

If $(u_1, u_2) \in P_{\tau_0} \cap B_{R_0}$ then $\Psi((u_1,u_2))$ has positive components due to the definition of Ψ and the fact that $u_1, u_2 \geq 0$. Hence

$$|||\Psi((u_1,u_2))||| = \sup_{t\in[0,\tau_0]} \left\{ \left\| T_t^1\left(f_1(\cdot)\right) + \int_0^t T_{t-s}^1\left(u_2^{1+\beta_1}(s,\cdot)\right) \mathrm{d}s \right\|_{L^\infty(\mathbb{R},\mathrm{d}x)} \right.$$
$$\left. + \left\| T_t^2\left(f_2(\cdot)\right) + \int_0^t T_{t-s}^2\left(u_1^{1+\beta_2}(s,\cdot)\right) \mathrm{d}s \right\|_{L^\infty(\mathbb{R},\mathrm{d}x)} \right\}$$
$$\leq \|f_1\|_{L^\infty(\mathbb{R},\mathrm{d}x)} + \|f_2\|_{L^\infty(\mathbb{R},\mathrm{d}x)} + \tau_0\left(R_0^{1+\beta_1} + R_0^{1+\beta_2}\right),$$

where we have used the contraction property of the operators T_t^i, $i = 1, 2$, to obtain the last inequality. It follows that $|||\Psi((u_1,u_2))||| \leq R_0$, i.e., Ψ is an operator from $P_{\tau_0} \cap B_{R_0}$ onto itself.

In order to prove the contraction property of Ψ we choose τ_0 as above in such a way that

$$\max_{i\in\{1,2\}} \left\{(1+\beta_i) R_0^{\beta_i}\right\} \tau_0 \in (0,1). \tag{5}$$

Let $(u_1,u_2), (\hat{u}_1, \hat{u}_2) \in P_{\tau_0} \cap B_{R_0}$. Using again the contraction property of the operators T_t^i, $i = 1, 2$, and the well-known inequality $|a^p - b^p| \leq p(a \vee b)^{p-1}|a-b|$, which holds for all $a, b > 0$ and $p \geq 1$, we obtain

$$|||\Psi((u_1,u_2)) - \Psi((\hat{u}_1,\hat{u}_2))|||$$
$$= \sup_{t\in[0,\tau_0]} \left\{ \left\| \int_0^t T_{t-s}^1\left(u_2^{1+\beta_1}(s,\cdot) - \hat{u}_2^{1+\beta_1}(s,\cdot)\right) \mathrm{d}s \right\|_{L^\infty(\mathbb{R},\mathrm{d}x)} \right.$$

$$
\left. + \left\| \int_0^t T^2_{t-s} \left(u_1^{1+\beta_2}(s,\cdot) - \hat{u}_1^{1+\beta_2}(s,\cdot) \right) \mathrm{d}s \right\|_{L^\infty(\mathbb{R},\mathrm{d}x)} \right\}
$$

$$
\begin{aligned}
&\leq \sup_{t\in[0,\tau_0]} \int_0^t \left\| u_2^{1+\beta_1}(s,\cdot) - \hat{u}_2^{1+\beta_1}(s,\cdot) \right\|_{L^\infty(\mathbb{R},\mathrm{d}x)} \mathrm{d}s \\
&\quad + \sup_{t\in[0,\tau_0]} \int_0^t \left\| u_1^{1+\beta_2}(s,\cdot) - \hat{u}_1^{1+\beta_2}(s,\cdot) \right\|_{L^\infty(\mathbb{R},\mathrm{d}x)} \mathrm{d}s \\
&\leq (1+\beta_1) R_0^{\beta_1} \int_0^{\tau_0} \left\| u_2(s,\cdot) - \hat{u}_2(s,\cdot) \right\|_{L^\infty(\mathbb{R},\mathrm{d}x)} \mathrm{d}s \\
&\quad + (1+\beta_2) R_0^{\beta_2} \int_0^{\tau_0} \left\| u_1(s,\cdot) - \hat{u}_1(s,\cdot) \right\|_{L^\infty(\mathbb{R},\mathrm{d}x)} \mathrm{d}s \\
&\leq \max_{i\in\{1,2\}} \left\{ (1+\beta_i) R_0^{\beta_i} \right\} \tau_0 \left|\left|\left| (u_1,u_2) - (\hat{u}_1,\hat{u}_2) \right|\right|\right| .
\end{aligned}
$$

From the last inequality we conclude, due to (5), that Ψ is a contraction in $P_{\tau_0} \cap B_{R_0}$. It follows from the Banach fixed-point theorem that Ψ has a unique fixed point in $P_{\tau_0} \cap B_{R_0}$, which is the unique mild solution of system (2). □

3 A Condition for Blowup in Finite Time

Our main result is the following

Theorem 2 *Let $\varphi_i \in L^2(\mathbb{R}, \mathrm{d}x) \cap C^2(\mathbb{R})$ be a strictly positive function and assume that the initial value f_i admits the representation*

$$
f_i(x) := \frac{h_i(x)}{\varphi_i(x)} \geq 0, \quad x \in \mathbb{R}, \tag{6}
$$

for some positive nontrivial $h_i \in L^2(\mathbb{R}, \mathrm{d}x)$, $i = 1, 2$. Suppose in addition that there exist strictly positive constants k_1, k_2 such that

$$
k_1 \leq \frac{\varphi_1(x)}{\varphi_2(x)} \leq k_2, \quad x \in \mathbb{R}. \tag{7}
$$

1. *Assume that $\beta_1 = \beta_2$. Then any non-trivial positive mild solution (u_1, u_2) of system (2) blows up in finite time.*

2. *Assume that* $\beta_1 > \beta_2$. *Let* $A_0 := \left(\frac{1+\beta_2}{1+\beta_1}\right)^{\frac{1+\beta_2}{\beta_1-\beta_2}} \frac{\beta_1-\beta_2}{1+\beta_1}$ *and suppose that*

$$\int_{\mathbb{R}} f_1(x)\,\mu_1(\mathrm{d}x) + \int_{\mathbb{R}} f_2(x)\,\mu_2(\mathrm{d}x) > 2^{\frac{\beta_2}{1+\beta_2}} A_0^{\frac{1}{1+\beta_2}}. \tag{8}$$

Then any mild solution (u_1, u_2) *of system (2) blows up in finite time.*

Proof Let (u_1, u_2) be a mild solution of system (2). We denote

$$w_i(t,x) := \varphi_i(x)\,u_i(t,x), \quad t \geq 0, \quad x \in \mathbb{R}.$$

Multiplying both sides of (3) by φ_i yields

$$w_i(t,x) = \varphi_i(x)\,T_t^i\left(\frac{h_i}{\varphi_i}(x)\right) + \int_0^t \varphi_i(x)\,T_{t-s}^i\left(w_{3-i}^{1+\beta_i}(s,x)\,\varphi_{3-i}^{-(1+\beta_i)}(x)\right)\mathrm{d}s. \tag{9}$$

Since the function $g_i(x) := \varphi_i^2(x)$ satisfies the differential equation

$$\frac{1}{2}\frac{\partial^2}{\partial x^2} g_i(x) - \frac{\partial}{\partial x}\left(g_i(x)\frac{\varphi_i'(x)}{\varphi_i(x)}\right) = 0, \quad x \in \mathbb{R},$$

it follows that $\mu_i(\mathrm{d}x) = \varphi_i^2(x)\,\mathrm{d}x$ is invariant for the semigroup $\{T_t^i,\ t \geq 0\}$. Let us write $\mathbb{E}^i[f] := \int_{\mathbb{R}} f(x)\,\varphi_i(x)\,\mathrm{d}x$. Due to (9) this implies that

$$\mathbb{E}^i[w_i(t,\cdot)] = \mathbb{E}^i[h_i(\cdot)] + \int_0^t \mathbb{E}^i\left[w_{3-i}^{1+\beta_i}(s,\cdot)\,\varphi_i(\cdot)\,\varphi_{3-i}^{-(1+\beta_i)}(\cdot)\right]\mathrm{d}s. \tag{10}$$

Define $a := \min\left\{k_1^2, \frac{1}{k_2^2}\right\}$. From assumption (7) we get $\frac{\varphi_i^2(x)}{\varphi_{3-i}^2(x)} \geq a$ for all $x \in \mathbb{R}$ and $i = 1, 2$. Therefore

$$\begin{aligned}
&\mathbb{E}^i\left[w_{3-i}^{1+\beta_i}(s,\cdot)\,\varphi_i(\cdot)\,\varphi_{3-i}^{-(1+\beta_i)}(\cdot)\right] \\
&= \int_{\mathbb{R}}\left(\frac{w_{3-i}(s,x)}{\varphi_{3-i}(x)}\right)^{1+\beta_i}\varphi_i^2(x)\,\mathrm{d}x \\
&\geq a\,\|\varphi_{3-i}\|_{L^2(\mathbb{R},\mathrm{d}x)}^2 \int_{\mathbb{R}}\left(\frac{w_{3-i}(s,x)}{\varphi_{3-i}(x)}\right)^{1+\beta_i}\frac{\varphi_{3-i}^2(x)}{\|\varphi_{3-i}\|_{L^2(\mathbb{R},\mathrm{d}x)}^2}\,\mathrm{d}x
\end{aligned}$$

$$\geq a \frac{\|\varphi_{3-i}\|^2_{L^2(\mathbb{R},dx)}}{\|\varphi_{3-i}\|^{2+2\beta_i}_{L^2(\mathbb{R},dx)}} \left(\int_{\mathbb{R}} \frac{w_{3-i}(s,x)}{\varphi_{3-i}(x)} \varphi^2_{3-i}(x)\,dx \right)^{1+\beta_i}$$

$$= a\,\|\varphi_{3-i}\|^{-2\beta_i}_{L^2(\mathbb{R},dx)} \left(\mathbb{E}^{3-i}\left[w_{3-i}(s,\cdot) \right] \right)^{1+\beta_i}, \tag{11}$$

where we have used Jensen's inequality to obtain the last inequality. Plugging (11) into (10) renders

$$\mathbb{E}^i\left[w_i(t,\cdot)\right] \geq \mathbb{E}^i\left[h_i(\cdot)\right] + a\,\|\varphi_{3-i}\|^{-2\beta_i}_{L^2(\mathbb{R},dx)} \int_0^t \left(\mathbb{E}^{3-i}\left[w_{3-i}(s,\cdot) \right] \right)^{1+\beta_i} ds. \tag{12}$$

Let $y_i(t)$ be the solution of the system

$$y_i'(t) = a\,\|\varphi_{3-i}\|^{-2\beta_i}_{L^2(\mathbb{R},dx)}\, y^{1+\beta_i}_{3-i}(t), \quad t > 0,$$

$$y_i(0) = \mathbb{E}^i\left[h_i(\cdot)\right], \quad i = 1, 2.$$

Putting $b := a \min\left\{ \|\varphi_1\|^{-2\beta_2}_{L^2(\mathbb{R},dx)}, \|\varphi_2\|^{-2\beta_1}_{L^2(\mathbb{R},dx)} \right\}$ we get the system of differential inequalities

$$y_i'(t) \geq b y^{1+\beta_i}_{3-i}(t), \quad t > 0,$$

$$y_i(0) = \mathbb{E}^i\left[h_i(\cdot)\right], \quad i = 1, 2.$$

Let $(z_1(t), z_2(t))$ be the solution of the system of ordinary differential equations

$$z_i'(t) = b z_j^{1+\beta_i}(t), \quad t > 0,$$

$$z_i(0) = \mathbb{E}^i\left[h_i(\cdot)\right], \quad i = 1, 2, \quad j = 3 - i.$$

By the Picard-Lindelöf theorem, this system with $(z_1(0), z_2(0)) = (0,0)$ has a unique local solution $(w_1(t), w_2(t)) \equiv (0,0)$ for all $t \in [0,\tau)$, for some $\tau \in (0,\infty]$. In our case $\mathbb{E}^i\left[h_i(\cdot)\right] \geq 0$. Therefore by a classical comparison theorem, $z_1(t), z_2(t) \geq 0$ for all $t \in [0,\tau)$.

Consider the new function

$$E(t) := z_1(t) + z_2(t), \quad t \geq 0.$$

We deal separately with the two cases in the statement of the theorem:

1. Case $\beta_1 = \beta_2$. Using the fact that

$$x^{1+\beta_1} + y^{1+\beta_1} \geq 2^{-\beta_1}(x+y)^{1+\beta_1}, \quad x \geq 0, \quad y \geq 0, \tag{13}$$

we get

$$\begin{aligned} E'(t) &= z_1'(t) + z_2'(t) \\ &= b\left(z_1^{1+\beta_1}(t) + z_2^{1+\beta_1}(t)\right) \\ &\geq 2^{-\beta_1} b E^{1+\beta_1}(t), \quad t > 0, \\ E(0) &= \mathbb{E}^1[h_1(\cdot)] + \mathbb{E}^2[h_2(\cdot)]. \end{aligned}$$

Let $I(t)$ be the solution of the ordinary differential equation

$$\begin{aligned} I'(t) &= 2^{-\beta_1} b I^{1+\beta_1}(t), \quad t > 0, \\ I(0) &= \mathbb{E}^1[h_1(\cdot)] + \mathbb{E}^2[h_2(\cdot)]. \end{aligned}$$

Since I is a subsolution of E (see [13], Lemma 1.2.) and I explodes at time

$$T^* = \frac{2^{\beta_1}}{b\beta_1 \left(\mathbb{E}^1[h_1(\cdot)] + \mathbb{E}^2[h_2(\cdot)]\right)^{\beta_1}} \in (0, \infty),$$

it follows that E explodes at some time $t_E \leq T^*$, and therefore, by a classical comparison theorem we get that

$$\begin{aligned} \mathbb{E}^1[w_1(t,\cdot)] &= \|u_1(t,\cdot)\|_{L^1(\mathbb{R},\mu_1)} = \infty \quad \text{or} \\ \mathbb{E}^2[w_2(t,\cdot)] &= \|u_2(t,\cdot)\|_{L^1(\mathbb{R},\mu_2)} = \infty \end{aligned}$$

for all $t \geq T^*$. Since $\|u_i(t,\cdot)\|_{L^1(\mathbb{R},\mu_i)} \leq \|u_i(t,\cdot)\|_{L^\infty(\mathbb{R},dx)} \|\varphi_i\|^2_{L^2(\mathbb{R},dx)}$ for all $t \in [0,\infty)$, $i = 1, 2$, we conclude that the mild solution (u_1, u_2) of system (2) blows up in finite time.

2. Case $\beta_1 > \beta_2$. Recall that for all $x, y \geq 0$, $\delta > 0$ and $p, q \in (1, \infty)$ such that $p^{-1} + q^{-1} = 1$ we have Young's inequality

$$xy \leq \frac{\delta^{-p} x^p}{p} + \frac{\delta^q y^q}{q}. \tag{14}$$

From the definition of A_0 it follows that

$$z_2^{1+\beta_1}(t) \geq z_2^{1+\beta_2}(t) - A_0, \quad \text{for all } t \geq 0.$$

In fact, it suffices to choose in (14)

$$x = 1, \quad y = z_2^{1+\beta_2}(t), \quad \delta = \left(\frac{1+\beta_1}{1+\beta_2}\right)^{\frac{1+\beta_2}{1+\beta_1}} \quad \text{and} \quad q = \frac{1+\beta_1}{1+\beta_2}.$$

Therefore we have

$$E'(t) \geq b\left(z_1^{1+\beta_2}(t) + z_2^{1+\beta_2}(t) - A_0\right).$$

Using again inequality (13) we conclude that

$$z_1^{1+\beta_2}(t) + z_2^{1+\beta_2}(t) \geq 2^{-\beta_2} E^{1+\beta_2}(t),$$

hence

$$E'(t) \geq b\left(2^{-\beta_2} E^{1+\beta_2}(t) - A_0\right).$$

Let $I(t)$ solve the ordinary differential equation

$$I'(t) = b\left(2^{-\beta_2} I^{1+\beta_2}(t) - A_0\right), \quad t > 0,$$
$$I(0) = \mathbb{E}^1[h_1(\cdot)] + \mathbb{E}^2[h_2(\cdot)].$$

It follows from the same comparison theorem as above that I is a subsolution of E. Using separation of variables we get, for $t \in (0, \infty)$,

$$t = \int_{E(0)}^{I(t)} \frac{dx}{b\left(2^{-\beta_2} x^{1+\beta_2} - A_0\right)} \leq \int_{E(0)}^{\infty} \frac{dx}{b\left(2^{-\beta_2} x^{1+\beta_2} - A_0\right)} =: T^*. \qquad (15)$$

But the hypothesis (8) implies that $T^* < \infty$. Hence (15) cannot hold for sufficiently large t, which yields that I explodes at a finite time $T^{**} \in (0, T^*]$. Therefore E explodes no later than T^* as well. From here we proceed as in the case $\beta_1 = \beta_2$ to conclude that the mild solution (u_1, u_2) of system (2) blows up in finite time also in this case.

□

The following result is an immediate consequence of the previous theorem. Recall that $E(0) = \int_{\mathbb{R}} f_1 \, d\mu_1 + \int_{\mathbb{R}} f_2 \, d\mu_2$ and

$$A_0 = \left(\frac{1+\beta_2}{1+\beta_1}\right)^{\frac{1+\beta_2}{\beta_1-\beta_2}} \frac{\beta_1 - \beta_2}{1+\beta_1}, \qquad b = \min\left\{k_1^2, \frac{1}{k_2^2}\right\} \min_{i\in\{1,2\}} \left\{\|\varphi_i\|_{L^2(\mathbb{R},dx)}^{-2\beta_i}\right\}.$$

Corollary 3 *Under the assumptions of Theorem 2, if $\beta_1 = \beta_2$ then the explosion time of any non-trivial positive solution of (2) is bounded above by*

$$T^* = \frac{2^{\beta_1}}{b\beta_1 (E(0))^{\beta_1}}.$$

If $\beta_1 > \beta_2$ *and* (8) *holds, then the time of explosion of* (2) *is bounded above by*

$$T^* = \int_{E(0)}^{\infty} \frac{\mathrm{d}x}{b\left(2^{-\beta_2}x^{1+\beta_2} - A_0\right)}.$$

Remark Theorem 2 and Corollary 3 remain valid when $\beta_2 > \beta_1$, with the obvious changes in the correspondent statements.

Acknowledgements The authors are grateful to an anonymous referee for her/his valuable comments. The second-named author gratefully acknowledges partial support from CONACyT (Mexico), Grant No. 257867.

References

1. K. Deng, H.A. Levine, The role of critical exponents in blow-up theorems: the sequel. J. Math. Anal. Appl. **243**, 85–126 (2000)
2. M. Dozzi, E.T. Kolkovska, J.A. López-Mimbela, Exponential functionals of Brownian motion and explosion times of a system of semilinear SPDEs. Stoch. Anal. Appl. **31**(6), 975–991 (2013)
3. M. Escobedo, M. Herrero, Boundedness and blow up for a semilinear reaction-diffusion system. J. Differ. Equ. **89**(1), 176–202 (1991)
4. V.A. Galaktionov, S.P. Kurdyumov, A.A. Samarskii, A parabolic system of quasilinear equations. I. (Russian) Differentsial'nye Uravneniya **19**(12), 2123–2140 (1983)
5. V.A. Galaktionov, S.P. Kurdyumov, A.A. Samarskii, A parabolic system of quasilinear equations. II. (Russian) Differentsial'nye Uravneniya **21**(9), 1544–1559 (1985)
6. H.A. Levine, The role of critical exponents in blowup theorems. SIAM Rev. **32**(2), 262–288 (1990)
7. J.A. López-Mimbela, A. Pérez, Global and nonglobal solutions of a system of nonautonomous semilinear equations with ultracontractive Lévy generators. J. Math. Anal. Appl. **423**(1), 720–733 (2015)
8. J.A. López-Mimbela, N. Privault, Large time behavior of reaction-diffusion equations with Bessel generators. J. Math. Anal. Appl. **383**(2), 560–572 (2011)
9. J.A. López-Mimbela, A. Wakolbinger, A probabilistic proof of non-explosion of a non-linear PDE system. J. Appl. Probab. **37**(3), 635–641 (2000)
10. M. Nagasawa, T. Sirao, Probabilistic treatment of the blowing up of solutions for a nonlinear integral equation. Trans. Am. Math. Soc. **139**, 301–310 (1969)
11. A. Pérez, Global existence and blow-up for nonautonomous systems with non-local symmetric generators and Dirichlet conditions. Differ. Equ. Appl. **7**(2), 263–276 (2015)
12. A. Pérez, J. Villa-Morales, Blow-up for a system with time-dependent generators. ALEA Lat. Am. J. Probab. Math. Stat. **7**, 207–215 (2010)
13. G. Teschl, *Ordinary Differential Equations and Dynamical Systems*. Graduate Studies in Mathematics, vol. 140 (American Mathematical Society, Providence, 2012)
14. Y. Uda, The critical exponent for a weakly coupled system of the generalized Fujita type reaction-diffusion equations. Z. Ang. Math. Phys. **46**(3), 366–383 (1995)
15. J. Villa-Morales, Blow up of mild solutions of a system of partial differential equations with distinct fractional diffusions. Electron. J. Differ. Equ. **2014**(41), 9 pp. (2014)

A Note on the Small-Time Behaviour of the Largest Block Size of Beta n-Coalescents

Arno Siri-Jégousse and Linglong Yuan

Abstract We study the largest block size of Beta n-coalescents at small times as n tends to infinity, using the paintbox construction of Beta-coalescents and the link between continuous-state branching processes and Beta-coalescents established in Birkner et al. (Electron J Probab 10(9):303–325, 2005) and Berestycki et al. (Ann Inst H Poincaré Probab Stat 44(2):214–238, 2008). As a corollary, a limit result on the largest block size at the coalescence time of the individual/block $\{1\}$ is provided.

Keywords Beta-coalescent · Kingman's paintbox construction · Continuous-state branching processes · Largest block size · Block-counting process

2010 Mathematics Subject Classification 60J25, 60F05, 92D15

1 Introduction and Main Results

Beta n-coalescents form a class of partition-valued coagulating Markov chains. This family was introduced by Schweinsberg [20] following pioneer works of Pitman [17], Sagitov [18] and Möhle and Sagitov [16]. Formally, a Beta n-coalescent $(\Pi^{(n)}(t), t \geq 0)$ is a continuous-time Markov chain with values in partitions of $[n] := \{1, 2, \ldots, n\}$ starting at $\Pi^{(n)}(0) = \{\{1\}, \{2\}, \ldots, \{n\}\}$. As n-coalescents can

A. Siri-Jégousse (✉)
IIMAS, Universidad Nacional Autónoma de México, Mexico City, Mexico
e-mail: arno@sigma.iimas.unam.mx

L. Yuan
Department of Mathematical Sciences, Xi'an Jiaotong-Liverpool University, Suzhou, People's Republic of China

© Springer International Publishing AG, part of Springer Nature 2018

D. Hernández-Hernández et al. (eds.), *XII Symposium of Probability and Stochastic Processes*, Progress in Probability 73,
https://doi.org/10.1007/978-3-319-77643-9_8

be used as models for the genealogy of a sample of n individuals, we refer to $[n]$ as the set of (labels of) individuals. Its dynamics are determined by a parameter $\alpha \in (0, 2)$: when $\Pi^{(n)}$ has b blocks, any k-tuple of them merges into one block at rate

$$\lambda_{b,k} := \frac{\beta(k-\alpha, b-k+\alpha)}{\beta(\alpha, 2-\alpha)} \tag{1}$$

where $\beta(a, b) = \Gamma(a)\Gamma(b)/\Gamma(a+b)$ is the Beta function. In this paper, we are only interested in the case $\alpha \in (1, 2)$.

Equation (1) induces exchangeability and consistency of these processes. Exchangeability means that if we permute the labels of individuals, the law of $\Pi^{(n)}$ stays unchanged. Consistency refers to that for any couple of integers $n < m$, the projection of $\Pi^{(m)}$ on $[n]$ has the same law as $\Pi^{(n)}$. By Kolmogorov's extension theorem [17], we can construct the so-called Beta-coalescent process $(\Pi(t), t \geq 0)$ taking values in partitions of $\mathbb{N}$ such that the projection of Π on $[n]$ is equal in distribution to $\Pi^{(n)}$. When $\alpha \in (1, 2)$ the Beta-coalescent has proper frequency (i.e., almost surely for any $t > 0$, Π has no singletons, see [17]) and comes down from infinity (i.e., almost surely for any $t > 0$, Π has a finite number of blocks, see [19]).

Berestycki et al. [2] provided many results on the behaviour of functionals of $\Pi(t)$ as t tends to 0, such as the number of blocks, the ranked sequence of asymptotic frequencies of those blocks and the asymptotic frequency of the largest block. For the latter, they establish the following result in Proposition 1.6:

Proposition 1.1 *Let $X(t)$ be the asymptotic frequency of the largest block of Π at time t, then*

$$(\alpha\Gamma(\alpha)\Gamma(2-\alpha))^{\frac{1}{\alpha}} t^{-\frac{1}{\alpha}} X(t) \xrightarrow{d} X, \text{ as } t \text{ goes to } 0 \tag{2}$$

where X is a Fréchet random variable with parameter α, i.e., $\mathbb{P}(X \leq x) = e^{-x^{-\alpha}}$, for any $x \geq 0$, and "$\xrightarrow{d}$" stands for the convergence in law.

This is a result in the infinite coalescent for $t \to 0$. Often, especially when used as a genealogy model, we are actually more interested in the n-coalescents and their asymptotic behaviour, since we can then interpret results in terms of the finite models (as in [7–9, 12, 13, 15, 21, 22]). Proposition 1.1 would in this sense be first taking $n \to \infty$, then $t \to 0$, while we would like a simultaneous limit $(t_n, n) \to (0, \infty)$. In this case, we could look at specific, interpretable/interesting small times t_n.

Such time is the external branch length of individual 1 (studied in [9], and with further extensions given recently in [22] and [24]), denoted by $T_1^{(n)}$ and defined by

$$T_1^{(n)} := \sup\{t, \{1\} \in \Pi^{(n)}(t)\}.$$

This can be seen as seeing the coalescent from the eyes of individual 1 and measuring its "distance" to the rest of the sample or its *genetic uniqueness* [6]. Here individual 1 represents a randomly chosen individual of the sample thanks to exchangeability. Observe that, since the Beta-coalescent has proper frequency when $\alpha \in (1, 2)$, this variable vanishes as we let n tend to infinity. We are now curious how the block structure of the coalescent looks like at this specific time (asymptotically).

One possible tool for this study is the minimal clade size, studied in [22] for $\alpha \in (1, 2)$ (see also [11] for $\alpha = 1$ and [5] for $\alpha = 2$). This is the size of the block containing 1 at time $T_1^{(n)}$. The size of the minimal clade gives the information of how many individuals share the genealogy with individual 1 after he merges. It was shown in [22] that the minimal clade size converges in law, without any renormalization, to a heavy-tailed random variable of index $(\alpha - 1)^2$.

Now we would like to compare this minimal clade size to the size of the largest block at time $T_1^{(n)}$, denoted by $\tilde{W}^{(n)}$. This comparison gives a first picture of the inhomogeneity of the block structure of the Beta n-coalescent at small times. To study $\tilde{W}^{(n)}$, we first consider the size of the largest block at any time t, denoted by $W^{(n)}(t)$. Hence, we have

$$\tilde{W}^{(n)} = W^{(n)}(T_1^{(n)}).$$

We obtain an asymptotic result for $W^{(n)}$ at the $n^{1-\alpha}t$ scale.

Theorem 1.2 *For a Beta n-coalescent with* $1 < \alpha < 2$, *as n tends to infinity*

$$(\alpha\Gamma(\alpha)\Gamma(2-\alpha))^{\frac{1}{\alpha}}(nt)^{-\frac{1}{\alpha}}W^{(n)}(n^{1-\alpha}t) \xrightarrow{d} X, \tag{3}$$

where X is a Fréchet random variable with parameter α.

Rewriting (3) as

$$\alpha\Gamma(\alpha)\Gamma(2-\alpha))^{\frac{1}{\alpha}}(n^{1-\alpha}t)^{-\frac{1}{\alpha}}\frac{W^{(n)}(n^{1-\alpha}t)}{n} \xrightarrow{d} X,$$

the reader can observe the similarity with (2).

To study the behaviour of $\tilde{W}^{(n)}$, we shall consider the restriction of $\Pi^{(n)}$ on $\{2, \dots, n\}$, denoted by $\Pi^{(n,2)} = (\Pi^{(n,2)}(t), t \geq 0)$. By consistency, the latter is equal in law to $\Pi^{(n-1)}$ modulo notations of the labels of individuals. Then $\tilde{W}^{(n)}$ is actually the largest block size of $\Pi^{(n,2)}(T_1^{(n)})$ plus 1, if $\{1\}$ coalesces with the largest block of $\Pi^{(n,2)}(T_1^{(n)})$ or plus 0 otherwise.

It has been established in the proof of Theorem 5.2 of [9] that conditional on $\Pi^{(n,2)}$, $n^{\alpha-1}T_1^{(n)}$ converges in law to a random variable T. More precisely,

$$\mathbb{P}(n^{\alpha-1}T_1^{(n)} \geq t|\Pi^{(n,2)}) \xrightarrow{d} \mathbb{P}(T \geq t) = (1 + \frac{t}{\alpha\Gamma(\alpha)})^{-\frac{\alpha}{\alpha-1}}. \tag{4}$$

This shows that in the decomposition of $\tilde{W}^{(n)} = W^{(n)}(T_1^{(n)})$, the terms $(W^{(n)}(n^{1-\alpha}t), t \geq 0)$ and $n^{\alpha-1}T_1^{(n)}$ are asymptotically independent. Combining (4) together with Theorem 1.2, we can describe the limit of $\tilde{W}^{(n)}$ as a mixture.

Corollary 1.3 *As n tends to infinity,*

$$\frac{\tilde{W}^{(n)}}{n^{\frac{1}{\alpha}}} \xrightarrow{d} \tilde{W}, \tag{5}$$

where $\tilde{W}$ is a positive random variable such that for any $x \geq 0$,

$$\mathbb{P}(\tilde{W} \leq x) = \int_0^\infty \frac{\exp(-x^{-\alpha}\frac{t}{\alpha\Gamma(\alpha)\Gamma(2-\alpha)})}{(\alpha-1)\Gamma(\alpha)}(1+\frac{t}{\alpha\Gamma(\alpha)})^{-\frac{2\alpha-1}{\alpha-1}}dt.$$

This note is organised as follows. In Sect. 2, we introduce the main tools such as the construction of Beta-coalescents via continuous-state branching processes and the paintbox construction of exchangeable coalescents. Section 3 is devoted to the proofs of Theorem 1.2.

2 Preliminaries

2.1 Ranked Coalescent and Paintbox Construction

Assume all along the rest of the paper that $1 < \alpha < 2$. Let $\Pi = (\Pi(t), t \geq 0)$ be the Beta-coalescent and denote by $K = (K(t), t > 0)$ the block-counting process of Π. In words, $K(t)$ stands for the number of blocks of $\Pi(t)$. It is known that Π is coming down from infinity: for any $t > 0$, $K(t)$ is finite almost surely [19]. Also recall that for any $t \geq 0$, $\Pi(t)$ is an exchangeable random partition of $\mathbb{N}$. This means that if we permute finitely many integers in $\Pi(t)$, the law of $\Pi(t)$ is unchanged. Applying Kingman's paintbox theorem on exchangeable random partitions [14], almost surely for every block $B \in \Pi(t)$, the following limit, called the *asymptotic frequency* of B, exists:

$$\lim_{m\to\infty} \frac{1}{m}\sum_{i=1}^{m} \mathbf{1}_{\{i\in B\}}.$$

Furthermore, when $t > 0$, the sum of all asymptotic frequencies equals 1 since Π is of proper frequency [17]. Hence, one can reorder all the asymptotic frequencies in a non-increasing way to define a sequence $\Theta(t) = \{\theta_1(t), \theta_2(t), \cdots, \theta_{K(t)}(t)\}$ where $\theta_1(t) \geq \theta_2(t) \geq \cdots \geq \theta_{K(t)}(t)$ and $\sum_{i=1}^{K(t)} \theta_i(t) = 1$. At time $t = 0$, every block is a singleton and then has asymptotic frequency 0. Hence one can naturally set $\Theta(0) = \{0, 0, \ldots\}$. Then the process $\Theta = (\Theta(t), t \geq 0)$ is well defined. We call it the *ranked coalescent*.

Given $\Theta(t)$ for some $t > 0$, one can recover the distribution of $\Pi(t)$ using again Kingman's paintbox theorem. Let us at first divide $[0, 1]$ into $K(t)$ subintervals such that their lengths are equal one to one to the values of elements of $\Theta(t)$. Then we throw individuals $1, 2, \cdots$ uniformly and independently into $[0, 1]$. Finally, all individuals within one interval form a block and this procedure provides a random exchangeable partition which has the same law as $\Pi(t)$. Thanks to the consistency property, the restricted partition $\Pi^{(n)}(t)$ can be obtained using the same procedure but throwing n particles instead of infinitely many.

2.2 Beta-Coalescents and Stable Continuous-State Branching Processes

To prove Theorem 1.2, we will use classical relations between Beta-coalescents and continuous-state branching processes (CSBPs) developed in [4] (see also Section 2 of [2]). We give a short summary to provide a minimal set of tools. A continuous-state branching process $(Z(t), t \geq 0)$ is a $[0, \infty]$-valued Markov process (in continuous time) whose transition semigroup $p_t(x, \cdot)$ satisfies the branching property

$$p_t(x + y, \cdot) = p_t(x, \cdot) * p_t(y, \cdot), \quad \text{for all } x, y \geq 0.$$

For each $t \geq 0$, there exists a function $u_t : [0, \infty) \to \mathbb{R}$ such that

$$\mathbb{E}[e^{-\lambda Z(t)} | Z(0) = a] = e^{-a u_t(\lambda)}. \tag{6}$$

If, almost surely, the process has no instantaneous jump to infinity, the function u_t satisfies the following differential equation

$$\frac{\partial u_t(\lambda)}{\partial t} = -\Psi(u_t(\lambda)),$$

where $\Psi : [0, \infty) \longrightarrow \mathbb{R}$ is a function of the form

$$\Psi(u) = \gamma u + \beta u^2 + \int_0^\infty (e^{-xu} - 1 + xu\mathbf{1}_{\{x \leq 1\}})\pi(dx),$$

where $\gamma \in \mathbb{R}, \beta \geq 0$ and π is a Lévy measure on $(0, \infty)$ satisfying $\int_0^\infty (1 \wedge x^2)\pi(dx) < \infty$. The function Ψ is called the *branching mechanism* of the CSBP.

As explained in [3], a CSBP can be extended to a two-parameter random process $(Z(t, a), t \geq 0, a \geq 0)$ with $Z(0, a) = a$. For fixed t, $(Z(t, a), a \geq 0)$ turns out to be a subordinator with Laplace exponent $\lambda \mapsto u_t(\lambda)$ thanks to (6).

There exists a measure-valued process $(M_t, t \geq 0)$ taking values in the set of finite measures on $[0, 1]$ which characterises $(Z(t, a), t \geq 0, 0 \leq a \leq 1)$. More precisely, $(M_t([0, a]), t \geq 0, 0 \leq a \leq 1)$ has the same finite-dimensional distributions as $(Z(t, a), t \geq 0, 0 \leq a \leq 1)$. Hence $(M_t([0, a]), 0 \leq a \leq 1)$ is a subordinator with Laplace exponent $\lambda \mapsto u_t(\lambda)$ and $Z(t, 1) = M_t([0, 1])$ is a CSBP with branching mechanism Ψ started at $M_0([0, 1]) = 1$. In particular, if the branching mechanism is $\Psi(\lambda) = \lambda^\alpha$, its Lévy measure is given by $\pi(dx) = \frac{\alpha(\alpha-1)}{\Gamma(2-\alpha)} x^{-1-\alpha} dx$ and, for all $t > 0$, M_t consists only of a finite number of atoms. For the construction of $(M_t([0, a]), t \geq 0, 0 \leq a \leq 1)$, we refer to [1, 4, 10].

A deep relation has been revealed in [4] between the Beta-coalescent and the CSBP with branching mechanism $\Psi(\lambda) = \lambda^\alpha$. It is described by the following two lemmas which are respectively Lemma 2.1 and 2.2 of [2]. To save notations, from now on, $(Z(t), t \geq 0)$ will always denote a continuous-state branching process $(Z(t, 1), t \geq 0)$.

Lemma 2.1 *Assume that $(Z(t), t \geq 0)$ is a CSBP with branching mechanism $\Psi(\lambda) = \lambda^\alpha$ and let $(M_t, t \geq 0)$ be its associated measure-valued process. If $(\Pi(t), t \geq 0)$ is a Beta-coalescent and $(\Theta(t), t \geq 0)$ is the associated ranked coalescent, then for all $t > 0$, the distribution of $\Theta(t)$ is the same as the distribution of the sizes of the atoms of the measure $\frac{M_{R^{-1}(t)}}{Z(R^{-1}(t))}$, ranked in decreasing order. Here $R(t) = (\alpha - 1)\alpha\Gamma(\alpha) \int_0^t Z(s)^{1-\alpha} ds$ and $R^{-1}(t) = \inf\{s : R(s) > t\}$.*

Let μ denote the Slack's probability distribution on $[0, \infty)$ (see [23]) characterised by its Laplace transform

$$\mathcal{L}_\mu(\lambda) = \int_0^\infty e^{-\lambda x} \mu(dx) = 1 - (1 + \lambda^{1-\alpha})^{-\frac{1}{\alpha-1}}, \quad \lambda \geq 0. \tag{7}$$

Lemma 2.2 *Assume $\Psi(\lambda) = \lambda^\alpha$. For any $t \geq 0$, let $D(t)$ be the number of atoms of M_t, and let $J(t) = (J_1(t), \cdots, J_{D(t)}(t))$ be the sizes of the atoms of M_t, ranked in decreasing order. Then $D(t)$ is Poisson with mean $\gamma(t) = ((\alpha - 1)t)^{-\frac{1}{\alpha-1}}$. Moreover, conditional on $D(t) = k$, the distribution of $J(t)$ is the same as the distribution of $(\gamma(t)^{-1}X_1, \cdots, \gamma(t)^{-1}X_k)$ where $X_1, \cdots, X_k$ are obtained by picking k i.i.d. random variables with distribution μ and then ranking them in decreasing order.*

Remark 2.1 From the relation between $(M_t, t \geq 0)$ and $(Z(t, a), t \geq 0, 0 \leq a \leq 1)$ and also the fact that for all $t > 0$, M_t has a finite number of atoms $D(t)$, we can deduce that for a given $t > 0$, there exist $0 \leq a_1, \cdots, a_{D(t)} \leq 1$ such that $\{Z(t, a_1) - Z(t, a_1-), \cdots, Z(t, a_{D(t)}) - Z(t, a_{D(t)}-)\}$ are exactly the sizes of the atoms of M_t. Markov property of $(Z(t, a), t \geq 0, 0 \leq a \leq 1)$ implies that for $s \geq t$, discontinuity points of the subordinator $(Z(s, a), 0 \leq a \leq 1)$ must be part (or all) of the points $a_1, \cdots, a_{D(t)}$. Therefore, $t \mapsto D(t)$ is almost surely non-increasing.

3 Proofs

In this section, we aim to prove Theorem 1.2 and Corollary 1.3. From now on, we will use the notations $t_n = n^{1-\alpha}t$ and $t'_n = \frac{t_n}{(\alpha-1)\alpha\Gamma(\alpha)}$. Lemma 2.1 entails that $\Theta(t_n)$ has the same law as $\frac{M_{R^{-1}(t_n)}}{Z(R^{-1}(t_n))}$. Moreover, Lemma 4.2 of [2] states that $\frac{R^{-1}(t_n)}{t'_n} \xrightarrow{P} 1$, as n goes to ∞. From this arises the idea of approximating the block sizes of the coalescent at time t_n by the atoms of the renormalized measure-valued process at time t'_n. The advantage of this approximation is that the time is no longer random. This idea will be executed through three steps. First, we will study the size of the largest atom of the rescaled measure M/Z at deterministic time t'_n, using tools of the theory of CSBPs. Second we show that the paintbox construction of an exchangeable partition can also be provided by using a different paintbox and by modifying it according to the differences between the paintboxes. In the third step, we use this construction to approximate the partition $\Pi^{(n)}$ at time t_n from partitions built from the rescaled atoms of M/Z at time $(1 \pm \varepsilon)t'_n$ for small ε.

3.1 *The Largest Atom Size of M/Z at a Fixed Time*

We start with a technical lemma associated to the measure μ. We write $a_n \sim b_n$ if $\lim_{n\to\infty} \frac{a_n}{b_n} = 1$. Recall from Equation (33) of [2] that

$$\mu([x,\infty)) \sim \frac{x^{-\alpha}}{\Gamma(2-\alpha)} \tag{8}$$

when x goes to ∞.

Lemma 3.1 *Let $k > 0$ and X be a random variable distributed according to μ. Define $\mathcal{X}$ such that conditional on X, $\mathcal{X}$ is a Poisson variable with parameter $\frac{X}{k}$. Then for any $x > 0$,*

$$\lim_{n\to\infty} n\mathbb{P}(\mathcal{X} \geq xn^{\frac{1}{\alpha}}) = \frac{(kx)^{-\alpha}}{\Gamma(2-\alpha)}.$$

Proof Let $M = \lfloor xn^{\frac{1}{\alpha}} \rfloor$. We start the proof with two claims. First, using Stirling's formula for $M!$ and a change of variable, we get that for any $0 < \beta < 1$,

$$\begin{aligned}\int_0^{M\beta} e^{-t}\frac{t^M}{M!}dt &= \int_0^{M\beta} e^{M-t}\left(\frac{t}{M}\right)^M (2\pi M)^{-\frac{1}{2}}(1+O(M^{-1}))dt \\ &= \int_0^{\beta} e^{M(1-t+\ln t)}(\frac{M}{2\pi})^{\frac{1}{2}}(1+O(M^{-1}))dt \\ &= O(e^{M(1-\beta+\ln\beta)}M^{\frac{1}{2}}).\end{aligned} \tag{9}$$

The last equality is due to the fact that $1 - t + \ln t$ is negative and increasing for $t \in (0, 1)$. Second, if $\beta > 1$, then

$$\int_{M\beta}^{\infty} e^{-t} \frac{t^M}{M!} dt = \int_{\beta}^{\infty} e^{M(1-t+\ln t)} (\frac{M}{2\pi})^{\frac{1}{2}} (1 + O(M^{-1})) dt.$$

Notice that $1 - t + \ln t$ is strictly decreasing and concave over $[\beta, \infty]$. Then there exists a positive number ε such that $1 - t + \ln t \leq -\varepsilon t$ for any $t \geq \beta$. Therefore,

$$\int_{M\beta}^{\infty} e^{-t} \frac{t^M}{M!} dt \leq \int_{\beta}^{\infty} e^{-\varepsilon M t} (\frac{M}{2\pi})^{1/2} (1 + O(M^{-1})) dt = O(e^{-\varepsilon M\beta} M^{-1/2}). \tag{10}$$

Now we can turn to $\mathcal{X}$. Thanks to successive integrations by parts,

$$\mathbb{P}(\mathcal{X} \geq M + 1) = \mathbb{E}[\int_0^{\frac{X}{k}} e^{-t} \frac{t^M}{M!} dt]. \tag{11}$$

Let $0 < \beta_1 < 1$ and $\beta_2 > 1$, then we have

$$\mathbb{P}(\mathcal{X} \geq M + 1) = I_1 + I_2 + I_3,$$

where

$$I_1 = \mathbb{E}[\int_0^{\frac{X}{k}} e^{-t} \frac{t^M}{M!} dt \mathbf{1}_{\{X < kM\beta_1\}}],$$

$$I_2 = \mathbb{E}[\int_0^{\frac{X}{k}} e^{-t} \frac{t^M}{M!} dt \mathbf{1}_{\{kM\beta_1 \leq X \leq kM\beta_2\}}],$$

$$I_3 = \mathbb{E}[\int_0^{\frac{X}{k}} e^{-t} \frac{t^M}{M!} dt \mathbf{1}_{\{X > kM\beta_2\}}].$$

Now let n tend to infinity. By (9), we get

$$0 \leq nI_1 \leq n\mathbb{P}(X < kM\beta_1) \int_0^{M\beta_1} e^{-t} \frac{t^M}{M!} dt \longrightarrow 0, \; n \to \infty. \tag{12}$$

It is easy to verify that $\int_0^{\infty} e^{-t} \frac{t^M}{M!} dt = 1$ for any integer $M \geq 0$. Then using together (8) and (10), we obtain

$$\lim_{n\to\infty} nI_3 = \lim_{n\to\infty} n\mathbb{P}(X > kM\beta_2) = \frac{(kx\beta_2)^{-\alpha}}{\Gamma(2-\alpha)}. \tag{13}$$

In the same way, we have

$$0 \le nI_2 \le n\mathbb{P}(kM\beta_1 \le X \le kM\beta_2) \longrightarrow \frac{(kx\beta_1)^{-\alpha}}{\Gamma(2-\alpha)} - \frac{(kx\beta_2)^{-\alpha}}{\Gamma(2-\alpha)},\ n \to \infty. \tag{14}$$

If β_1 and β_2 are close enough to 1, nI_2 can be bounded by an arbitrarily small positive number for n large enough. The proof is finished by combining (12), (13) and (14). □

Fix $t > 0$. If $D(t) \neq 0$, let $\bar{J}_i(t) = \frac{J_i(t)}{Z(t)}$ for $1 \le i \le D(t)$. Let $\{d_1(t), \cdots, d_{D(t)}(t)\}$ be an interval partition of $[0, 1]$ such that the Lebesgue measure of $d_i(t)$ is $\bar{J}_i(t)$. Build a partition of $[n]$ thanks to a paintbox associated with $\{d_1(t), \cdots, d_{D(t)}(t)\}$. Let $N_i(t)$ be the number of integers in the i-th interval and $N(t) = \max\{N_i(t) : 1 \le i \le D(t)\}$. This random variable stands for the size of the largest block of a partition of $[n]$ obtained by a paintbox construction from the atoms of M/Z at time t.

Lemma 3.2 *Let $x > 0$. Then*

1)

$$\lim_{n\to\infty} \mathbb{P}(N(t_n') \le xn^{\frac{1}{\alpha}}) = \exp(-\frac{tx^{-\alpha}}{\alpha\Gamma(\alpha)\Gamma(2-\alpha)}).$$

2) *Let $0 < y < x$. Then*

$$\lim_{n\to\infty} \mathbb{P}(\exists i : J_i(t_n') < n^{\frac{1-\alpha}{\alpha}} y, N_i(t_n') \ge xn^{\frac{1}{\alpha}}) = 0. \tag{15}$$

Proof

1) Let us throw a Poisson number of integers on $[0, 1]$ with parameter $nZ(t_n')$. Then, conditional on $\{J_i(t_n') : 1 \le i \le D(t_n')\}$, the number of integers falling in $d_i(t_n')$, denoted by $\mathcal{N}_i$, is a Poisson variable with parameter $nJ_i(t_n')$ and $\{\mathcal{N}_i : 1 \le i \le D(t_n')\}$ forms a family of (conditional) independent random variables. Let $\mathcal{N}$ be the maximum of all $\mathcal{N}_i$'s. Then, using Lemmas 3.1 and 2.2, as n tends to infinity,

$$\begin{aligned}\mathbb{P}(\mathcal{N} \le xn^{\frac{1}{\alpha}}) &= \mathbb{E}[\Pi_{i=1}^{D(t_n')}\mathbb{P}(\mathcal{N}_i \le xn^{\frac{1}{\alpha}})] \\ &\longrightarrow \exp(-\gamma(\frac{t}{(\alpha-1)\alpha\Gamma(\alpha)})^{1-\alpha}\frac{x^{-\alpha}}{\Gamma(2-\alpha)}) \\ &= \exp(-\frac{tx^{-\alpha}}{\alpha\Gamma(\alpha)\Gamma(2-\alpha)}).\end{aligned}$$

Lemma 2.2 implies that $Z(t_n')$ tends in probability to 1 as n goes to infinity. Hence N and $\mathcal{N}$ are close in the limit and standard comparison techniques allow to conclude.

2) As $Z(t_n')$ converges to 1, it is easy to show that (15) is equivalent to

$$\lim_{n\to\infty} \mathbb{P}(\exists i : J_i(t_n') < n^{\frac{1-\alpha}{\alpha}} y, \mathcal{N}_i \geq x n^{\frac{1}{\alpha}}) = 0.$$

Let $\tilde{\mathcal{N}} = \max\{\mathcal{N}_i : J_i(t_n') < n^{\frac{1-\alpha}{\alpha}} y\}$. It is necessary and sufficient to show that $\lim_{n\to\infty} \mathbb{P}(\tilde{\mathcal{N}} \geq x n^{\frac{1}{\alpha}}) = 0$. Notice that conditional on $J_i(t_n')$, $\mathcal{N}_i$ is a Poisson variable with parameter $nJ_i(t_n')$. Let $\{P_1(yn^{\frac{1}{\alpha}}), P_2(yn^{\frac{1}{\alpha}}), \cdots\}$ be a sequence of i.i.d. Poisson variables with parameter $yn^{\frac{1}{\alpha}}$ and also independent of $D(t_n')$. Then

$$\begin{aligned}\mathbb{P}(\tilde{N} \geq x n^{\frac{1}{\alpha}}) &\leq \mathbb{P}\left(\max\{P_i(yn^{\frac{1}{\alpha}}) : 1 \leq i \leq D(t_n')\} \geq x n^{\frac{1}{\alpha}}\right) \\ &= 1 - \mathbb{E}[(\mathbb{P}(P_1(yn^{\frac{1}{\alpha}}) < x n^{\frac{1}{\alpha}}))^{D(t_n')}].\end{aligned}$$

Using (11) and (9), one gets

$$\mathbb{P}(P_1(yn^{\frac{1}{\alpha}}) < x n^{\frac{1}{\alpha}}) = 1 - o(\frac{1}{n}).$$

Meanwhile, Lemma 2.2 tells that $\frac{D(t_n')}{n}$ converges in probability to $\gamma(\frac{t}{((\alpha-1)\alpha\Gamma(\alpha)})$ as n goes to infinity. Hence the proof is finished.

□

Remark 3.1 The key points to prove (15) is that $Z(t_n')$ converges to 1 in probability and $\frac{D(t_n')}{n}$ is asymptotically bounded by a positive value from above. The distribution of $\{J_i(t_n')\}_{1\leq i\leq D(t_n')}$ is not necessary to know. Actually (15) remains true if t_n' is random and conditions on $Z(t_n')$ and $D(t_n')$ are still satisfied. This fact will be used in the proof of Theorem 1.2.

3.2 Alternative Paintbox Construction

Let $(A_1, \cdots, A_k)$ and $(B_1, \cdots, B_k)$ be two partitions of [0, 1] with $k \geq 1$. We throw n particles uniformly and independently on [0, 1] and group those within the same intervals of $(B_1, \cdots, B_k)$, which gives a sequence of k numbers $(N_{B_1}, \cdots, N_{B_k})$ such that N_{B_i} is the number of particles located in B_i. We can obtain the law of this sequence in another way using $(A_1, \cdots, A_k)$. Throw n particles uniformly and independently on [0, 1]. Let $I := \{i : 1 \leq i \leq n, l(A_i) \leq l(B_i)\}$, where $l(\cdot)$ denotes the Lebesgue measure. If a particle falls in A_i with $i \in I$, then move this particle to B_i. If a particle falls in A_i with $i \in I^c$, then associate to this particle an independent Bernoulli variable with parameter $\frac{l(B_i)}{l(A_i)}$. If the Bernoulli variable gives 1, then the particle is put into B_i. Otherwise, this particle will be put into B_j for $j \in I$ with probability

$$\frac{l(B_j) - l(A_j)}{\sum_{h \in I}(l(B_h) - l(A_h))}. \tag{16}$$

We denote by $N^A_{B_i}$ the new amount of particles in B_i. We have the following result.

Lemma 3.3 *The following identity in law holds.*

$$(N^A_{B_1}, \cdots, N^A_{B_k}) \stackrel{(d)}{=} (N_{B_1}, \cdots, N_{B_k}).$$

Proof Notice that only the Lebesgue measure of each element of $(A_1, \cdots, A_k)$ and $(B_1, \cdots, B_k)$ matters. So one can always assume that $[0, 1]$ is divided in a way that A_i is contained in B_i for $i \in I$ and B_i is contained in A_i for $i \in I^c$. Then if a particle is located in A_i for $i \in I$, it is also located in B_i. But if a particle is located in A_i for $i \in I^c$, with probability $\frac{l(B_i)}{l(A_i)}$ it is located in B_i. Assume that this particle is not located in B_i, then it must be in $\cup_{h \in I} B_h \backslash A_h$. Using the uniformity of the throws, this particle falls in B_j with probability (16). □

3.3 *Proof of Theorem 1.2*

Let us first recall some technical results from [2]. Let $\varepsilon > 0$, $t > 0$ and recall t_n and t'_n. Let $t_- = (1 - \varepsilon)t'_n$ and $t_+ = (1 + \varepsilon)t'_n$. Define the event $B_{1,t} := \{t_- \le R^{-1}(t_n) \le t_+\}$. It can be found in Lemma 4.2 of [2] that there exists a constant C_{17} such that

$$\mathbb{P}(B_{1,t}) \ge 1 - C_{17} t_n \varepsilon^{-\alpha}. \tag{17}$$

Also from Lemma 5.1 of [2], there exists a constant C_{18} such that for all $a > 0$, $t > 0$ and $\eta > 0$,

$$\mathbb{P}(\sup_{0 \le s \le t} |Z(s, a) - a| > \eta) \le C_{18}(a + \eta)t\eta^{-\alpha}. \tag{18}$$

Thus, if we define $B_{2,t} := \{1 - n^{\frac{1-\alpha}{2\alpha}} \le Z(s) \le 1 + n^{\frac{1-\alpha}{2\alpha}},\ \forall s \in [t_-, t_+]\}$, we obtain that

$$\mathbb{P}(B_{2,t}) \ge 1 - C_{19} t(1 + \varepsilon)(1 + n^{\frac{1-\alpha}{2\alpha}})n^{\frac{1-\alpha}{2}} \tag{19}$$

where $C_{19} = C_{18}/(\alpha - 1)\alpha\Gamma(\alpha)$.

Fix any $s \geq 0$ and let π be the random partition of $[n]$ obtained from a paintbox associated with $\frac{M_{R^{-1}(s)}}{Z(R^{-1}(s))}$. Then $\pi \overset{d}{=} \Pi^{(n)}(s)$. Observe that if $R^{-1}(s) \geq t_-$, we can as well at first build a partition from a paintbox associated with $\frac{M_{t_-}}{Z(t_-)}$ and then use Lemma 3.3 to obtain that associated with $\frac{M_{R^{-1}(s)}}{Z(R^{-1}(s))}$ which has the same law as π.

By Markov and branching properties of CSBPs, for any $s \geq t_-$, we can consider the CSBP as the sum of $D(t_-)$ independent CSBP's which we denote by $m_i(s) = Z_i(s - t_-, J_i(t_-))$. Notice that $m_i(s)$ can be 0 while $J_i(t_-)$ is always positive. Let us then build a partition $V^{(n)}(s) = (V_1^{(n)}(s), V_2^{(n)}(s), \ldots, V_{D(t_-)}^{(n)}(s))$ of $[n]$ from a paintbox associated with $(\frac{m_i(s)}{Z(s)}, 1 \leq i \leq D(t_-))$. Let $I_i^{(n)}(s)$ be the number of particles in $V_i^{(n)}(s)$. and $I_+^{(n)}(s) = \sup\{I_i^{(n)}(s), 1 \leq i \leq D(t_-)\}$. Fix $x > 0$ and define $B_{3,t} = \{\exists k : I_k^{(n)}(t_-) \geq xn^{\frac{1}{\alpha}}, J_k(t_-) \geq n^{\frac{2(1-\alpha)}{\alpha}}, \sup_{t_- \leq s \leq t_+} |m_k(s) - J_k(t_-)| \leq \varepsilon J_k(t_-)\}$.

On the event $B_{3,t}$, we have that $I_+^{(n)}(t_-) \geq xn^{\frac{1}{\alpha}}$. Conditional on $B_{1,t}$ we can also build the partition $V^{(n)}(R^{-1}(t_n))$ from a paintbox associated to the partition $Z(t_-)^{-1}(J_1(t_-), \ldots, J_{D(t_-)}(t_-))$ and Lemma 3.3. Let $B(m, p)$ be a binomial variable with parameters $m \geq 2$ and $0 \leq p \leq 1$. Lemma 3.3 implies that

$$\begin{aligned}
&\mathbb{P}\left(I_+^{(n)}(R^{-1}(t_n)) \geq (1-2\varepsilon)xn^{\frac{1}{\alpha}} | B_{1,t} \cap B_{2,t} \cap B_{3,t}\right) \\
\geq &\mathbb{P}\left(B\left(\lceil xn^{\frac{1}{\alpha}}\rceil, \frac{m_k(R^{-1}(t_n))Z(t_-)}{J_k(t_-)Z(R^{-1}(t_n))} \wedge 1\right) \geq (1-2\varepsilon)xn^{\frac{1}{\alpha}} | B_{1,t} \cap B_{2,t} \cap B_{3,t}\right) \\
\geq &\mathbb{P}\left(B\left(\lceil xn^{\frac{1}{\alpha}}\rceil, (1-\varepsilon)\frac{1-n^{\frac{1-\alpha}{2\alpha}}}{1+n^{\frac{1-\alpha}{2\alpha}}}\right) \geq (1-2\varepsilon)xn^{\frac{1}{\alpha}}\right) \\
= &\mathbb{P}\left((xn^{\frac{1}{\alpha}})^{-1}B\left(\lceil xn^{\frac{1}{\alpha}}\rceil, (1-\varepsilon)\frac{1-n^{\frac{1-\alpha}{2\alpha}}}{1+n^{\frac{1-\alpha}{2\alpha}}}\right) \geq (1-\varepsilon)-\varepsilon\right).
\end{aligned}$$

A law of large numbers argument implies that

$$\mathbb{P}\left(I_+^{(n)}(R^{-1}(t_n)) \geq (1-2\varepsilon)xn^{\frac{1}{\alpha}} | B_{1,t} \cap B_{2,t} \cap B_{3,t}\right) \geq 1-\varepsilon \tag{20}$$

for n large enough. Now observe from (18) that

$$\begin{aligned}
\mathbb{P}(B_{3,t}) = &\mathbb{P}(\exists k : I_k^{(n)}(t_-) \geq xn^{\frac{1}{\alpha}}, J_k(t_-) \geq n^{\frac{2(1-\alpha)}{\alpha}}) \\
&\times \mathbb{P}(\sup_{t_- \leq s \leq t_+} |m_k(s) - J_k(t_-)| \leq \varepsilon J_k(t_-) | \exists k : I_k^{(n)}(t_-) \geq xn^{\frac{1}{\alpha}}, \\
&J_k(t_-) \geq n^{\frac{2(1-\alpha)}{\alpha}})
\end{aligned}$$

$$\geq \mathbb{P}(\exists k : I_k^{(n)}(t_-) \geq x n^{\frac{1}{\alpha}}, J_k(t_-) \geq n^{\frac{2(1-\alpha)}{\alpha}})$$
$$(1 - 2tC_{19} n^{\frac{(1-\alpha)(2-\alpha)}{\alpha}} (1+\varepsilon)\varepsilon^{1-\alpha}).$$

By Lemma 3.2, we obtain that

$$\mathbb{P}(\exists k : I_k^{(n)}(t_-) \geq x n^{\frac{1}{\alpha}}, J_k(t_-) \geq n^{\frac{2(1-\alpha)}{\alpha}})$$
$$\sim \mathbb{P}(\exists k : I_k^{(n)}(t_-) \geq x n^{\frac{1}{\alpha}}) = \mathbb{P}(I_+^{(n)}(t_-) \geq x n^{\frac{1}{\alpha}})$$
$$\sim 1 - \exp(-(1-\varepsilon)\frac{t x^{-\alpha}}{\alpha\Gamma(\alpha)\Gamma(2-\alpha)}).$$

In consequence,

$$\liminf_{n\to\infty} \mathbb{P}(B_{3,t}) \geq 1 - \exp(-(1-\varepsilon)\frac{t x^{-\alpha}}{\alpha\Gamma(\alpha)\Gamma(2-\alpha)})$$

when n tends to ∞. Then, thanks to (17) and (19), we deduce that

$$\liminf_{n\to\infty} \mathbb{P}(B_{1,t} \cap B_{2,t} \cap B_{3,t}) \geq 1 - \exp(-(1-\varepsilon)\frac{t x^{-\alpha}}{\alpha\Gamma(\alpha)\Gamma(2-\alpha)}).$$

Combining the latter with (20), we obtain

$$\liminf_{n\to\infty} \mathbb{P}\left(I_+^{(n)}(R^{-1}(t_n)) \geq (1-2\varepsilon) x n^{\frac{1}{\alpha}}\right) \geq 1 - \exp(-(1-\varepsilon)\frac{t x^{-\alpha}}{\alpha\Gamma(\alpha)\Gamma(2-\alpha)}). \tag{21}$$

Next, we seek to find an upper bound for $\mathbb{P}\left(I_+^{(n)}(R^{-1}(t_n)) \geq x n^{\frac{1}{\alpha}}\right)$. Conditional on $B_{1,t}$, we construct $V^{(n)}(t_+)$ from $V^{(n)}(R^{-1}(t_n))$ using the method in Lemma 3.3. Let

$$B_{4,t} = B_{1,t} \cap \{\exists k : I_k^{(n)}(R^{-1}(t_n)) \geq x n^{\frac{1}{\alpha}}, m_k(R^{-1}(t_n)) \geq n^{\frac{2(1-\alpha)}{\alpha}},$$
$$\sup_{R^{-1}(t_n) \leq s \leq t_+} \frac{|m_k(s) - m_k(R^{-1}(t_n))|}{m_k(R^{-1}(t_n))} \leq \varepsilon\}.$$

Similarly as for the lower bound,

$$\begin{aligned}
&\mathbb{P}\left(I_+^{(n)}(t_+) \geq (1-2\varepsilon)xn^{\frac{1}{\alpha}} | B_{2,t} \cap B_{4,t}\right) \\
\geq &\mathbb{P}\left(B\left(\lceil xn^{\frac{1}{\alpha}}\rceil, \frac{Z(R^{-1}(t_n))m_k(t_+)}{Z(t_+)m_k(R^{-1}(t_n))} \wedge 1\right) \geq (1-2\varepsilon)xn^{\frac{1}{\alpha}} | B_{2,t} \cap B_{4,t}\right) \\
\geq &\mathbb{P}\left(B\left(\lceil xn^{\frac{1}{\alpha}}\rceil, (1-\varepsilon)\frac{1-n^{(1-\alpha)/\alpha}}{1+n^{(1-\alpha)/\alpha}}\right) \geq (1-2\varepsilon)xn^{\frac{1}{\alpha}}\right) \longrightarrow 1. \qquad (22)
\end{aligned}$$

Using the strong Markov property of the CSBP and (18), we have

$$\mathbb{P}(B_{4,t}) = \mathbb{P}(B_{1,t} \cap \{\exists k : I_k^{(n)}(R^{-1}(t_n)) \geq xn^{\frac{1}{\alpha}}, m_k(R^{-1}(t_n)) \geq n^{\frac{2(1-\alpha)}{\alpha}}\}) \qquad (23)$$

$$\times (1 - 2tC_{19}n^{\frac{(1-\alpha)(2-\alpha)}{\alpha}}(1+\varepsilon)\varepsilon^{1-\alpha}) \qquad (24)$$

Notice that using (18), in the sense of convergence of probability

$$\lim_{n\to\infty} \sup_{t_- \leq s \leq t_+} Z(s) = \lim_{n\to\infty} \inf_{t_- \leq s \leq t_+} Z(s) = 1$$

Together with (17), we get the following convergence in probability

$$\lim_{n\to\infty} Z(R^{-1}(t_n)) = 1.$$

Recall Remark 2.1 where it is deduced that $t \mapsto D(t)$ is non-increasing. Thus, on the event $B_{1,t}$, we have $D(t_-) \leq D(R^{-1}(t_n)) \leq D(t_+)$. It is then easy to see that $\frac{D(R^{-1}(t_n))}{n}$ is asymptotically bounded from above by a certain positive number. Now we can apply Remark 3.1 and get

$$\mathbb{P}(B_{4,t}) = \mathbb{P}(\exists k : I_k^{(n)}(R^{-1}(t_n)) \geq xn^{\frac{1}{\alpha}}) + o(1) = \mathbb{P}(I_+^{(n)}(R^{-1}(t_n)) \geq xn^{\frac{1}{\alpha}}) + o(1). \qquad (25)$$

Using (22), (19) and (25), we get that

$$\begin{aligned}
&\limsup_{n\longrightarrow\infty} \mathbb{P}(I_+^{(n)}(R^{-1}(t_n)) \geq xn^{\frac{1}{\alpha}}) \\
\leq &\lim_{n\longrightarrow\infty} \mathbb{P}(I_+^{(n)}(t_+) \geq (1-2\varepsilon)xn^{\frac{1}{\alpha}}) \\
= &1 - \exp(-(x(1-2\varepsilon))^{-\alpha}\frac{t(1+\varepsilon)}{\alpha\Gamma(\alpha)\Gamma(2-\alpha)}). \qquad (26)
\end{aligned}$$

Since ε can be arbitrarily small, (21) and (26) allow to conclude.

Acknowledgements The authors would like to thank Fabian Freund for some very valuable comments. Arno Siri-Jégousse would like to welcome Raphaël Clodic Griffon, born the day the final version was sent. ASJ is supported by CONACyT Grant CB-2014/243068.

References

1. J. Berestycki, N. Berestycki, J. Schweinsberg, Beta-coalescents and continuous stable random trees. Ann. Probab. **35**(5), 1835–1887 (2007)
2. J. Berestycki, N. Berestycki, J. Schweinsberg, Small-time behavior of Beta-coalescents. Ann. Inst. H. Poincaré Probab. Stat. **44**(2), 214–238 (2008)
3. J. Bertoin, J.-F. Le Gall, The Bolthausen-Sznitman coalescent and the genealogy of continuous-state branching processes. Probab. Theory Relat. Fields **117**(2), 249–266 (2000)
4. M. Birkner, J. Blath, M. Capaldo, A.M. Etheridge, M. Möhle, J. Schweinsberg, A. Wakolbinger, Alpha-stable branching and Beta-coalescents. Electron. J. Probab. **10**(9), 303–325 (2005)
5. M.G.B. Blum, O. François, Minimal clade size and external branch length under the neutral coalescent. Adv. Appl. Probab. **37**(3), 647–662 (2005)
6. A. Caliebe, R. Neininger, M. Krawczak, U. Rösler, On the length distribution of external branches in coalescence trees: genetic diversity within species. Theor. Popul. Biol. **72**(2), 245–252 (2007)
7. I. Dahmer, G. Kersting, A. Wakolbinger, The total external branch length of Beta-coalescents. Comb. Probab. Comput. **23**, 1–18 (2014)
8. J.-F. Delmas, J.-S. Dhersin, A. Siri-Jégousse, Asymptotic results on the length of coalescent trees. Ann. Appl. Probab. **18**(3), 997–1025 (2008)
9. J.-S. Dhersin, F. Freund, A. Siri-Jégousse, L. Yuan, On the length of an external branch in the beta-coalescent. Stoch. Process. Appl. **123**, 1691–1715 (2013)
10. P. Donnelly, T.G. Kurtz, Particle representations for measure-valued population models. Ann. Probab. **27**(1), 166–205 (1999)
11. F. Freund, A. Siri-Jégousse, Minimal clade size in the Bolthausen-Sznitman coalescent. J. Appl. Probab. **51**(3), 657–668 (2014)
12. G. Kersting, The asymptotic distribution of the length of beta-coalescent trees. Ann. Appl. Probab. **22**(5), 2086–2107 (2012)
13. G. Kersting, J. Schweinsberg, A. Wakolbinger, The evolving beta coalescent. Electron. J. Probab. **19**(64), 1–27 (2014)
14. J.F.C. Kingman, The coalescent. Stoch. Process. Appl. **13**(3), 235–248 (1982)
15. L. Miller, H. Pitters, Small-time behaviour and hydrodynamic limits of beta coalescents. Preprint on Arxiv. https://arxiv.org/pdf/1611.06280.pdf
16. M. Möhle, S. Sagitov, A classification of coalescent processes for haploid exchangeable population models. Ann. Probab. **29**(4)(500), 1547–1562 (2001)
17. J. Pitman, Coalescents with multiple collisions. Ann. Probab. **27**(4), 1870–1902 (1999)
18. S. Sagitov, The general coalescent with asynchronous mergers of ancestral lines. J. Appl. Probab. **36**(4), 1116–1125 (1999)
19. J. Schweinsberg, A necessary and sufficient condition for the Λ-coalescent to come down from infinity. Electron. Commun. Probab. **5**, 1–11 (2000)
20. J. Schweinsberg, Coalescent processes obtained from supercritical Galton-Watson processes. Stoch. Process. Appl. **106**(1), 107–139 (2003)

21. B. Şengül, Asymptotic number of caterpillars of regularly varying Λ-coalescents that come down from infinity. Electron. Commun. Probab. **22** (2017)
22. A. Siri-Jégousse, L. Yuan, Asymptotics of the minimal clade size and related functionals of certain Beta-coalescents. Acta Appl. Math. **142**(1), 127–148 (2016)
23. R. Slack, A branching process with mean one and possibly infinite variance. Probab. Theory Relat. Fields **9**(2), 139–145 (1968)
24. L. Yuan, On the measure division construction of Λ-coalescents. Markov Process. Relat. Fields **20**, 229–264 (2014)

The manufacturer's authorised representative in the EU is Springer Nature Customer Service Centre GmbH, Europaplatz 3, 69115 Heidelberg, Germany. If you have any concerns regarding our products, please contact ProductSafety@springernature.com

Printed and bound by CPI Group (UK) Ltd, Croydon, CR0 4YY

30/09/2026

02225654-0007